The Flora of Essex County, Massachusetts

By Stuart Kimball Harris

Peabody Museum

SALEM, MASSACHUSETTS

1975

CONTENTS

PREFACE

A great deal has transpired in the field of systematic botany since John Robinson published the *Flora of Essex County, Massachusetts* in 1880. The fifth edition of Gray's *Manual of Botany* from which he worked included a total of 2426 species of vascular plants, relatively few of which were divided into varieties. The eighth edition, which appeared in 1950, includes 5523 species and enough varieties and forms to make a total of 8340 different taxa. The genus *Antennaria* has grown from a single species to 32, *Carex* from 132 species to 267, *Panicum* from 19 to 76, and *Rubus* from 11 to 205. In Robinson's day the broad interpretation of specific lines made it possible to correctly identify most species in the field so that sight records made by competent botanists could be accepted and the preservation of specimens to verify such reports did not seem important. Thus there seem to be no specimens to support some of the plants recorded in Robinson's *Flora* although usually there is one specimen collected somewhere in the County. With the much narrower concept of the species used at present specimens are an absolute necessity. Since a single species of Robinson's time may include several modern species it is impossible to make use of Robinson's records unless specimens are available for examination.

Fortunately the Golden Age of plant collecting occurred during the past seventy-five years in New England and much more material is now at hand for study than was preserved in Robinson's era. Each new edition of Gray's Manual seemed to stimulate a new spurt of collecting. The New England Botanical Club was founded in 1895 and a club herbarium was soon established with the lofty aim of including a specimen of every species growing in every township in New England. While the goal will never be reached the herbarium now contains over 215,500 specimens with Essex County well represented. Many amateur botanists built up large private herbaria. The advent of the automobile was a mixed blessing. Any point in the County could be reached in a short time so that out of the way areas were made much more accessible. The roadside flower was in much less jeopardy from the driver of a speeding automobile than from the driver of a horse and buggy. However the improvement in transportation has led to a mass movement from city life and the boom in country homes has destroyed many previously undisturbed areas.

Field botany in New England has now sunk to a very low ebb, the peak being passed soon after the end of the first World War. The cream of the rarities has been found. The trend towards smaller and more compact homes did not allow space for the keeping of private herbaria and most of those previously amassed have now been turned over to institutions. The increased tempo of living does not encourage the undisturbed leisure needed for critical study of specimens. Few amateurs are still left who thoroughly know the local flora and almost none are coming along to fill their places. Soon the student of local flora will have gone the way of the Labrador Duck, the Passenger Pigeon and the Heath Hen.

Stuart Kimball Harris

In 1951 I asked Stuart Harris to write a modern *Flora of Essex County* for the museum to publish. He did an immense amount of work for many years, both in the field and in the office, but unfortunately the Introduction and certain other parts were not completed finished when he died in mid-1969. Many people contributed parts for those missing: Dr. Robert Goodale, Mrs. Chandler Robbins II, Dr. Hurd C. Willett, and above all Richard J. Eaton who wrote two of them and gave me invaluable advice. Charles F. Allen II, Museum Assistant, drew the End Maps. The list of those whom Harris would have mentioned for help given is unknown, it would certainly have included Ralph C. Bean as well as curators at the Harvard Botanical Museum and other institutions.

I am much indebted to the aforementioned and to Professors Albion R. Hodgdon and Wesley N. Tiffeny for information and to Mrs. Harris for her help.

Dorothy Eastman Snyder, Editor
Consultant in Natural History

FOREWORD

When the Trustees of the Peabody Museum decided to sponsor the publication of a modern *Flora of Essex County,* they found in Stuart Harris a uniquely qualified author. He was a professional botanist whose activities had centered on the New England scene for many years. His home was within a half-hour's drive of any town in the County and an hour at most from many thousands of herbarium specimens collected from there by his predecessors and housed under one roof in Cambridge. Being energetic in the field and well-nigh tireless at his presses and typewriter, he accepted with enthusiasm this chance to exploit his interest in the vegetation of an area which harbors many species of plants of saline, sweet and acid soils, - an area where the northern White Fringed Orchis *(Habenaria blephariglottis)* grows in the shadow of the southern Sweet Bay *(Magnolia virginiana),* where Moosewood or Striped Mapel *(Acer pensylvanicum)* occurs rather abundantly within a few rods of the sea-cliffs at Manchester and Hobble-bush *(Viburnum alnifolium)* not far away in the same town, both characteristic of more northern upland woods. A brief account of the procedures he adopted in gathering material for a new flora of the County is to be found in a biographical sketch of the author. It gives a hint of his boundless energy, his distaste of superficiality, his obvious willingness to burn the midnight oil.

During many years of close association with Stuart in the affairs of the New England Botanical Club, I counted him as a friend who commanded deep respect for his professional competence, modesty and quiet humor. With the brashness of a somewhat opinionated amateur, I occasionally questioned one of his identifications. In a spirit of reviewing his own taxonomic opinion, he would go in silence to the cases and return with a suite of specimens which usually confirmed his original judgement to our mutual satisfaction. On the rare occasions when Jupiter had obviously nodded, he would admit his error with sincere thanks.

The diversity of his interests is impressive. He liked to travel and to sketch. Weaving and silversmithing were hobbies in which he attained considerable skill and which provided an outlet for his flair for design. A glance at the titles of his published papers reveals an interest in such diverse subjects as lichens, meterological phenomena on Mt. Washington, mushrooms, the botanical lore of the Navajo Indians, the Alcan (Alaska) Highway Survey, ornithology, the natural history of the Great Smoky Mountains, the arctic-alpine flora of the White Mountains of northern New Hampshire.

I have a vivid picture in my mind's eye of Stuart doing his weekly stint of proofreading for the Botanical Club's journal, *Rhodora,* in its bailiwick on the fourth floor of what we continue incorrectly, to call the Gray Herbarium in Cambridge. Tipping back in his chair, feet on table and unlighted pipe gripped rather precariously between his front teeth, he was following the drone of his companion reading from a manuscript. Now and again he stops him in mid-sentence to query some possible deviation from the author's text. Suddenly he pulls out his watch, rolls up the galleys and announces: "Time to go fishing." A contemplative hour had beckoned, and he was off to a well-deserved interlude between a busy day and a busy evening at home. What he had in mind to do after dinner, he did not divulge. He might start working on a silver mug planned for a grandchild, or perhaps sit at his loom for an hour. Certain it is that he would turn to his botanical worktable as a nightcap.

This book is a scholarly work such as would be expected from a botanist writing for people with botanical tastes.

March 1972

Richard Jefferson Eaton

GEOLOGY IN RELATION TO THE FLORA

Essex County forms the northeast corner of Massachusetts, its boundaries lying between Lat. 42° 25' - 42° 55' north, and Long. 70° 35' - 71° 15' west. Roughly diamond shaped the southernmost area is in Saugus and Nahant 7 miles north of Boston, the most northerly point in Salisbury on the New Hampshire border. Cape Ann is on the east, while the Atlantic Ocean bounds the county from northeast to southeast. Northwest is the New Hampshire state line, with the city of Lawrence on the western angle. Essex County is roughly 30 miles across and encompasses 500 square miles. Of the larger cities and towns the 1970 census shows Lynn with over 90,000 inhabitants; Lawrence almost 67,000; Peabody, Haverhill and Salem over 40,000; Beverly, Methuen, Gloucester and Saugus with more than 25,000. The county was settled very early in the 17th century, with population centers along the coast because ships and shipping were the main industry, and farming inland. As the industrial revolution developed in America manufacturing centers grew up, factory hands moved in and farmers moved west. But except for the places mentioned and where railways and highways were built much of the county retained its rural aspect until after World War II. Then a belt road around Boston, Route 128 was completed and many small electronic plants were built; a good deal of farmland was taken for Route 128 and also turned into housing developments as homes for electronic workers and people overflowing from Boston as a result of the post war population explosion. Now Essex County is essentially suburban and has very little farmland left. Practically every road in the county is now hot topped and this has had a serious effect in lowering the water table.

Essex County has a remarkable diversity of landscapes within its boundaries, due to its geologic history and largely the result of continental glaciation. The county lies in that part of the United States which has been thoroughly eroded during geologic time so that most evidences of geologic history are gone. The bed rock seen today is very ancient – mostly Cambrian-granites, diorites, gabbros and volcanics covered with glacial till, drift, clay and huge erratic boulders. There are very few sedimentary rocks in the county and most of them are strongly metamorphosed. A full and detailed description of the soils of Essex County is available in *The Soil Description of Essex County, Massachusetts* which appeared as #14, Series 1925, U.S.D.A. Bureau of Chemistry and Soils.

There are evidences of past glaciation all over the county: kames, eskers, many small glacial lakes, kettle holes, sphagnum bogs, fresh water swamps, glacial grooves and scratches, erratic boulders, sand dunes, glacial drift, till and moraines both lateral and terminal. The seashore is a typical drowned coastline. Cape Ann, once a mountain, is a solid granite peninsula jutting out into the ocean and covered with a terminal moraine. Almost an island, cut off from the mainland by the Annisquam River it is connected only by a marsh through a channel which has been dredged; two bridges now link the Cape to the rest of the county. From Gloucester to Nahant the coast is rocky with good harbors (drowned river valleys) at Rockport, Gloucester - one of the best on the eastern seaboard - Manchester, Salem, Beverly and Marblehead. There is little sand in this area and the only beaches of any size are at Swampscott, Lynn and Nahant.

From Cape Ann to the New Hampshire border sandy beaches, sand dunes and salt marshes prevail with drumlins and off shore bars. The harbors are shallow and poor and exist only at the mouth of rivers, the entire shore line is much indented and if stretched out would reach 100 miles.

The Merrimac is the largest river in the county, draining a huge area starting in the foothills of the White Mts., flowing south through New Hampshire and then northwesterly across the top of the county parallel to the state line. It is a large mature river by the time it reaches Essex County and the sea, with a wide valley cut through glacial deposits. The Parker, Ipswich and Essex Rivers are small by comparison draining only the county and each emptying into a salt water marsh with a large estuarine area, the largest salt marsh area in the state lying north of Boston. South of Cape Ann there are numerous small brooks but the only rivers are the Danvers and Saugus, both of which are tidal for the greater part of their length, with upper reaches which have recently been altered by road construction. The Shawsheen is the only important inland river, it rises near Lowell outside the county and is a tributary of the Merrimack. The Ipswich rises in Middleton and also drains the center of the county.

There are no great mountains or cliffs in Essex County, the predominant hills are drumlins, elliptical hills of unassorted till left by the continental glacier. These line up with their longitudinal axes running in a northwest - southeast direction all across the county. The highest point, a drumlin of 420 feet is Holt's Hill in Andover, Boston Hill near by is almost as high (385). Most of the islands in the Newburyport to Ipswich areas are drumlins and many more are incorporated in the marsh and dune areas. They are lacking only in the Cape Ann to Nahant region where the hills are all solid granite and under 250 feet in height. Once the bare drumlin was a characteristic of Essex County, but now there are very few cows grazing so trees and shrubs cover almost all the hills.

Fresh water swamps abound in the lowlands as well as numerous lakes and small ponds - legacies of the ice sheet - places where ice blocks melted.

There are at least two true quaking bogs, one in Andover at the base of Holt's Hill and the other near Lynn Woods abutting the Lynnfield town line. This bog shows its glacial origin in its position, lying on a rocky ridge 200 feet above sea level.

Essex County has been settled for so long that there are no extensive forests, nor many very old trees but it is surprisingly well wooded with second, third and fourth growths. There are mixed hardwoods and evergreens in Essex, Rowley and Ipswich. White Pine *(Pinus strobus)*, Pitch Pine *(Pinus rigida)* and Hemlock *(Tsuga canadensis)* as well as Red Cedar *(Juniperus virginiana)* are common. Black Spruce *(Picea mariana)* is found locally in bogs, with Larch *(Larix laricina)* and White Cedar *(Chamaecyparis thyoides)* in the sphagnum bogs and swamps.

All five native birches are found in the county. Gray Birch *(Betula lenta)* is the most common, River Birch *(B. nigra)* is found in dunes in Ipswich, along the Merrimack, in North Andover and is now widespread in the county. Beech *(Fagus grandifolia)*, oaks *(Quercus)* with seven native species: White *(Q. alba)*, Swamp-White *(Q. bicolor)*, Chinquapin *(Q. prinoides)*, Red *(Q. rubra)*, Scarlet *(Q. coccinea)*, Black *(Q. velutina)* and Scrub *(Q. ilicifolia)* are common or frequent, and the introduced English Oak *(Q. robur)* is fairly widespread.

The American Elm *(Ulmus americana)* is dying out as a result of Dutch Elm disease. Three hickories *(Carya)* may be found, the Shagbark *(C. ovata)* being the most common, the two Pignuts *(C. glabra* and *C. cordiformis)* are frequent or occasional. Butternuts *(Juglans cinerea)*, White and Red Ash *(Fraxinus americana, pennsylvanica)*, Hop Hornbeams *(Ostrya virginiana)*, Black and Choke Cherry *(Prunus serotina* and *virginiana)* grow throughout the county. The maple family *(Acer)* is represented by 9 species, only 3: Red Maple *(Acer rubrum)*, the introduced Silver Maple *(A. saccharinum)* and Box-elder *(A. negunda)* are at all common. Sassafras *(Sassafras albidum)* and Tupelo *(Nyassa sylvatica)* are fairly common.

Essex County lies in the Transition Zone. However some trees and plants of the Canadian Zone, for instance the Paper-Birch *(Betula papyrifera)*, Hobblebush *(Viburnum alnifolium)*, Black Spruce *(Picea mariana)* and Larch *(Larix laricina)* are found as well as stragglers from the Austral Zone such as the Laurel-Magnolia *(Magnolia virginiana)* growing in the Magnolia section of Gloucester, and not found again north of Long Island.

The transitional position of Essex County adds to the interest of its several types of environments: rocky shore, sandy shore, salt marsh, sand dunes, mixed evergreen and deciduous woods, broad river valleys, fresh water swamps, sphagnum bogs, inland ponds and the great boulder fields of the terminal moraine of the glacier. The fact that it also lies on the flyway of one of the great bird migration routes of the new world makes Essex County unique. In spite of the immense changes of the last part of the twentieth century the natural history of Essex County continues to be of the greatest interest.

Sarah Fraser Robbins

CLIMATE

Long-term climatic features of southeastern New England, particularly coastal sections such as Essex County are probably well represented by Boston records. Northern and more western sections, particularly in the Green and White Mountains are less well represented.

Essex County being much like Boston, weather can be summarized from the latter's figures and data as:

A. TEMPERATURE:

A slow warming trend during the past century, most pronounced from about 1920 to 1955, with maximum warmth reached in the early 1950's. This showed a rise of some two degrees in annual mean temperatures during the winter months, and about a three degree rise in winter and early spring over the late 19th century temperature. Particularly warm early springs (March - mid April) were reached by 1950.

Since the middle '50's the temperature trend has been downwards, and particularly toward colder winters and colder early springs. This downward trend probably is not yet completed, and has now amounted to about 35-40% of the rise indicated as having occurred from the late 19th to the mid 20th century.

B. PRECIPITATION:

This element has not shown as clear long-term trends as temperature. Boston rainfall has averaged around 40 inches per year from 1820 - 1870, but a very wet period (56 inches) occurred from 1856-1871, continuing less wet to 1883, and again a brief wet period (52 inches) from 1953-1959. Wet periods apply to both winter and summer. From 1884 to 1952 Boston rainfall averaged only 38.5 inches, with only one year (1937) slightly surpassing the 50 inch mark.

The 1960's have been dry again, the driest four year period in 150 years being 1963-1966, 1965 being the driest year in 150, with a rainfall of only 23.71 inches.

Snowfall, and particularly late spring snow cover on the ground, tends to go with the cold periods, light snowfall and early disappearance of snow with the warm periods.

Hurd C. Willett, Professor Emeritus,
Department of Meteorology,
Massachusetts Institute of Technology

HISTORY OF BOTANICAL COLLECTING

It is a great privilege for me to add an historical chapter to Stuart Harris's "Flora of Essex County". If his untimely death had not prevented him from writing it himself, I feel sure that he would have done so in the tradition of John Robinson. When Robinson published a *Flora of Essex County* in 1880 he prefaced it with a "Sketch of Some of the Early Botanists" which is so complete that I have borrowed extensively from it. Robinson gives credit to Manasseh Cutler for the beginning of the study of Botany in Essex County. "Previously the plants had only been noticed by writers upon more general subjects of Natural History . . . . But from Cutler's time there has been a steady succession of botanists, chiefly amateurs, who have kept alive an interest in the subject." Cutler was born in Killingly, Connecticut May 3, 1742, graduated from Yale 1765, studied law and was admitted to the bar in 1767. He then changed his interests, studied for the ministry and settled in the Hamlet Parish (later Hamilton) in Ipswich, Massachusetts. He was a chaplain during the Revolution and on his return switched again and studied medicine which he thereafter practiced among his parishioners. In 1787 his efforts secured the passage of the ordinance by which freedom was declared in the Northwestern Territories and he organized the first band of pioneers who went from Hamilton by ox team to Marietta, Ohio. He himself drove the entire distance in a sulky. He then entered Congress where he served two terms and while in Philadelphia came to know Benjamin Franklin. Perhaps some of Franklin's scientific interest rubbed off on Cutler for in 1783 he wrote "An Account of the vegetable productions, naturally growing in this part of America botanically arranged", which was published in the first volume of the Memoirs of the American Academy of Arts and Sciences in 1785. He here described some three hundred and fifty species of flowering plants. It was his intention to extend this work. His manuscripts were left at his death in 1823 to Professor Edward Tuckerman ("who intends that their final destination shall be the library at Harvard"). After Cutler's death the study of botany was carried on by a group of very active and enthusiastic younger men, chiefly Osgood, Nichols, Oakes, and Pickering.

Dr. George Osgood, Jr. was born in Fairhaven, March 25, 1784. He studied medicine with his father and settled in Danvers in 1804. Through association with Cutler, Nichols and Oakes he became an enthusiastic botanist. He contributed to Bigelow's *Florula Bostoniensis* and in 1853 published a list of plants flowering in his area. He died May 26, 1863.

Dr. Andrew Nichols was born in Danvers, November 22, 1785 and studied medicine under Dr. Waterhouse. He was particularly interested in the botany of what is now Peabody and was one of the founders of the Essex County Natural History Society. He died March 31, 1853.

Robinson states that "William Oakes must be acknowledged as the most eminent botanist of Essex County birth." Oakes was born July 1, 1799 in Danvers New Mills, Mass., the son of Caleb and Mehitable (Pope) Oakes. He studied under Prof. Peck at Harvard who inspired him to devote his life to Natural History. For about two years after graduation he studied law under Leverett Saltonstall in Salem, then moved to Ipswich, married Sarah Farley, widow of William Treadwell, gave up the law and devoted the rest of his life to Natural History.

Oakes began collecting botanical specimens throughout New England, always duplicating his specimens to obtain all the variations and oddities. Walter Dean abstracted fifty letters from Oakes to Dr. J.W. Robbins with whom he was closely associated in his collecting. One letter reveals a truly open-minded skepticism. Beginning with a long dissertation on nomenclature and casting doubts on the validity of some claims he discusses doubtful plants. "The truth is, our plants are in a sad state and can only be determined from specimens in herbaria which themselves are not free from mistakes . . . I think you are . . . too much disposed to rely on less important distinctions and also that you construe the descriptions in the books too strictly. It is best to collect and preserve numerous specimens of each species and look about and collect extremes and oddities. In some genera and species of plants the forms of the parts vary greatly according to the stage of growth and accidental circumstances."

Oakes collected from virtually all of New England but was chiefly concerned with Essex County, Western Massachusetts, Connecticut, Vermont and the White Mountains of New Hampshire. He contributed a list of plants for Thompson's *History of Vermont.* His chief interest no doubt was his ambition to publish a comprehensive Natural History of the White Mountain region. He made innumerable trips to this area and spent a fortune in preparing the manuscript of *White Mountain Scenery.* His tragic death by drowning July 31, 1848 left this work to be published posthumously. It was beautifully illustrated with lithographs by Isaac Sprague. Asa Gray in his obituary states "Oakes' services to botany are not to be measured by the amount of his actual publications, but there are few botanists in this country who are not indebted to him for some part of their knowledge or for some of the finest specimens in their herbaria."

Dr. Charles Pickering, son of Timothy and Lurena (Cole) Pickering and grandson of Col. Timothy Pickering of Revolutionary fame, was born at Starucca Creek on the Susquehanna, Pennsylvania, in 1805. He was educated at Harvard in the class of 1823, graduating at the medical school in 1826. In 1838 he was appointed naturalist to the U.S. (Wilkes) Exploring Expedition; and to perfect his knowledge of animals and plants in foreign parts, he made very extensive journeys after his return from that expedition. He was the author of several works of great value which in their preparation required much untiring research; among them are *Geographical Distribution of Animals and Plants* and *Chronological History of Plants,* the latter work occupying the last sixteen years of his life.

During his college life Dr. Pickering spent much of his time at Wenham, at the homestead of his grandfather, Col. Pickering, and here he was in the habit of botanizing in company with William Oakes, a favorite locality being the "Great Swamp." It is but right that Essex County should claim a share of the honor of his name, for it was here that his attention was drawn to botany, and in the *Chronological History of Plants,* page 1063, we find the following entry "1824– 'In this year, after an excursion in 1823, with William Oakes diverting my attention from entomology, my first botanical discovery'. Dr. Pickering died at Boston, March 17, 1878."

Rev. John L. Russell, born in Salem, December 2, 1808, was in the 1828 Class at Harvard and graduated from the divinity school in 1831. He was especially interested in cryptogamic botany on which he frequently lectured.

According to Robinson, "The only attempt at an enumeration of county plants as such is that of Mr. Cyrus M. Tracy of Lynn."

Robinson mentions others who contributed to the subject, George D. Phippen, Calvin Pool, S. B. Buttrick, Dr. G. A. Perkins, Geo. F. A. Markoe, Rev. Mr. Chute, Dr. Henry P. Wheatland and S. P. Fowler, B. D. Greene, Wm. P. Richardson, S. Bass, Mrs. Alex Bray, Mrs. Charles Grover, Mrs. J. Babson, Mrs. Davis, Frank Lufkin, John H. Sears, W. P. Conant and "many others."

An account of Robinson written by Albert P. Morse appeared in *Rhodora,* Volume 31, December 1929. Morse relates that Robinson was born July 3, 1846 in Salem, Mass. and died there on April 9, 1925. The son of John and Lucy Pickering (Stone) Robinson, he graduated from private elementary and public high schools but did not go on to college. Later he regretted this and took summer courses at Harvard and studied the natural sciences on his own. After a clerkship in the commission house of Alden Speare, he became associated with the Peabody Academy of Science (Museum) in Salem. Here he was greatly influenced by Dr. Henry Wheatland, the moving spirit in the founding of the Essex Institute. Robinson was also much interested in history, rare coins and marine material. At age 19 he was named Curator of Civil War Relics, but at the same time he was drawn toward the natural sciences. He studied Gray's *Lessons in Botany* and Dana's *Geology* and any other textbooks he could lay his hands on and took a summer course in botany under Prof. George L. Goodale and a winter course in cryptogamic botany under Prof. W. G. Farlow.

In 1874 he was in charge of the herbarium of the Peabody Museum. He contributed over 1000 sheets of Essex County plants to the Museum and published a catalogue of the County Flora. He was encouraged and aided in this by C. G. Sprague, F. S. Collins, Caleb Cooke, John H. Sears and W. P. Conant. He also published *Ferns in Their Homes and Ours.* Robinson gave much attention to woody plants and so naturally came to know Prof. Charles S. Sargent. For two years he was assistant in charge of the herbarium at the Arnold Arboretum. He was also active in the Massachusetts Horticultural Society which made him an "Honorary Professor of Botany."

In June 1875 he was appointed Trustee of the Peabody Academy of Science, was its Treasurer and in charge of its Museum.

Robinson gave an interesting account of himself in the third person which Morse quotes –
"He never found a plant new to botanical literature nor rearranged the specific or generic limits of those already known. His interest in botany was in outdoor observation, and he detested 'closet-work'. Physiological botany and morphology not the dry descriptive indoor work, was what he enjoyed. He had a supreme contempt for the continual disarrangement and rearrangement of names; and in the confusion that the nomenclaturists seemingly gloated over in pages of discussions over 'rights of priority' by which established names were thrown aside for forgotten ones, – this thoroughly disgusted him. The study of the structure and development of plants seemed more worthwhile and the plants themselves in the field more interesting through microscopic studies, so much so that, later he cared little to collect them for herbarium preservation."

In 1875 after Dr. Packard's resignation, Prof. E.S. Morse was elected Director of The Peabody Museum. But as Morse soon left for Japan, Robinson was named Acting Director. He now had to manage all the other collections as well as the herbarium.

It was inevitable that the great enthusiasm of these earlier botanists as well as the increasing interest in horticulture would result in an ever growing number of persons who scoured the woods and fields and swamps of Essex County to collect or note hitherto unknown material. The founding of the New England Botanical Club in 1899 brought together both professional and amateur botanists to correlate and report their findings. Significant discoveries were published in the Club's magazine *Rhodora.* In the main these reports recorded new plants or stations or odd variations of well known species.

Harris quotes largely from this publication and the essential purpose of his *Flora of Essex County* is a listing of the plants and the localities where they are found. A glance at Harris' list of source material will show what a host of contributors he could draw on, 619 including the 35 listed by Robinson. Note should be made of the 53 references to *Rhodora.* It is timely that all this material, published since John Robinson's *Flora* should be brought up to date. The length of Harris' catalogue attests the need for recording not only the discovery of hitherto unreported species and varieties of native plants but also the introduction of those of foreign origin.

Robert Lincoln Goodale, M.D.

COLLECTORS

Abbott, Miss Florence I. n.d. Andover
Abbott, Mrs. S. E. 1898 Andover
Adams, V. P. 1880 Gloucester
Ahler, Miss Olga E. 1917 Lynnfield
Alcott, Rev. William Penn died 12 Oct. 1919
Allen, Glover M. 1912 Ipswich
Ames, Oakes died 28 Apr. 1950
Anderson, William Arthur died 20 Feb. 1949 Rockport
Andrade, J. 1957 North Andover
Andrews, William P. 1879 Swampscott
Annable, Mr. 1857 Hamilton
Arimoto, Shintaro n.d. Topsfield
Ashburn, N. 1945 Andover
Ashburn, P. 1945 Andover
Atherton, Daniel Y. 1960-1970 died April, 1971 Gloucester, Rockport

B., Mrs. C. n.d. Gloucester (Boynton?)
B., J. Y. 1884 Georgetown, Merrimac (CNSH)
Babson, Mrs. Ann J. 1863-1877
Bailey, U. n.d. Topsfield
Bailey, William Whitman died 20 Feb. 1914
Banvard, Rev. Joseph 1864 died 28 Sept. 1887
Bartlett, Mr. 1878 Marblehead
Bartlett, Harley Harris 1908 Lynn died
Bartlett, Miss Mira W. Haverhill
Barton, William G. 1879 Beverly
Bartow, Mrs. Edward 1888 Beverly
Bartram, Edwin B. 1908 Rockport
Bass, Dr. S. 1816-1819 Salem
Batchelder, Charles Foster died 7 Nov. 1954
Batchelder, Jacob 1858 Nahant
Bates, James A. n.d. Boxford, Georgetown
Beal, William 1887 Nahant
Bean, Ralph Carleton 1919-1954
Beeber, J. A. 1887 Plum Island
Beeber, Mrs. T. R. n.d. Swampscott (CNSH)
Benson, Mrs. n.d. Topsfield
Bickford, Mrs. n.d. Georgetown (CNSH)
Bigelow, Dr. Jacob (1787-1879)
Birrell, William 1935, 1940 Wenham
Bishop, Harlow 1928 Rockport
Blake, Rev. Joseph 1872-1886
Blake, Sidney Fay 1918, 1928 died 1959
Blankinship, J. W. 1895 Nahant
Blatchford, Mary 1894 Haverhill
Bolles, Rev. E. C. n.d. Salem
Booth, George W. 1959 Andover
Boott, Mrs. 1819 Andover
Boott, Francis 1816 Lynn
Boott, Dr. John Wright 1816 Ipswich (1788-1845)
Boott, William (1805-1887)
Bosson, Abraham n.d. Salem
Boutwell, Frederic S. 1907 Andover
Boutwell, Winthrop S. 1951 Andover
Boynton, Mrs. C. 1880 Gloucester
Boynton, Mrs. J. F. n.d. Boxford
Bradford, George P. 1858, 1859 Salem
Bragdon, H. W. 1943 North Andover
Brainard, Prof. n.d. Newburyport
Bray, Mrs. Alexander 1875 Gloucester, Manchester
Bray, Mrs. Maria H. 1887 Gloucester
Brinton, Dr. J. B. 1884 Gloucester
Brooks, Miss 1880 Salem
Brooks, Miss M. W. 1923 Essex County
Brown, Mr. n.d. Peabody
Brown, C. E. 1912
Brown, Lillian F. 1885, 1912 Boxford, Topsfield
Bubier, George M. 1909 Ipswich
Buck, Miss Alice n.d. Andover
Buck, Henry H. 1895-1898
Bucklin, Mrs. S. J. 1898 Andover
Bullard, Miss Louisa D. 1880-1882
Bullard, William M. 1908 Manchester
Burnett, Francis 1963 Manchester
Buttrick, Samuel Bartlett 1855-1878 died 11 Nov. 1887
Buxton, Mrs. H. H. 1925, 1926
Byers, Montgomery L. 1958 Boxford

Cabot, Miss n.d. Lawrence
Carleton, Miss Hattie C. 1885-1893
Carpenter, Miss Jane B. 1952 North Andover
Carter, Miss M. E. 1905 Rockport
Castelhans, Dr. Karl 1884 Newburyport
Chadwick, George W. 1882 Boxford
Chamberlain, Edward Blanchard 1899, 1902 Newburyport died 2 Feb. 1925
Chandler, Mrs. G. W. n.d. Andover
Chapman, Miss Jean M. Haverhill
Chapman, Miss J. E. 1918 Gloucester
Chapman, J. H. n.d. Marblehead
Chapman, J. W. n.d. Marblehead
Cheever, Austin Walter 1900-1933
Cheever, Clarence Alonzo 1900-1933 Died 31 Dec. 1946
Chivers, A. H. 1902 Amesbury
Church, George Lyle 1936-1942
Churchill, Judge Joseph Richmond, 1882-1914 Died 14 Feb. 1933

Chute, Rev. Ariel Parish 1856, 1858 Died 18 Dec. 1887
Clark, Dr. prior 1858 Lynn
Clark, Austin H. 1924, 1925 Essex, Manchester
Clark, M. n.d. Topsfield
Clark, Moseley 1932 Newburyport
Clark, S. S. 1916 Nahant
Clarke, Miss Cora H. 1905-1912 Gloucester
Clarkson, Edward Hale 1917 Merrimac Died 4 Sept. 1934
Clay, Annie n.d. Topsfield
Coe, Miss M. A. 1902 Plum Island
Cole, Arthur E. 1885-1956 Died Oct. 1962
Cole, E. N. 1861
Cole, Harrison R. Haverhill
Cole, M. H. 1881, 1884 Boxford
Collins, Frank Shipley 1890-1918 Died 25 May 1920
Collins, John A., Jr. 1900, 1932 Lawrence
Conant, A. W. n.d. Topsfield
Conant, Woodbury P. 1877-1918
Conley, G. T. n.d. Topsfield
Conley, Marion n.d. Topsfield
Conley, P. n.d. Topsfield
Conley, Mrs. S. n.d. Topsfield
Cook, Mr. 1857 Wenham
Cooke, Caleb 1872-1879 Beverly, Hamilton Died 5 June 1880
Crittenden, Miss Viola E. 1931 Beverly
Crosby, Mrs. Alpheus 1876 Salem
Crosby, Mrs. Martha Kingman 1861 Saugus
Cross, Henry J. 1850, 1859 Beverly
Cummin, J. W. n.d. Lynnfield
Currier, Mary E. n.d. Newburyport
Curtis, C. H. 1878 Gloucester
Curtis, C. L. 1908 Andover
Curtis, Mrs. E. S. 1945 Essex
Cushing, Caleb 1819
Cushing, Margaret 1927 Newburyport
Cushman, Joseph Augustine 1905-1908 Died 16 Apr. 1949
Cutler, Rev. Manasseh 1783, 1806 Born 1742, Died 1823

Dame, Lorin Low 1881 Gloucester Died 27 Jan. 1903
Davenport, George Edward 1874-1895 Died 29 Nov. 1907
Davis, Lucy 1875 Wenham
Dawson, Jackson T. n.d. Andover
Deane, G. C. 1914 Ipswich
Deane, Walter 1884-1926 Died 30 July 1930
Deering, Lila n.d. Topsfield
Dewey, Chester Salem Died 1867
Dixon, Prof. George H. 1877, 1878
Dodge, Mrs. A. n.d. Topsfield
Dodge, Miss B. 1885 Rowley
Dodge, Edward 1898 Ipswich, Plum Island
Dodge, Mrs. Paul A. 1918 Rowley
Dodge, Raynal 1890-1912 Died 20 Oct. 1918
Dodge, V. R. n.d. Topsfield
Dow, A. M. n.d. Topsfield
Dow, Bertha n.d. Topsfield
Dow, E. M. n.d. Topsfield
Dow, George Francis 1915, 1916 Topsfield
Dow, G. T. n.d. Topsfield
Dow, L. M. n.d. Topsfield
Downs, Mrs. S. M. (A. L.) 1890
Dudley, Kendall 1957 North Andover
Duncan, Stewart 1951, 1953

Eames, Arthur Johnson 1908 Lynn
Eaton, Alvah Augustus 1882-1904 Died 29 Sept. 1908
Eaton, Daniel Cady Died 29 June 1895
Eaton, Richard Jefferson 1925-1962
Edwards, William n.d. Salem
Emerson, George Barrell 1838 Topsfield Died 1881
Emerton, James Henry 1865-1879 Died 5 Dec. 1930
Emery, G. E. 1871 Lynn, Saugus
Emilio, S. Gilbert 1926, 1927 Hamilton, Saugus Died 1973
Emmerton, James Arthur 1856 Essex County Died 31 Dec. 1888

Farlow, William Gilson 1903, 1916 Died 3 June 1919
Farrington, George P. 1877 Salem
Farwell, H. C. 1912 Salem
Farwell, J. A. 1927 Newbury
Fassett, Norman Carter 1921-1924 Died 14 Sept. 1954
Faxon, Charles Edward 1877-1903 Died 6 Feb. 1918
Faxon, Edwin, 1878-1893 Died 12 June 1898
Faxon, Walter 1879, 1895 Danvers, Ipswich
Felt, George P. 1925 Essex County
Fenollosa, S. K. 1889 Swampscott
Fernald, Merritt Lyndon 1902-1932 Died 22 Sept. 1950
Files, Miss L. H. 1939 Salem
Fiske, Dr. 1880 Salem
Fitch, Miss C. M. 1883 Gloucester
Flint, George Foster Marblehead Died 23 June 1882
Floyd, Mrs. C. n.d. Topsfield
Floyd, Fred Gillan 1897, 1904 Ipswich, Plum Island, Nahant Died 7 Mar. 1941

Floyd, Olive n.d. Topsfield
Follansbe, P. B. n.d. Andover
Forbes, Fayette Frederick 1900-1913 Died 20 Nov. 1935
Fowler, Samuel Page 1850-1882 Died 15 Dec. 1888
Foye, Elmer Paul 1952-1958 Topsfield, Essex
Frame, B. n.d. Topsfield
French, Arthur prior 1891 Middleton
Frost, A. E. 1872-1884
Frye, Susan M. 1880-1882
Fuller, T. Otis 1893-1905
Furber, Jane M. 1911 Ipswich, Plum Island

Gates, Ruth C. 1928 Rockport (Mrs. Lyman B. Smith after 1928)
Gay, E. 1857 Peabody
Gilbert, E. J. C. n.d. Essex County
Gilbert, Grace 1894, 1896 Gloucester, Nahant
Gilbreth, M. E. 1889 Andover
Gleason, Mrs. George 1906 Topsfield
Glück, H. 1924 Newbury
Goldsmith, William Gleason 1901 Andover Died 7 Oct. 1910
Goldsmith, Prof. W. S. n.d. Andover, North Andover
Goodale, Prof, George Lincoln 1873 Hamilton Died 12 April 1923
Goodale, Joseph L. Ipswich Died 1957
Goodale, H. M. n.d. Topsfield
Goodale, Robert L. 1961 Ipswich
Goodale, Susan S. Ipswich 1968
Goodrich, Miss Carrie 1881, 1884 Essex County
Gould, F. M. n.d. Topsfield
Gould, Mrs. H. n.d. Topsfield
Gould, Mrs. M. n.d. Topsfield
Gowing, Mrs. Belle P. 1888 Andover
Graves, Elizabeth 1885 Beverly
Gray, Arthur F. n.d. Danvers
Gray, Asa 1847-1886 Hamilton, Beverly Died 30 Jan. 1888
Gray, Paul E. Haverhill
Grayce, Robert L. 1952 Gloucester, Rockport
Greene, Miss A. M. 1882 Methuen, North Andover
Greene, Mr. B. D. n.d. Essex County
Greene, E. W. 1886 North Andover
Greenough, J. T. 1891, 1895 Groveland, Ipswich
Gregory, J. J. H. 1883 Ipswich
Grigg, Frederick William 1911-1924 Died 16 Sept. 1952
Grover, Mrs. Charles (A.L.) 1875 Gloucester
Gutterson, Myron Edwards 1901-1909

H. n.d. Topsfield
Hall, Mrs. n.d. Groveland (CNSH)
Hall, A. F. 1925-1927 Marblehead, Salem
Harris, Hilda Frame 1959 Newburyport
Harris, Stuart Kimball 1931-1964 Died 30 July 1969
Hart, S. K. 1882 Manchester
Haskell, Mrs. A. J. n.d. North Andover
Havens, Samuel 1884 Methuen
Hayward, Mrs. N. P. 1860 Danvers
Hedge, Frederick H. n.d. Lawrence
Herrick, Mrs. William n.d. Topsfield
Hewitt, William 1901 Gloucester
Higginson, Rev. Francis 1629 Salem
Higginson, S. 1882 Gloucester
Hitchings, E. H. 1880-1891
Hoar, E. S. 1857, 1858 Gloucester, Nahant
Hobson, M. n.d. Topsfield
Hodgdon, Albion Reed 1945 Boxford
Hodges, Mrs. n.d. Topsfield
Hogan, D. F. 1927 Salem
Holder, Dr. John Bassett n.d. Lynn, Swampscott Died 28 Feb. 1888
Holt, Dr. Edward W. A. n.d. North Andover
Hood, E. J. n.d. Topsfield
Hooper, Nathaniel 1865 Essex County
Hopkins, Mrs. J. I. 1952 Essex
Hopkins, Milton 1932-1934
Hopkinson, Mrs. n.d. Marblehead (CNSH)
Horncliffe, Oliver C. 1902 Beverly
Horner, Charlotte Nichols Saunders 1870-1899
Horner, W. S. 1879-1886
Howe, Martha Jane (Mrs. Wm. B.) 1885 Boxford
Hoyt, W. D. 1915 Gloucester
Hubbard, F. Tracy 1911-1913
Hubbard, G. G. 1911 Ipswich
Hunnewell, Francis Wells, 2nd 1913-1936 Died 1964
Hunnewell, James Melville 1898-1902 Gloucester, Manchester
Huntington, J. Warren 1899-1909 Amesbury

Ives, Miss Jessie S. 1925-1927
Ives, John M. 1856-1860 Died 29 July 1883
Ives, John S., Jr. 1910 Salem

Jack, J. S. 1903 Topsfield
Jackman, C. n.d. Topsfield
Jackson, Miss Anna L. 1894 Beverly
Jackson, J., Jr. n.d. Salem
Jacobs, Allen n.d. Topsfield
Jenkins, Mrs. n.d. Topsfield
Jenks, Charles William 1880-1884

Jewell, E. Louise 1951 Boxford
Johnson, B. F. n.d. Lynn
Johnson, Charles W. 1912 Gloucester
Johnston, Elizabeth 1944 Gloucester
Jones, Florence H. 1884 Boxford

Kelsey, Seth Low 1931 Boxford
Kennedy, George Golding 1859-1914 Died 31 March 1918
Kennedy, Mrs. J. M. 1910 Danvers
Kettle, Mrs. 1859 Danvers
Keyes, C. 1886 Nahant
Kidder, Nathaniel Thayer 1884-1886 Died 14 Oct. 1938
Kilham 1891 Boxford
Kimball, H. 1890 Wenham
Kimball, L. G. 1875 Wenham
Kimball, Mrs. M. W. 1874-1885
Kimball, Walter 1877 Wenham
Kingman, C. C. 1895-1899
Kittredge, Miss H. A. 1897 North Andover
Kneeland, A. E. n.d. Topsfield
Kneeland, Phoebe Ellen n.d. Topsfield Died 1932
Knowlton, Clarence Hinkley 1903-1939 Died 10 May 1956

Lane, William Coolidge 1887, 1880 Beverly, Manchester
Larrabee, R. W. 1883 Nahant
Larrimee, George 1913 Peabody
Leavitt, Dr. Robert Greenleaf 1902 Newburyport Died 2 Oct. 1942
Lee, John C. n.d. Essex County
Lewis, Henry Stockton 1958-1961
Lindsay, N. Allen 1883-1889 Marblehead
Little, Dr. Henry 1823 Topsfield
Lloyd, H. H. 1884 Andover
Long, Bayard 1913
Long, H. n.d. Topsfield
Loring, C. G. 1876 Beverly
Lothrop, Mrs. Sturgis 1927 Manchester
Low, H. C. 1888 Danvers
Lowell, John Amory 1841-1857
Lufkin, Frank 1875-1891
Lufkin, H. S. 1861 Lynn
Lund, James 1880-1889
Lund, Miss M. B. n.d. Plum Island (CNSH)
Lunt, J. Richard 1911-1915

Mack, Miss H. C. n.d. Essex County
Mack, Dr. William n.d. Salem
Mackintosh, E. L. 1923, 1924
Mackintosh, Robert Bryant 1890-1938 Died 16 Jan. 1939
Mann, Horace 1849 Marblehead
Manning, J. W. 1883 Beverly
Manning, W. H. 1881-1886
Markoe, George F. H. 1860-1864
Mason, F. L. 1911 Andover
Matthews, Sara W. 1896 Ipswich
May, Dr. John B. circa 1908 Gloucester Died 1964
McCoy, L. L. 1951 North Andover
M'Grane, William 1877 Salem
McGregor, Frank John 1939-1958 Died 23 Feb. 1962
McKelvey, S. D. 1922 Gloucester
Merrill, J. W. n.d. Manchester
Merrill, William W. 1849-1877
Merrill, William, Jr. 1885 Boxford
Meyer, G. A. 1884 Andover
Miller, Henry J. 1954 North Andover
Milliken, R. S. 1912-1916 Plum Island, West Newbury
Minns, Susan n.d. Ipswich
Mitchell, William 1876 Andover (in Robinson, Gray says James)
Moore, Albert Hanford 1902-1906 Andover
Moore, Mrs. G. F. 1898 Andover
Moore, John William 1930 Lynnfield, Salem
Morissey, Eva n.d. Topsfield
Morong, Thomas 1872-1885
Morse, Albert Pitts 1912-1928 Died 29 Apr. 1936
Morse, E. C. 1885 Rowley
Morse, E. S. 1881, 1899 Gloucester, Marblehead
Morse, G. S. 1883 Boxford
Morss, Charles Henry 1872-1901 Died 15 Dec. 1930
Morss, R. D. 1901 North Andover
Moulton, Edward 1885-1889
Moulton, J. prior 1858 Lynn, Saugus
Munroe, Mrs. J. 1889 Lynnfield
Murdock, John, Jr. 1904, 1913 Died 29 Jan. 1915

Newhall, Asa T. prior 1858 Lynnfield
Newhall, E. K. 1899 Gloucester
Nichols, Dr. Andrew 1816-1852 Died 31 Mar. 1853 Born 22 Nov. 1785
Nichols, Andrew, Jr. 1886 Middleton
Nichols, Rev. H. P. 1873 Andover
Nichols, Miss Mary W. 1882-1891 Danvers
Nickerson 1884 Lawrence
Norris, Miss 1915 Middleton, Salem
Norris, Charles H. 1860 Manchester
Noyes, Mr. S. 1905 Georgetown
Nuttall, Mr. n.d. Gloucester, Salem

Nye, Dr. J. M. n.d. Lynn

Oakes, William 1817-1842 Died 31 July 1848
Born 1 July 1799
Ober, Frederick A. 1887 Ipswich
Ogden, Edith B. 1938 Wenham
Ogden, Eugene Cecil 1938 Newbury, Wenham
Oliver, James E. 1861 Lynn
O'Neil, J. R. 1925 Rowley
Ordway, D. Leighton 1937 Ipswich
Osborn, A. A. 1930 Peabody
Osgood, Alfred n.d. Newburyport
Osgood, Dr. George 1819-1863 Died 26 May 1863
Born 25 March 1784

P., C. 1853 Nahant (Pickering, Charles?)
Page, Miss A. L. 1856-1909
Page, Dr. C. G. 1835 Salem
Paine, Miss Harriet E. 1883-1885
Palfrey, H. G. 1890-1893 Haverhill
Palmer, Dr. Charles 1870, 1877 Ipswich
Palmer, Ernest Jesse 1942 Newbury
Parker, E. S. 1878 Groveland
Parkhurst, George Byam 1882 Boxford
Parkhurst, Miss Hattie E. 1882-1885
Parkhurst, John 1882, 1883 Boxford
Parsons, Miss Kate 1885 Nahant
Parsons, Judge Theophilus 1806 Gloucester
Patch, Harry 1879 Salem
Peabody, Mrs. Dr. n.d. Essex County
Peabody, Charles 1895 Salem
Peabody, Mrs. F. n.d. Topsfield
Peabody, Francis Howard 1892-1902 Died 22 Sept. 1905
Peabody, J. C. n.d. Newburyport
Peabody, J. P. 1886 Boxford
Peabody, M. R. n.d. Topsfield
Peabody, S. W. n.d. Topsfield
Peabody, W. C. 1846 Rockport
Pease, Arthur Stanley 1893-1943 Died 7 Jan. 1964
Peattie, Donald Culross 1921 Saugus Died 1964
Peirce, Mrs. Mary A. n.d. Topsfield Died 1929
Perkins, Dr. n.d. Salem (CNSH)
Perkins, Albert C. 1868, 1883 Ipswich, Rowley
Perkins, Dr. Anne E. 1927 Boxford
Perkins, C. E. 1880 Methuen
Perkins, Mrs. Esther A. R. n.d. Topsfield
Perkins, Dr. George A. 1877, 1879 Died 18 May 1895
Perkins, Dr. Henry C. 1824, 1864 Newburyport
Died 1 Feb. 1875
Perkins, Mrs. J. W. n.d. Topsfield

Perley, Miss 1884 Boxford
Perley, Henry E. 1885, 1886 Georgetown
Perley, Mary Ellen 1880-1915
Perley, Mary Paige 1881-1888
Perley, Sarah Peabody 1882 Boxford
Perley, Sidney 1884 Boxford
Phillips, John C. 1877-1910
Phillips, W. A. 1858 Swampscott
Phippen, George Delan 1837-1889 Died 26 Dec. 1895
Phippen, Joshua n.d. Salem
Pickering, Dr. Charles 1821-1860 Died 17 March 1878 Born 1805
Pickman, Dudley L. 1925 Danvers
Pierce, Archie n.d. Topsfield
Pierce, Mrs. E. n.d. Topsfield
Pierce, Mrs. M. I. n.d. Topsfield
Pieree, M. L. n.d. Topsfield
Pierson, C. L. 1914 Beverly
Pitman, B. C. 1857 Wenham
Plummer, Miss 1884, 1890 Newbury
Pool, Calvin 1875 Rockport
Poor, Mrs. Winifred n.d. Topsfield
Pope, Robert A. 1898 Newburyport, Salisbury
Porter, A. n.d. Topsfield
Porter, Blanche Merritt 1927-1950
Porter, Frederick 1866 Salem
Porter, Harriet A. 1898 Topsfield
Porter, H. S. 1901, 1902 Ipswich, Topsfield
Pranker, Edward 1859 Saugus
Preble, M. H. 1917 Nahant
Prince, Miss F. C. 1883-1902
Prince, J. C. 1895 Manchester
Proctor, T. E. 1900, 1901 Topsfield
Purdie, Henry Augustus 1888-1909 Died 29 Mar. 1911
Putnam, Rev. Mr. 1884 Rockport

R., J. L. 1884-1890
Rand, Edward Lothrop 1896, 1910 Gloucester, Marblehead Died 9 Oct. 1924
Raymond, John 1902 Beverly
Rayner, Mrs. n.d. Boxford
Rearden, Miss Harriet 1906 Georgetown
Redmond, John J. 1878 Beverly
Rehder, Prof. Alfred 1903 Lynn Died 21 July 1949
Rich, William Penn 1886-1906 Died 30 Nov. 1930
Rich, Mrs. W. P. 1897 Gloucester
Richardson, E. S. L. 1838 Essex County
Richardson, Dr. William Putnam 1835 Essex County
Died 27 March 1857

Ripley, Winfield Scott, Jr. 1914-1916 Gloucester, Saugus Died 8 Sept. 1932
Robbins, Dr. J. W. 1867 Gloucester, Wenham Died 1879
Roberts, H. H. n.d. Topsfield
Robinson, Benjamin Lincoln 1896, 1903 Gloucester, Newburyport Died 27 July 1935
Robinson, Elizabeth Robbins (Kemble) 1899 Manchester
Robinson, Prof., J. H. 1894 Gloucester
Robinson, John 1865-1904 Died 9 Apr. 1925
Robinson, Lucy P. 1879 Essex
Robinson, Mary K. 1874, 1890 Beverly, Swampscott
Rodman, Mary C. 1886-1889 Died 1919
Rogers, Miss A. F. 1898 Gloucester
Rogers, Edward S. 1876 Salem
Root, Oscar M. 1942-1958 Died Jan. 11, 1969
Ropes, Nathaniel 1875 Manchester, North Andover
Ropes, Willis H. 1919, 1925 Salem
Rossbach, George Bowyer 1928 Nahant
Rossi, Mary n.d. Topsfield
Rouleau, Ernest 1945 Rockport
Russell, Abigail 1945 North Andover
Russell, Rev. John Lewis 1826-1868 Died 7 June 1873 Born 2 Dec. 1808
Russell, O. 1859 Salem
Russell, Robert 1957 North Andover
Russell, Mrs. Robert Shaw (Margaret P.) 1917 Lynnfield

Safford, Wm. O. Salem (1897)
St. John, Harold 1911-1926
Sanford, Samuel Newton Folius 1924-1955 Died 17 Sept. 1958
Sargent, Charles Sprague 1900 Ipswich Died 23 March 1927
Sargent, G. K. III 1958 North Andover
Sargent, Miss H. C. 1885 North Andover
Saunders, Miss Mary T. 1868-1879
Schiff, David 1951 North Andover
Schneider, Mrs. E. S. 1907 Lawrence
Schubert, Bernice G. 1938, 1941 Newbury, West Newbury
Schweinfurth, Charles 1950, 1957
Sears, John Henry 1865-1905 Died 26 Feb. 1910
Sears, Nettie 1876, 1877 Essex County
Seymour, Arthur Bliss 1908 Hamilton
Seymour, Frank Conkling 1933-1938
Shafer, P.A. 1931 Rockport
Shattuck, Otis E. 1958, 1962 Boxford
Shaw 1945 Andover
Shaw, L. A. 1910 Beverly
Sheffield, Mrs. George 1926 Gloucester
Sheppard, S. A. D. 1863 Amesbury
Silsbee, William H. 1869 Beverly
Smart, Mr. 1880 Saugus
Smith, Alonzo 1884 Salem
Smith, Francis E., Jr. 1923-1938 Lynnfield, Rockport
Smith, Lyman Bradford 1918-1945
Smith, Ruth C. (Ruth C. Gates) 1928-1945 Rockport
Snyder, Dorothy E. 1950, 1951, 1971-3
Sornborger, J. D. 1918 Rowley
Spencer, Mr. 1882 Salem
Spofford, Carrie n.d. Boxford
Spofford, M. n.d. Topsfield
Sprague, C. J. 1880
Stevens, Francis J. 1884 Boxford
Stevens, H. C. 1927 Danvers
Steyermark, Julian Alfred 1931 Ipswich
Still, Mrs. n.d. Topsfield
Stockman, Charles C. 1953, 1954
Stone, Arthur R. 1884 Wenham
Stone, Frank 1879 Beverly, Salem Died 26 Aug. 1891
Stone, I. B. 1900, 1902 Manchester
Streeter, Gilbert L. 1850-1859
Stubbs, A. P. 1926 Saugus
Sutton, William 1874, 1876 Andover Died 18 Apr. 1882
Svenson, Henry Knute 1917-1938
Swan, Charles Walter 1880 1893 Died 1 Dec. 1921
Sweeney, Charles n.d. Topsfield
Sweeney, Lois n.d. Topsfield

Taylor, C. M. 1897 Gloucester
Taylor, Horace 1944 Gloucester
Taylor, Michael W. 1957 Boxford
Tenney, R. n.d. Georgetown (CNSH)
Thayer, Edward S. n.d. Topsfield
Thayer, J. n.d. Topsfield
Thomas, J. B. C. 1950 Andover
Thomas, Mrs. J. B. C. 1950 Andover
Thomas, J. F. 1952 North Andover
Thompson, David A., II. 1957 North Andover
Thompson, E. W. 1924 Andover
Thompson, Lovell 1917 Marblehead
Thoreau, Henry David 1858 Died 1862
Thurlow, Thomas C. 1882 Boxford
Tilton, Rev. George H. Essex County Died 19 July 1926
Todd, Miss M. n.d. Topsfield
Torrey, Prof. C. C. 1898 Andover

Torrey, George Safford 1914 Lynn
Torrey, R. n.d. Georgetown
Towne, Annie F. 1901 Topsfield
Towne, Benjamin B. 1901, 1904 Topsfield
Towne, George W. n.d. Middleton, Topsfield
Towne, Mrs. George W. n.d. Topsfield
Towne, Hiram N. 1885 Boxford
Towne, Harriet Rose 1912, 1918 Danvers, Topsfield
Towne, J. H. n.d. Topsfield
Towne, L. A. n.d. Topsfield
Towne, Mrs. Laura J. n.d. Topsfield
Towne, M. I. n.d. Topsfield
Towne, Serene Josephine n.d. Topsfield
Towne, V. M. n.d. Topsfield
Towne, Mrs. W. n.d. Topsfield
Townsend, Charles W. 1904-1922 Ipswich
Tracy, Cyrus M. 1844-1891 Died 28 Sept. 1891
True, Eben 1885, 1897 Salisbury
True, R. H. 1932 Essex
Tucker, Jonathan n.d. Salem
Tuckerman, Edward n.d. Manchester Died 1886
Twisden, Elizabeth J. 1885 Boxford
Twitchell, Miss 1903 Andover
Twitchell, Lawson B. 1883 Boxford

U., Mrs. G. P. n.d. Haverhill (CNSH)
U., S. P. n.d. Boxford, Haverhill
Upton, Miss L. H. 1875 Essex

Vanderbilt, W. H. n.d. Salem

Wade, Francis C. 1955-1964
Wadleigh, Samuel H. 1866, 1884 Georgetown, Topsfield
Walcott, Charles F. 1917 Marblehead
Walton, Mason A. 1891, 1901 Gloucester
Ward, Mrs. Martha E. 1915, 1918 Lynn, Swampscott
Ware, B. P. n.d. Marblehead
Ware, Robert Allison 1905-1911 Ipswich, Topsfield Died 16 Nov. 1946
Waters, David 1880-1884 Essex, Salem
Watson, Miss C. A. 1875 North Andover
Watson, Sereno 1877 Lawrence
Weatherby, Charles Alfred 1908-1938 Died 21 June 1949
Weatherby, Una F. 1930 Gloucester
Weinland, Dr. D. F. 1857 Manchester
Weir, Miss 1875 Rockport
Wells, T. W. n.d.
Wendte, L. A. 1898 Newburyport
Wentworth, L. A. 1892-1919
Wetherall, Cyril 1950, 1954
Wheat, Thomas H. 1968-1971 Gloucester, Rockport
Wheatland, Dr. Henry 1828-1878 Died 27 Feb. 1893
Wheeler, Mrs. Alonzo 1876, 1880 Gloucester, Rockport
Wheeler, Leston Ansil 1922 Methuen
Whinton, Herman F., Jr. 1958 Boxford
Whipple, A. J. 1883 Salem
Whipple, Mrs. C. E. n.d. Danvers, Topsfield
Whipple, G. n.d. Topsfield
Whipple, William 1877 Salem
White, Donald 1913
White, Edson Walter 1913 Salisbury
Whitman, Bailey W. 1900 Gloucester
Whittier, John Greenleaf n.d. Amesbury, Salisbury
Wiegand, Karl McKay 1908, 1911 Ipswich Died 12 March 1942
Wies, D. W. 1946 Marblehead
Willard, Mr. 1850 North Andover
Williams, Emile Francis 1892-1906 Died 19 Dec. 1929
Williams, Mrs. F. R. 1937 Peabody
Williams, M. E. n.d. Topsfield
Williams, Mary n.d. Topsfield
Willson, Miss Alice 1890 Boxford
Wilmarth, Mrs. H. D. 1877 Boxford
Wilson, Mrs. Joseph T. (Sophila A.) 1896, 1898 Nahant Died 29 Dec. 1925
Wilson, Hugh 1875 Salem
Winslow, Evelyn James 1910 Died 9 Feb. 1949
Witherby, J. B. 1941 Rockport
Woodbury, Barbara n.d. Topsfield
Woodbury, Edward B. n.d. Topsfield
Woodbury, Ruth 1966 Peabody
Wright, Charles 1882 Danvers

Young, Herbert A. 1878-1882

Zink, George Enos 1943-1951

Stuart K. Harris

SPECIES EXCLUSAE

N.B.: "P.A.S." referred to by Robinson is now the Peabody Museum of Salem.

Asplenium filix-foemina Bernh., var. *rhaeticum* Moore
"Banks and roadsides. Common" Robinson

Camptosorus rhizophyllus (L.) Link.
Georgetown C.N.S. Horner, Middleton

Polypodium virginianum L., f. *cambricoides* F.W. Gray
"A specimen corresponding to this variety was found some years since, at Andover, by Mr. Jackson T. Dawson." Robinson

Ginkgo biloba L.
Salem, A. F. Hall (1925, 1927). A.P. Morse (1927). From cultivated trees.

Taxus baccata L.
Salem, Sears (1885) introduced.

Abies balsamea (L.) Mill.
Andover woods, Robinson (1884); Danvers, Sears (1885) cultivated.

Abies fraseri Pursh
Boxford, H. 8357 (1953) cultivated.

Picea alba L.
Wenham, Robinson (1884) cultivated.

Pseudotsuga taxifolia Britt.
Ipswich, H. 13447 (1957) cultivated.

Pinus nigra Arnold
Salem, Little Misery Island, H. 13284 (1957) probably planted.

Pinus banksiana Lamb.
Haverhill, H. 7688 (1951); Newburyport, H. 9481 (1954); Marblehead, H. 12604 (1956). All probably planted.

Taxodium distichum (L.) Richard
Salem, Robinson (1877) introduced.

Echinodorus tenellus (Mart.) Buchenau
Both Raynal Dodge and Sears made marginal notes in their copies of Robinson's Flora that they found this in the Downfall area of Byfield, but no specimens seen.

Eragrostis trichodes (Nutt.) Wood (E. tenuis Gray)
" "In the same situations with the preceeding *(E. capillaris)*, in the environs of Salem, and as far as Philadelphia" (Dr. Chas. Pickering, Chron. Hist. Pl. p. 810). Probably noticed at the same date (about 1824). There is no other record of this species in Essex County." Robinson.

Bouteloa gracilis (HBK.) Lag. *(B. oligostachya* Torr.)
"Near old carpet factory, Tapleyville 1880 (J.H.S. & J.R.) Probably introduced from the west with wool." Robinson.

Zizania aquatica L., var. *angustifolia* Hitch.
"The report of this plant from Andover, Mass. (Fassett 1927, p. 228), may be incorrect, for the herbarium sheet, from the herbarium of G.U. Hay of St. John, New Brunswick, does not bear the name Massachusetts. There is an Andover, N. B., on the upper St. John River, near the Maine border." Fassett p. 88.

Paspalum setaceum Michx.
"Fields. Common." Robinson

Echinochloa walteri (Pursh) Nash (*Panicum crus-galli* L., var. *hispidum* (Gray's Manual))
"Similar situations *(Panicum crus-galli)*. Common. (Nat. from Eur.)" Robinson.

Setaria verticillata (L.) Beauv.
"A street grass in Salem, etc. (Adv. from Eu.)" Robinson

Scirpus lineatus Michx.
"Essex County, about 1824 (memo. Dr. Chas. Pickering). Not in the county herbarium P.A.S." Robinson

Scirpus rubricosus Fern. (*S. eriophorum* Michx.) =*S. cyperinus*?
"Meadows. Very common and variable. Flowering from June to September." Robinson

Carex rosea Schkuhr.
"Meadows, etc. Common." Robinson

Carex adjusta Boott =*C. straminea* or *C. leporina* or *C. argyrantha*
"West Newbury (W.P. Conant)." Robinson

Carex salina Wahlenb.
"Near Newburyport (memo. Prof. Brainard). Both this species and the last *(C. maritima)* are found in many places near the coast of Massachusetts." Robinson

Carex virescens Muhl.
"Copses, etc. Frequent." Robinson

Carex albursina Sheldon (*C. laxiflora* Lam., var *latifolia* Boott)
"Chebacco woods, etc. Frequent." Robinson

Carex laxiflora Lam., var. *intermedia* Boott
"Salem, West Newbury (W.P. Conant)" Robinson

Carex styloflexa Buckl. (*C. laxiflora* Lam., var. *styloflexa* Boott)
"Groveland, etc. (W.P. Conant)" Robinson

Carex pseudo-cyperus L.
"Middleton, Aug. 1880 (J.R.). Not previously noticed." Robinson

Carex oligosperma Michx.
"Essex county (Oakes) in herb. P.A.S. This specimen may have been collected outside the limits of the county." Robinson

Lilium superbum L.
" "Danvers" (J.H. Sears). The only locality yet reported for the County." Robinson

Polygonatum canaliculatum (Muhl.) Pursh (*P. giganteum* Dietrick)
" 'Ship Rock' G.D. Phippen (Proc. E. I. Vol. II, p. 204); 'Haverhill' Mrs. S. M. Downs. Not represented in the collection of the P.A.S., and possibly only large forms of the last (*P. biflorum* Ell.)." Robinson

Dioscorea villosa L.
Topsfield, H. 9592 (1954); planted

Habenaria orbiculata (Pursh) Torr.
" 'In Danvers, Dr. Nichols' (Bigelow's Fl. Bost. 2d. ed. 1824), 'In the deep woods at Gloucester' (Mrs. Bray and Mrs. Grover) where the only specimen found was a very fine one." Robinson
Bray specimen is *H. macrophylla*

Habenaria ciliaris (L.) R.Br.
" 'Lynnfield, 1856' (Proc. E. I. Vol. II, p. 47). This proves to be a mistake, another plant being intended, as probably is also the *'Orchis ciliaris'* of Dr. Osgood's list." Robinson

Isotria verticillata (Willd.) Raf.
"North Andover (Prof. Goldsmith), Haverhill (E.H. Hitchings). More frequent south of Boston." Robinson. *Andover Townsman* 20 #37 14 June 1907 reports the findings of the species in Andover by Frederic S. Boutwell on 9 June. See p. 107, added later by S.K.H.

Corallorhiza odontorhiza (Willd.) Nutt.
" 'Rare. Rather plentiful on the east side of Edward's Swamp' (Tracy). Not represented in the P.A.S. County herbarium." Robinson

Aplectrum hyemale (Muhl.) Torr.
"Reported by Mr. G.D. Phippen as collected at a field meeting of the Essex Institute, held at West Lynn, Aug. 26, 1857 (Proc. E. I. Vol. II. p. 225); also stated by Mrs. Horner to have been found at Georgetown some years since. Not represented in the P.A.S. county herbarium." Robinson

Salix viminalis L.

"Danvers, 1853 (Dr. Osgood's list). Reported in Salem and one or two other places, but it can hardly be called a thoroughly naturalized plant. Bigelow (Fl. Bost. 2d. ed. 1824) speaks of this species as growing "In swamps at Danvers and elsewhere." As he does not mention *Salix sericea,* it may be this species that is referred to, as it is frequent in swamps in the vincity of Danvers, the leaves somewhat resembling those of *S. viminalis.* (Adv. from Eu.)" Robinson

Juglans regia L.

"Cultivated in Salem, to a small extent, where the fruit perfects well. The *J. regia* noticed in Saugus (Proc. E. I., Vol. II, p. 389), proves to be *J. nigra.*" Robinson

Betula pumila L.

" 'Shore of North Andover pond' and 'on a railroad track in Groveland' (S.P. Fowler, Proc. E. I. Vol. II, pp. 402 and 409), Rockport (Pool in "Pigeon Cove and vicinity"). Specimens brought to the P.A.S., in 1879, from the shore of Pentucket Pond, Georgetown, as this species, are but small specimens of *B. nigra.* Not represented in the county herbarium. Very doubtful as a native of this region." Robinson

Polygonum glaucum Nutt. *(P. maritimum* L.)

"Beverly, Lynn, Ipswich, etc." Robinson

Claytonia virginica L.

"This plant is not represented in the county collection at the Peabody Academy of Science; and, although found abundantly north and west of us, is as yet only reported at 'Haverhill, North Parish,' (Mrs. Downs); and in 'Lynnfield by Mr. Chute' (memo. G.D. Phippen)."

Claytonia caroliniana Mx. is the species evidently intended by those who collected the specimens, as that is the one found in this region. Both are, however, included in one species by some authors." Robinson

Spergularia media (L.) Presl.

"All salt marshes. It is quite difficult to separate these plants in a satisfactory manner; but besides the two species already noticed, another, or a variety, is common; possibly," Robinson

Cerastium nutans Raf.

"Cerastium nutans Raf., which finds a place in some local printed lists, certainly must be an error." Robinson

Silene caroliniana Walt., var. *pensylvanica* (Michx.) Fern. (*S. pensylvanica* Michx.)

"Reported at Danvers, Andover, and occasionally in other parts of the county." Robinson

Cimicifuga racemosa (L.) Nutt.

"Collected along the embankment of the railroad near West Gloucester some years since (memo. Mr. G.D. Phippen). It is probable that the plants were introduced although they were at the time quite abundant." Robinson

Podophyllum peltatum L.

"Extensively introduced in old gardens, and said by some to be a native of the county, but this is doubtful." Robinson

Argemone mexicana L.

"On the road to Flax Pond, Lynn, 1879 (Herbert A. Young). Introduced from tropical America. Not permanently established." Robinson

Alyssum montanum L.

"Danvers, 1880 (J.H. Sears). (Nat. from Eu.) Not previously reported." Robinson
Specimen is *A. alyssoides* L. S.K.H.

Sisymbrium arvense

"Found by Rev. J.L. Russell, with other weeds, persisting for many years, in Salem" Robinson
Specimen is only New England collection of *Rorippa sessiliflora* (Nutt.) Hitch. S.K.H.

Cardamine hirsuta L.
"Brooks and other wet places. Common. Sold in the market as Water Cress and often supposed to be *N. officinalis."* Robinson Specimens are *C. pensylvanica* Muhl. S.K.H.
Arabis laevigata (Muhl.) Poir.
"Beverly (John C. Phillips)" Robinson Specimen is *A. drummondi* Gray. S.K.H.
Sedum telephium L.
"In Wenham and Danvers, this has become one of the worst weeds in grass land, and as every bit left to itself will soon take root, it is very difficult to exterminate." Robinson. Specimens are all *S. purpurem* (L.) Link S.K.H.
Crataegus oxyacantha L.
"Near Pranker's mills, Saugus' (Tracy); Derby estate, Salem. Cultivated and established in some places. (Adv. from Eu.)" Robinson
Crataegus tomentosa L. (Black Thorn)
"In the region of the last *(C. coccinea).* The form found in the county seems to be var. *punctata* the leaves, however, vary considerably." Robinson
Prunus americana Marsh.
"Represented in the P.A.S. county herbarium by a specimen collected by the late Mr. Oakes, the precise locality being unknown." Robinson.
Oakes specimen is *P. nigra* Ait. S.K.H.
Prunus pumila L.
" 'Outlet of mineral spring pond' (memo. Dr. Chas. Pickering); 'Andover, June 1873' (memo. Rev. H.P. Nichols). Not represented in the county collection at the P.A.S." Robinson
Probably *P. susquehanae* Willd. S.K.H.
Lespedeza virginica (L.) Britt. (*L. reticulata* Pers.)
"Georgetown (Mrs. Horner); Danvers (J. H. Sears). Not rare." Robinson
Probably *L. intermedia* (S. Wats.) Britt. S.K.H.
Acer nigrum Michx. f. (*A. saccharinum* Wang., var. *nigrum*)
"is now and then met with under cultivation." Robinson
Ampelopsis brevipedunculata (Maxim,) Trautv. (*Vitis heterophylla* Thunb.)
"Found escaped at Nahant, 1878, by Miss Mary T. Saunders" Robinson
Althaea officinalis L.
" 'Marsh, Salisbury, Sept.' (Mrs. Downs). Not in collection at the Peabody Academy of Science." Robinson
Hibiscus syriacus L.
"Much cultivated and escaping by seeds." Robinson
Opuntia hemifusa Raf. (*O. vulgaris* Mill.)
"Near 'Kernwood' in Salem, some thirty years ago (Hugh Wilson)." Robinson
Eleagnus angustifolia L.
Boxford, H. 18061 (1958); cultivated.
Oenothera fruticosa L.
"Beverly Farms (John C. Phillips). Perhaps introduced." Robinson
Hedera helix L.
"The English Ivy is not sufficiently hardy to become spontaneous, although frequently planted out as a climber. Only noticed in flower when unprotected in Salem (Dr. Fiske), Sept., 1880." Robinson
Rhododendron nudiflorum (L.) Torr.
" 'Banks of the Merrimac; West Parish, Andover' (Mrs. S.M. Downs). As this species is abundant in New Hampshire, it ought to be quite frequently met with north of the Merrimac." Robinson.
Convolvulus spithamaeus L.
"Reported from two localities but not represented in the P.A.S. collection. The identification is a little uncertain." Robinson

Stachys riddellii House (*S. cordata* Riddell)
"Specimens collected at Danvers, by Mr. Sears, were referred here by Prof. Watson." Robinson
Specimen is apparently *S. palustris* L., var. *homotricha* Fern. S.K.H.

Stachys hyssopifolia Michx.
" 'Appeared mysteriously on the wall at Paradise, Salem, 1824' (Dr. Chas. Pickering)." Robinson

Petunia axillaris BSP. (*P. nyctaginiflora* Juss.)
"On a waste heap, Nahant, Aug., 1878; Wenham, by the roadside, etc. Not permanently established, although very common in cultivation." Robinson
Nahant specimen seems to be *P. violacea*. S.K.H.

Scrophularia nodosa L.
" 'Boston street, between Washington street and the paper factory, Lynn' (Tracy); Beverly Farms (G.D. Phippen)." Robinson
Probably *S. lanceolata* Pursh S.K.H.

Mimulus moschatus Dougl.
Boxford, Elizabeth J. Twisden (1885) "cultivated".

Viburnum nudum L.
"Danvers (S.P. Fowler); Essex, Haverhill, Amesbury, etc. Not so common as the last *(V. lentago)*." Robinson
Is *V. cassinoides* L. S.K.H.

Dipsacus sylvestris Huds.
"Danvers, 1853 (Dr. Osgood's list) and noticed later by Mr. Sears, Rockport (Mrs. Wheeler)." Robinson

Lobelia siphilitica L.
"Reported at Andover, but as it is not mentioned otherwise, this may have been a mistake. It is probable that this species is known here only in cultivation." Robinson

Solidago stricta Ait.
"Wenham swamp, 1824 (Dr. Chas. Pickering); 'Bowler swamp and similar localities in Lynn' (Tracy). Not in the P.A.S. county herbarium. It is very possible there is some mistake and that this species does not grow here, as no specimens have since been found and forms of *S. neglecta* are sometimes mistaken for it." Robinson

Aster infirmus Michx.
" 'Essex County' (Dr. Chas. Pickering). Robinson

Stuart K. Harris

STATISTICAL SUMMARY

	Genera		Species		Varieties		Forms and Hybrids	
DIVISION PTERIDOPHYTA	Nat.	Int.	Nat.	Int.	Nat.	Int.	Nat.	Int.
Fam. Equisetaceae	1		6		1		9	
Lycopodiaceae	1		7		4			
Selaginellaceae	1		2					
Isoëtaceae	1		2					
Ophioglossaceae	2		7				3	
Osmundaceae	1		3				2	
Schizaeaceae	1		1					
Polypodiaceae	14		27		5		11	
Marsileaceae	1		1					
DIVISION SPERMATOPHYTA								
SUBDIVISION GYMNOSPERMEAE								
Fam. Taxaceae	1		1					
Pinaceae	7		10	3	1			
SUBDIVISION ANGIOSPERMAE								
CLASS MONOCOTYLEDONEAE								
Fam. Typhaceae	1		2				1	
Sparganiaceae	1		8					
Zosteraceae	4		22		7		1	
Najadaceae	1		2					
Juncaginaceae	2		2					
Alismataceae	3		8		1		3	
Hydrocharitaceae	2		3					
Gramineae	40	18	115	54	27	4	16	
Cyperaceae	11		145	4	15		9	
Araceae	5		6				5	
Lemnaceae	3		4					
Xyridaceae	1		3					
Eriocaulaceae	1		2					
Commelinaceae		2		2				
Pontederiaceae	2		2				3	
Juncaceae	2		18		8		3	
Liliaceae	13	5	21	11	1		[1]+2	
Amaryllidaceae	1	1	1	1				
Iridaceae	2		6	2				
Orchidaceae	12		29		3		1	

CLASS DICOTYLEDONEAE SUBCLASS ARCHICHLAMYDEAE	Genera		Species		Varieties		Forms and Hybirds	
	Nat.	**Int.**	**Nat.**	**Int.**	**Nat.**	**Int.**	**Nat.**	**Int.**
Fam. Salicaceae	2		17	6	2	1	2	
Myricaceae	2		3		1			
Juglandaceae	2		6		1			
Corylaceae	5		11		3		4	
Fagaceae	3		10	3		1		
Ulmaceae	2		3				4	
Moraceae		3		4				
Cannabinaceae		2		3				
Urticaceae	5		5	2	1			
Santalaceae	1		1					
Loranthaceae	1		1					
Aristolochiaceae	1		1					
Polygonaceae	3	1	23	17	5	2	3	
Chenopodiaceae	5	4	15	12	3	2		
Amaranthaceae	2		5	4				
Nyctaginaceae	1			2				
Phytolaccaceae	1		1					
Aizoaceae		2		2				
Portulacaceae		2		3				
Caryophyllaceae	7	8	14	25	1		1	
Ceratophyllaceae	1		2					
Nymphaeaceae	3	2	3	2	1		1	
Ranunculaceae	10	3	28	8	3	4	4	
Berberidaceae	2		1	2				
Magnoliaceae	1	1	1	2				
Lauraceae	2		2					
Papaveraceae	3	4	4	5				
Capparidaceae	1		1					
Cruciferae	7	20	13	39	2	2		2
Resedaceae		1		2				
Sarraceniaceae	1		1					
Droseraceae	1		2					
Crassulaceae	1	2	1	6				
Saxifragaceae	7	2	10	5			1	
Hamamelidaceae	1		1			1		
Platanaceae	1		1					
Rosaceae	14	5	63	27	7	1	3	

	Genera		Species		Varieties		Forms and Hybirds	
	Nat.	Int.	Nat.	Int.	Nat.	Int.	Nat.	Int.
Leguminosae	12	13	26	39	1 +4	4		2
Linaceae	1		2	1				
Oxalidaceae	1		3	2				3
Tropaeolaceae		1		1				
Geraniaceae	1	1	3	[1]+5				
Rutaceae		4		4				
Simaroubaceae		1		1				
Polygalaceae	1		5		2			
Euphorbiaceae	2	1	6	5				
Callitrichaceae	1		3	1				
Anacardiaceae	1	1	7	1	2			
Aquifoliaceae	2		5		1		1	
Celastraceae	1	1	1	2				
Aceraceae	1		5	4	1			
Hippocastanaceae		1		1				
Balsaminaceae	1		1	1				
Rhamnaceae	1	1	2	2				
Vitaceae	2		5	1			1	
Tiliaceae	1		2	1				
Malvaceae	1	4	1	9				1
Guttiferae	1		9	2	1		1	
Elatinaceae	1		2					
Cistaceae	3		9		1			
Violaceae	1		17	2			[1] +4	
Thymelaeaceae	1	1	1	1				
Lythraceae	3	1	3	2	1	1		
Nyssaceae	1		1					
Melastomataceae	1		1					
Onagraceae	4	1	14	4			1	
Haloragaceae	2		6				2	
Araliaceae	2		4					
Umbelliferae	12	11	16	11			1	
Cornaceae	1		8					
SUBCLASS METACHLAMYDEAE								
Fam. Clethraceae	1		1					
Pyrolaceae	4		9		1			
Ericaceae	12	1	24	1	2		4	

	Genera		Species		Varieties		Forms and Hybirds	
	Nat.	Int.	Nat.	Int.	Nat.	Int.	Nat.	Int.
Diapensiaceae		1		1				
Primulaceae	5	2	10	5	1			
Plumbaginaceae	1		1					
Oleaceae	1	2	3	3	2			
Gentianaceae	4	1	6	2			2	
Apocynaceae	1	1	3	1				
Asclepiadaceae	1	1	8	1				
Convolvulaceae	2	1	3	5		2	1	
Polemoniaceae		1		1				
Boraginaceae	2	8	3	13				
Verbenaceae	1		2	3				
Labiatae	10	18	20	30	4	3	3	1
Solanaceae	2	7	3	22		1		
Scrophulariaceae	14	6	26	23	4		5	2
Bignoniaceae		2		2				
Martyniaceae		1		2				
Orobanchaceae	2		2					
Lentibulariaceae	1		8					
Phrymaceae	1		1					
Plantaginaceae	1		3	4	1	1		1
Rubiaceae	4	1	15	1			2	
Caprifoliaceae	6	1	13	6				
Valerianaceae		1		1				
Dipsacaceae	1	2	2	2				
Cucurbitaceae	3	2	2	4				
Campanulaceae	3	1	8	2				
Compositae	28	30	127	82	24	6	19	10
Totals	Native			Adventive				
				GRAND TOTAL TAXA				1767

SUMMARY BY MINOR GROUPS

	Native	Introduced	Total
Families	116	15	131
Genera	411	222	633
Species	1189	578	1767
Additional Varieties	156	36	192
Additional forms and hybrids	141	28	169

REFERENCES

Alcott, William P.	1895	Botanical Notes. *Bull. Essex Inst.* 27:92-97. footnote p. 91.
Atwood, Luther	1898	Plants and Trees in Lynn Woods. 10th *Ann. Park Comm. Report* 25-43.
Bailey, W. Whitman	1900	The *Figassya* Hardy Plant in New England. *Rhodora* 2:234.
Baldwin, Henry	1884	*The Orchids of New England.* John Wiley & Sons, N.Y. pp. 158.
Bartlett, Harley Harris	1906	The Salt-marsh *Iva* of New England. *Rhodora* 8:25, 26.
	1907	The Retrograde Color Varieties of *Gratiola aurea. Rhodora* 9:122-124.
Bartlett, Mira W.	1924	Ferns (A List of the Native Ferns Found Growing in Haverhill) Bull. Haverhill Public Library 8:17-19.
	1925	A Partial List of the Wild Flowers Found Growing in the Thirty-two Square Miles Comprised in the City of Haverhill. Bull. Haverhill Public Library 8:61-76.
Bartram, Edwin B.	1909	A *Teucrium* New to Massachusetts. *Rhodora* 11:148.
Bean, Ralph C.	1953	Two Unusual Plants in Essex County, Massachusetts. *Rhodora* 55:348.
Bigelow, Jacob	1814	*Florula Bostoniensis* Cummings & Hilliard pp. 268.
	1824	ed. 2. Cummings & Hilliard pp. 423.
	1840	ed. 3. Charles C. Little & James Brown pp. 468.
Birrell, William	1945	*Thermopsis mollis* in Eastern Massachusetts. *Rhodora* 47:91.
Blake, S. F.	1912	The Forms of *Peltandra virginica. Rhodora* 14:102-106.
	1929	A New Variety of *Bidens Eatoni. Rhodora* 31:100.
Boivin, Bernard	1944	American *Thalictra* and Their Old World Allies. *Rhodora* 46; 337-377, 391-445, 453-487.
Butters, Frederick K.	1917	Taxonomic and Geographic Studies in North American Ferns. *Rhodora* 19:169-216.
Buttrick, Samuel Bartlett	1862	Calendar of Spring for 1855 and 1856. Proc. Essex Inst. 2:172, 173.
	1862	List of Plants found in Salem & Within seven miles of Salem, 1857. Proc. Essex Inst. 2:233-242.
	1873	List of Plants found in Chebacco region, 12 Aug. 1873. Proc. Essex Inst. 5:139, 140.
	1873	List of Plants found in Danvers, 5 Sept. 1873. Proc. Essex Inst. 5:161.
Chamberlain, E. B.	1903	An Interesting Form of *Leersia oryzoides. Rhodora* 5:118.
Chase, Agnes	1904	The North American Allies of *Scirpus lacustris. Rhodora* 6:65-71.
Churchill, J. R.	1907	Preliminary Lists of New England Plants. XVII. *Rhodora* 7:33-38.

	1856	*Collectanea Botanica,* or Notices of Rarer Plants found in Essex County, Massachusetts. Proc. Essex Inst. 1:270-273.
Collins, John A., Jr.	1901	Two Wool-waste Plants at Lawrence, Massachusetts. *Rhodora* 3:92.
Colman, Miss Harriet W.	1902	The Trees of Newburyport. City Improvement Society pp. 32.
Cutler, Manasseh	1785	An account of some of the vegetable Productions, naturally growing in this Part of America, botanically arranged. Mem. Amer. Acad. 1:396-493.
Cutler, William P. and J. P. Cutler	1888	Life, Journals & Correspondence of Rev. Manasseh Cutler, L. L. D. Cincinnati. Robert Clarke & Co. 2 vol.
Davenport, George Edward	1881	The Flora of Essex County, Massachusetts. John Robinson (review) Bot. Gaz. 6:187, 188.
	1894	Two new ferns from New England. Bot. Gaz. 19: 192-497.
	1896	New or Little-known Plants. Garden & Forest 9:444-446.
Day, Mary A.	1899	The Local Floras of New England. *Rhodora* 1:174-176.
	1900	The Local Floras of New England (Addenda). *Rhodora* 2:74.
Deane, Walter	1901	Notes on the *Ericaceae* of New England. *Rhodora* 3:193-198.
	1901	Albino Fruit of *Vacciniums* in New England. *Rhodora* 3:263-266.
	1901	Notes on the *Umbelliferae* of New England. *Rhodora* 3:209-213.
	1904	Preliminary Lists of New England Plants - XVII *Rhodora* 6:151-161.
	1915	A Correction concerning *Sagittaria teres* Watson and *Potamogeton lucens* L. *Rhodora* 17:142.
Dodge, Raynal	1896	Ferns and Fern Allies of New England. Binghamton, N.Y. Willard N. Clute & Co. pp. 52.
Drew, W. B.	1936	The North American Representatives of *Rununculus,* section *Batrachium. Rhodora* 38:1-47.
Eaton, A. A.	1897	*Lycopodium alopecuroides* in Massachusetts. Fern. Bull. 5:3-5.
Eaton, Richard J.	1928	The Present Status of *Magnolia virginiana* in Massachusetts. *Rhodora* 30:207, 208.
	1931	Peculiar Aspects of the New England Distribution of *Arceuthobium pusillum. Rhodora* 33:92-101.
	1933	Another Massachusetts Station for *Calluna vulgaris. Rhodora* 35:294.
Emerton, James H.	1871	Plants Collected in South Salem & Swampscott, 1 May 1871. Bull. Essex Inst. 3:58, 59.
Emery, George E.	1871	Climbing Fern. Am. Nat. 5:115, 116.
Farlow, W. G.	1916	*Montia sibirica* in Massachusetts. *Rhodora* 18:240.
Fassett, N. C.	1922	*Lophotocarpus* on the North-Eastern River-Estuaries. *Rhodora* 24:71-73.

1924 A Study of the Genus *Zizania*. *Rhodora* 26:153-160.
1925 *Bidens hyperborea* and its Varieties. *Rhodora* 27:166-171.
1951 *Callitriche* in the New World. *Rhodora* 53:137-155, 161-194, 209-222.

Fernald, M. L.

1899 Some Undescribed and Little-known Varieties of *Aster* and *Solidago*. *Rhodora* 1:187-191.
1903 A New *Bidens* from the Merrimac Valley. *Rhodora* 5:90-92.
1904 Preliminary Lists of New England Plants XIII Juncaceae. *Rhodora* 6:34-41.
1906 *Paronychia argyrocoma* and Its New England Representative. *Rhodora* 8:101-104.
1909 *Salix pedicellaris* and Its Variations. *Rhodora* 11:157-162.
1910 Notes from the Phaenogamic Herbarium of the New England Botanical Club, - I. *Rhodora* 12: 185-192.
1911 The Variations of *Lathyrus palustris* in Eastern America. *Rhodora* 13:47-52.
1912 A Second Station for *Cyperus Grayii* in Essex County, Massachusetts. *Rhodora* 14:22.
1912 Two Rare *Junci* of Eastern Massachusetts. *Rhodora* 14:55, 56.
1913 The Indigenous Varieties of *Prunella vulgaris* of North America. *Rhodora* 15:179-186.
1915 The American Variations of *Lycopodium annotinum*. *Rhodora* 17:123-125.
1915 Michaux's *Panicum muricatum*. *Rhodora* 17:105-107.
1916 The Genus *Puccinellia* in Eastern North America. *Rhodora* 18:1-23.
1916 The Genus *Sabatia* in New England. *Rhodora* 18:145-152.
1917 The Genus *Elatine* in Eastern North America. *Rhodora* 19:10-15.
1917 New Species, Varieties and Forms of *Saxifraga*. *Rhodora* 19:141-144.
1917 A New *Vitis* from New England. *Rhodora* 19:144-147.
1918 The Specific Identity of *Bidens hyperborea* and *B. colpophila*. *Rhodora* 20:146-150.
1919 The Identity of *Angelica lucida*. *Rhodora* 21:144-147.
1919 *Panicum* section *Capillaria* in New England. *Rhodora* 21:110-114.
1919 Two New *Myriophyllums* and a Species New to the United States. *Rhodora* 21:120-124.
1922 The Northern Variety of *Asperella hystrix*. *Rhodora* 24:229-231.
1923 The Identities of the Sand Cherries of Eastern North America. *Rhodora* 25:69-74.
1925 The Maritime Plantains of North America. *Rhodora* 27:93-104.
1931 *Potentilla canadensis* and *P. simplex*. *Rhodora* 33:180-191.
1932 *Lathyrus japonicus* versus *L. maritima*. *Rhodora* 34:177-187.
1933 *Calamagrostis epigejos*, var. *georgica* in America. *Rhodora* 35:64, 65.

1933 Recent Discoveries in the Newfoundland Flora *Agropyron* section *Goularda* in Eastern North America. *Rhodora* 35:161-182.

1933 Recent Discoveries in the Newfoundland Flora *Taraxacum* in Eastern North America. *Rhodora* 35:369-386.

1938 New Species, Varieties and Transfers. *Rhodora* 40:331-358.

1941 Another Century of Additions to the Flora of Virginia. *Rhodora* 43:485-553, 559-630, 635-657.

1943 Notes on *Danthonia. Rhodora* 45:239-246

1943 Studies in North American Species of *Scirpus. Rhodora* 45:279-296.

1945 Botanical Specialties of the Seward Forest and Adjacent areas of Southeastern Virginia. *Rhodora* 47:91-142, 149-182, 191-204.

1945 Eastern North American Representatives of *Alnus incana. Rhodora* 47:333-361.

1945 Transfers In and Animadversions on *Artemisia. Rhodora* 47:247-

1946 Identifications & Reidentifications of North American Plants. *Rhodora* 48: 137-162, 184-197, 207-216.

1949 Studies of Eastern American Plants. Pt. II. # 1. Blackberries, old and new. *Rhodora* 51:43-51.

Fernald, M.L. & C. H. Bissell 1910 The North American Variations of *Lycopodium clavatum. Rhodora* 12:50-55.

Fernard, M.L. & A.E. Brackett 1929 The Representatives of *Eleocharis palustris* in North America. *Rhodora* 31:57-77.

Fernald, M.L. & C.A. Weatherby 1921 *Equisetum fluviatile* or *E. limosum? Rhodora* 23:43-47.

Fernald, M.L. & K.M. Wiegand 1910 The North American Variations of *Juncus effusus. Rhodora* 12:81-93.

1913 *Calamagrostis Pickeringii* Gray. var. *debilis* (Kearney) n. comb. *Rhodora* 15:135, 136.

Fogg, John M., Jr. 1928 The Clandestine Form of *Leersia oryzoides. Rhodora* 30:81-85.

Fuller, T. Otis 1890 *Magnolia glauca* in Massachusetts. *Gard. & Forest* 3:23.

Gaiser, L.O. 1946 The Genus *Liatris. Rhodora* 48: 165-183, 216-263, 273-326, 331-382, 393-412.

Gray, Asa 1868 "*Saxifraga virginiensis* was found full double at Danvers, Mass., three years ago." *Am. Nat.* 2:484.

1876 *Calluna vulgaris,* the Ling or Heather, rediscovered in Massachusetts. Am. Nat. 10:489, 490.

1876 *Sedum reflexum* L. *Am. Nat.* 10:553.

Grayce, Robert L. 1952 Cape Ann Forests: a Review. *Essex Inst. Historical Collections* 88:207-218.

Greene, Edward L. 1901 A New Northern *Eupatorium. Rhodora* 3:83, 84.

Greenman, J.M. 1901 The Genus *Senecio* in New England. *Rhodora* 3:3-7.

Gutterson, M.E. 1902 *Cuphea procumbens* at Andover, Massachusetts. *Rhodora* 4:247, 248.

Harris, Stuart K. 1933 Notes on the Flora of Haverhill, Massachusetts. *Rhodora* 35:261, 262.

1958 *Cabomba caroliniana* in Essex County, Massachusetts. *Rhodora* 60:116.

1958 *Nelumbo lutea* in Essex County, Massachusetts. *Rhodora* 60:116.

1958 A New Color Form of *Solidago sempervirens. Rhodora* 60:261, 262.

1958 *Rorippa sessiliflora* in Essex County, Massachusetts. *Rhodora* 60:260, 261.

1959 *Elsholtzia ciliata* in Essex County, Massachusetts. *Rhodora* 61:63.

1959 The Northernmost Station of *Magnolia virginiana,* Its History and Present Status. *Rhodora* 61:65-70.

1959 *Peucedanum palustre,* an Interesting Addition to the Flora of Essex County, Massachusetts. *Rhodora* 61:181.

1959 Two *Crucifers* New to Essex County, Massachusetts. *Rhodora* 61:187.

1959 Two Grasses New to Essex County, Massachusetts. *Rhodora* 61:247, 248.

Harris, S.K. & R.C. Bean 1951 *Wolffia columbiana* in Methuen, Massachusetts. *Rhodora* 53:272.

Haverhill Agassiz Association 1900 A list of the wild flowers and ferns found in Winnekeni Park. Annual Report of the Park Commissioners of the City of Haverhill, Mass. for the year ending Dec. 31, 1900.

Higginson, Rev. Francis 1792 New Englands Plantation. Mass. Hist. Coll. 1:121.

Hopkins, Milton 1937 *Arabis* in Eastern & Central North America. *Rhodora* 39: 63-98, 106-148, 155-186.

Horner, C.N.S. 1876 List of Plants of Georgetown and Its Vicinity. *Georgetown Advocate* 2: #20 Feb. 26, #22 March 11, #23 March 18, #25 April 1, #29 April 29, #32 May 20.

1885 Notes on the Flora of South Georgetown. Bull. Essex Inst. 5:107-110.

1887 Notes on some Introduced Plants in Eastern Massachusetts. Bull. Torrey Club 14:219.

Hubbard, F. Tracy 1912 Some *Panicums* of Essex County, Massachusetts. *Rhodora* 14:36-40.

1913 Further Notes on the *Panicums* of Essex County, Massachusetts. *Rhodora* 15:36-38.

Huntington, J.W. 1901 *Hottonia inflata:* in Amesbury Massachusetts. Does it occur farther North? *Rhodora* 3:216.

Jack, J.G. 1889 *Magnolia glauca* in its Most Northern Home. Gard. & Forest 2:363, 364.

Kennedy, G.G. 1916 Some Historical Data Regarding the Sweet Bay and Its Station on Cape Ann. *Rhodora* 18:205-212.

Knowlton, C.H.	1907	Newly Observed Stations for Massachusetts Plants. *Rhodora* 9:11-15.
	1914	New England Distribution of *Ilex opaca* and *Ilex glabra*. *Rhodora* 16:163-165.
	1921	Herbarium of Rev. W. P. Alcott, *Rhodora* 23:47.
	1929	Index to the Flora of the Boston District. *Rhodora* 31:41-43.
	1940	Two Introduced Plants. *Rhodora* 42:524, 525.
Lamson-Scribner, F. & E.D. Merrill	1901	The New England Species of the Genus *Panicum*. *Rhodora* 3:93-129.
Leavitt, R.G.	1905	On Translocation of Characters in Plants. *Rhodora* 7:13-19, 21-31.
Linder, D.H.	1922	Some Varieties of *Panicum virgatum*. *Rhodora* 24:11-16.
	1946	List of Pteridophytes and flowering plant life of the Brooks School region (North Andover, Andover & Boxford), Essex County. Brooks School 1946. pp. 14.
Macbride, J. Francis	1916	The Correct Name of an Introduced *Symphytum*. *Rhodora* 18:23-25.
Markoe, George F.H.	1864	List of Plants seen in Vicinity of W. Gloucester. *Proc. Essex Inst.* 3:24-27.
	1866	List of Plants Seen in Gloucester. *Proc. Essex Inst.* 4:1iii.
McGregor, Frank J.	1945	Rediscovery of *Paronychia argyrocoma* var. *albimontana* at Newburyport, Massachusetts. *Rhodora* 47:90, 91.
McVaugh, Rogers	1936	Studies in the Taxonomy & Distribution of the Eastern North American Species of *Lobelia*. *Rhodora* 38:241-263, 276-298, 305-329, 346-362.
Moore, Albert Hanford	1910	William Gleason Goldsmith. *Rhodora* 12:227, 228.
Moore, A.H. & A.S. Pease	1902	List of Plants introduced into Andover, Mass. in 1902. Privately printed.
Morse, A.P.	1918	*Amaranthus Powellii* and *Digitalis lanata* in New England. *Rhodora* 20:203.
	1924	*Lepidium latifolium* in New England. *Rhodora* 26:197, 198.
	1929	John Robinson, Botanist, of Salem, Massachusetts. *Rhodora* 31:245-254.
	1931	The American Lotus at West Peabody, Massachusetts. *Rhodora* 33:230.
Moulton, Edward	1902	Flora of This Vicinity (Newburyport). *Newburyport Daily Standard* 14 March - 7 November. 32 articles.
New England Botanical Club		Reports on the Flora of the Boston District. I. *Rhodora* 9:81-86, 1907; II. *Rhodora* 10:59-64, 1908; III. *Rhodora* 10:128-131, 1908; IV. *Rhodora* 11:75-80, 1909; V. *Rhodora* 11:204-209, 1909; VI. *Rhodora* 12:3-7, 1910; VII. *Rhodora* 12:95-99, 1910; VIII. *Rhodora* 13:27-29, 1911; IX. *Rhodora* 13: 72, 73, 1911; X. *Rhodora* 13:82-85, 1911; XI. *Rhodora* 13:104, 105, 1911; XII. *Rhodora* 13:232-235, 1911; XIII. *Rhodora* 13:248-251, 1911; XIV. *Rhodora* 14:76-78, 1912; XV. *Rhodora* 14:107-113, 1912; XVI. *Rhodora* 15:54-59,

		1913; XVII. *Rhodora* 15:122-132, 1913; XVIII. *Rhodora* 15:144-151, 1913; XIX. *Rhodora* 16:106-113, 1914; XX. *Rhodora* 17:169-180, 1915; XXI. *Rhodora* 17:202, 203, 1915; XXII. *Rhodora* 18:86-90, 1916; XXIII. *Rhodora* 18:165-168, 1916; XXIV. *Rhodora* 18:213-221, 1916; XXV. *Rhodora* 18:248-252, 1916; XXVI. *Rhodora* 20:15-18, 1918; XXVII. *Rhodora* 20:55-59, 1918; XXVIII. *Rhodora* 20:164-171, 1918; XXIX. *Rhodora* 20:208, 209, 1918; XXX. *Rhodora* 21:78-83, 1919; XXXI. *Rhodora* 21:125-128, 1919; XXXII. *Rhodora* 22:72-75, 1920; XXXIII. *Rhodora* 22:123-127, 1920; XXXIV. *Rhodora* 23:113-118, 1921; XXXV. *Rhodora* 24:92-95, 1922; XXXVI. *Rhodora* 24:152-156, 1922; XXXVII. *Rhodora* 24:231-236, 1922; XXXVIII. *Rhodora* 25:25-31, 1923; XXXIX. *Rhodora* 25:37-43, 1923; XL. *Rhodora* 25:60-67, 1923; XLI. *Rhodora* 25:185-187, 1923; XLII. *Rhodora* 26:13-19, 1924; XLIII. *Rhodora* 26:34-36, 1924; XLIV. *Rhodora* 26:55-60, 1924; XLV. *Rhodora* 26:82-88, 1924; XLVI. *Rhodora* 26:107-111, 1924; XLVII. *Rhodora* 26:130-132, 1924; XLVIII. *Rhodora* 26:149-152, 1924; XLIX. *Rhodora* 26:166-168, 1924.
		Reports on the Flora of Massachusetts. I. *Rhodora* 30:12-19, 1928; II. *Rhodora* 35:351-359, 1933; III. *Rhodora* 44:213-220, 1942 IV. *Rhodora* 49:257-277, 1947.
Oakes, William	1841	Notice of some rare Plants of New England, with descriptions of some new species. *Hovey's Mag.* 7:178-186.
	1841	On the advance of Spring in the eastern part of Massachusetts. *Hovey's Mag.* 7:201-208.
	1847	Notice of Some of the Plants of New England. *Hovey's Mag.* 13:217-220.
Ogden, E.C.	1943	The Broad-leaved Species of *Potamogeton* of North America North of Mexico. *Rhodora* 45:57-105, 119-163, 171-214.
O'Neill, Hugh	1940	*Cyperus polystachyos* var. *filicinus. Rhodora* 42:84-86.
Osgood, George M.D.	1853	List of Plants, with date of blooming, in vicinity of Salem. *Salem Gazette.*
Paine, Harriet E.	1883	Groveland Plants Not Reported by Mr. Robinson in County Flora. *Bull. Essex Inst.* 15:134.
	1883	Plants shown at the Meeting in Groveland, Mass., Aug. 1883. *Bull. Essex Inst.* 15:133.
Peabody, Elizabeth P.	1849	Vegetation about Salem, Mass. pp. 224-, 245 in *Aesthetic Papers.* Putnam. N.Y.
Pease, Arthur Stanley	1901	Some Wild Flowers of Andover. School Dept.
	1902	*Hieracium praealtum* at Andover, Massachusetts. *Rhodora* 4:195.
	1902	Two new stations for *Arceuthobium. Rhodora* 4:249.
	1903	*Erodium malacoides* at Lawrence, Massachusetts. *Rhodora* 5:39.

	1903	The Genus *Trisetum* in Andover, Massachusetts. *Rhodora* 5:289, 290.
	1914	A form of *Potentilla tridentata. Rhodora* 16:194, 195.
Pease, A.S. & A.H. Moore	1906	Peculiarities of *Botrychium lanceolatum* in America. *Rhodora* 8:229.
(Perley, Mary Ellen)	1876	A list of the plants of Boxford (East), 304 species being listed. *Georgetown Advocate* 2: #22 March 11, #24, March 25.
Phillips, J.C.	1911	Ten Years of observations on the migration of *Anatidae* at Wenham Lake, Massachusetts. *Auk.* 28:188-200.
Phippen, George D.	1870	"Dark Lane", with Allusions to other Localities of Wild Plants in Salem. *Bull. Essex Inst.* 2:97-104.
Pickering, Charles	1877	Chronological History of Plants. Boston. Little, Brown & Co. pp. 1222.
Pool, Calvin W.	1873	List of Trees, Shrubs & Plants of Cape Ann. In Pigeon Cove and Vicinity by H.C. Leonard. F.A. Searle. Boston.
Rehder, Alfred	1908	The New England Species of *Psedera. Rhodora* 10:24-29.
Rich, William P.	1900	The Heather in New England. *Rhodora* 2:54, 55.
Robinson, B.L.	1900	Variations of *Ilex verticillata. Rhodora* 2:104-106.
	1902	Further Stations for *Veronica chamaedrys. Rhodora* 4:152.
	1902	The New England *Polygonums* of the Section *Avicularia. Rhodora* 4:65-73.
Robinson, John	1872	List of plants found in bloom about Salem on 20 May 1872 & chart of dates of flowering 1856-1871. *Bull. Essex Inst.* 4:76-78.
	1873	List of ferns found in Danvers. *Bull. Essex Inst.* 5:155, 156.
	1875	Addenda to the Ferns of Essex County. *Bull. Essex Inst.* 7:147.
	1875	Ferns of Essex County, Massachusetts. *Bull. Essex Inst.* 7:44-54.
	1877	*Salix candida* in Essex County. *Am. Nat.* 11:432.
	1878	The Pine Its Life & Importance in Essex County. *Bull. Essex Inst.* 10.
	1879	Notes on the Native and extensively cultivated Woody Plants of Essex County, Massachusetts. *Bull. Essex Inst.* 11:72-106.
	1879	Notes on the Woody Plants of Essex County, Massachusetts. Essex Inst., Salem. pp. 38. From *Bull. Essex Inst.* 11.
	1880	Date of Flowering of Trees and Shrubs in Eastern Massachusetts, 1880. *Trans. Mass. Hort. Soc.* 161-173.
	1880	The Flora of Essex County, Massachusetts. Essex Institute, Salem. pp. 200.
	1880	Notes on the Flora of Essex County, Massachusetts, with sketches of the early Botanists, and a list of the Publications on these Subjects. *Bull. Essex Inst.* 12:81-99.
	1881	Date of Flowering of Trees and Shrubs in Eastern Massachusetts, 1881. *Trans. Mass. Hort. Soc.* 348-458.
	1881	Our Trees in Winter. *Bull. Essex Inst.* 13:38-51.

1884 The Progress of Botany in Essex County During the Last Half Century, Especially as Influenced by the Essex County Natural History Society and the Essex Institute 1834-1884. *Bull. Essex Inst.* 16:122-132.

1884 Native Woods of Essex County, Massachusetts. Reports of P.A.S. 103-137, 1884 (1885).

1888 *Aquilegia Canadensis* L., var. *flaviflora* (Tenney) Britton. *Bull. Torrey Club* 15:165, 166.

1891 Our trees. Essex Institute, Salem. pp. 120.

1902 Concerning the Plants Mentioned in Young's Chronicles. *Rhodora* 4:81, 82.

1926 Plants Growing the First Season in an Uncovered Cellar. *Rhodora* 28:69-74.

1881 Robinson's Flora of Essex County, Massachusetts (review). *Am. Nat.* 15:129.

Root, Oscar M. 1948 Trees at Brooks. Archbishop. June 11-17.

Rossbach, Ruth P. 1940 *Spergularia* in North and South America. *Rhodora* 42:57-83, 105-143, 158-193, 201-213.

Russell, John Lewis 1839 Notice of Rare Plants; with a Description of a Curious Variety of *Cladonia uncialis*. *Jour. Essex County Nat. Hist. Soc.* 1:#2. 96-100.

1841 Notice of some Plants found in this vicinity. *Hovey's Mag. Hort.* 7:130, 131.

1849 Some Plants of Danvers and Wenham, Essex County, Massachusetts. *Hovey's Mag. Hort.* 15:289-296.

1850 Notice of some Plants of Lynnfield, Danvers, Manchester, etc., etc., Essex County, Massachusetts. *Hovey's Mag. Hort.* 16:97-107.

1852 Further Notice of Rare Plants. *Jour. Essex County Nat. Hist. Soc.* 1, #3, 125, 126.

1856 Notice of three varieties of native Columbine (*Aquilegia canadensis* L.) *Proc. Essex Inst.* 1:268.

1858 Rambles Among the Plants of Essex County, Massachusetts. *Hovey's Mag. Hort.* 24:356-365.

1862 (A discussion of some spring wildflowers in minutes of meeting of 24 April 1857). *Proc. Essex Inst.* 2:146-149.

1862 Notice of Josselyn's New England Rarities, (presented at meeting of 9 Jan. 1857) *Proc. Essex Inst.* 2:94-115.

1869 Double Early Saxifrage. *Am. Nat.* 2:327.

1869 The Double Saxifrage found in Danvers, Massachusetts. *Am. Nat.* 2:610.

St. John, Harold 1916 A revision of the North American Species of *Potamogeton* of the Section *Coleophylli*. *Rhodora* 18:121-138.

1917 *Arenaria lateriflora* and its Varieties in North America. *Rhodora* 19:259-262.

1920 The Genus *Elodea* in New England. *Rhodora* 22:17-29.

Sargent, C.S. 1903 Recently Recognized Species of *Crataegus* in Eastern Canada and New England. I. *Rhodora* 5:52-66.

Saunders, Mary T. 1895 The Flora of Colonial Days. *Bull. Essex Inst.* 27:74-88.

Schneider, Mrs. E.S.	1907	*Acanthospermum australe* at Lawrence, Massachusetts. *Rhodora* 9:26.
Sears, J.H.	1868	Flowering of *Hepatica triloba. Am. Nat.* 2:161.
	1881	List of plants collected at Marblehead Neck, 3 Aug. 1881. *Bull. Essex Inst.* 13:149.
	1881	List of plants collected in Rowley, 7 July 1881. *Bull. Essex Inst.* 13:119.
	1881	List of plants collected at Saugus, 30 Aug. 1881. *Bull. Essex Inst.* 13:153, 154.
	1882	Paper on dates of appearance of local spring flowers and list of plants seen in bloom this season, 1 May 1882. *Bull. Essex Inst.* 14:74-76.
	1883	List of plants found near Dodge's Mill, Rowley, 29 May 1883. *Bull. Essex Inst.* 15:112, 113.
	1883	Weeds of Essex County. *Bull. Essex Inst.* 15:93-104.
	1884	List of plants found in bloom about Salem on 8 Dec. and 14 Dec. 1884. *Bull. Essex Inst.* 16:170.
	1886	List of native and introduced plants observed in flower in the vicinity of Salem, during the spring of 1886, on or before May 1. *Bull. Essex Inst.* 18:95-98.
	1908	A Southern Flora and Fauna of Post-Pleistocene Age in Essex County, Massachusetts. *Rhodora* 10:42-46.
Sheppard, S.A.D. & G.F.H. Markoe	1863	Plants Collected at Salisbury and Amesbury, June 25, 1863. *Proc. Essex Inst.* 3:272.
Smith, Lyman B.	1929	A New Form of *Erythronium americanum. Rhodora* 31:36.
Snyder, Dorothy E.	1950	*Dalea alopecuroides* on Plum Island, Essex County, Massachusetts. *Rhodora* 52:299.
Svenson, H.K.	1929	Monographic Studies in the Genus *Eleocharis. Rhodora* 31:121-135, 152-163, 167-191, 199-299.
	1932	Monographic Studies in the Genus *Eleocharis* - II. *Rhodora* 34:193-203, 215-227.
	1937	Monographic Studies in the Genus *Eleocharis* - IV. *Rhodora* 39:210-231, 236-273.
Townsend, Charles W.	1905	The Birds of Essex County, Massachusetts. *Memoirs of The Nuttall Ornithological Club,* Cambridge, Massachusetts.
	1906	The Birds of Essex County, 1906 and Supplement to, 1920. (Introduction)
Tracy, C.M.	1858	Studies of the Essex Flora. Stevenson & Nichols, Lynn. pp. 88.
	1862	Noticeable Traits of the Flora of Nahant. *Proc. Essex Inst.* 2:272-277.
	1867	A Variety of the Ox-eye Daisy. *Am. Nat.* 1:496.
	1869	Variation in the *Sarracenia. Am. Nat.* 3:327.
	1873	A Blue *Anagallis. Am. Nat.* 7:309, 310.
Tuckerman, Edward	1843	Descriptions of several New Plants of New England. *Hovey's Mag. Hort.* 9:142-145.

Upton, Miss L.H.	1875	List of Plants Found In Flower at Essex, 3 June 1875. *Bull. Essex Inst.* 7:109.
Ward, Martha E.	1919	*Galax aphylla* introduced in Massachusetts. *Rhodora* 21:24.
Weatherby, C.A.	1927	The Group of *Acalypha virginica* in Eastern North America. *Rhodora* 29:193-204.
	1932	An Addition to the Wool-waste Flora of Eastern Massachusetts. *Rhodora* 34:214, 215.
Wentworth, L.A.	1903	Two Plants New to the Flora of Lynn, Massachusetts. *Rhodora* 5:256, 257.
Wheeler, Louis C.	1940	*Euphorbia* Subgenus *Chamaesycae* in Canada and the United States Exclusive of Southern Florida. *Rhodora* 43:97-154, 168-205, 223-286.
Whittier, John Greenleaf	1863	Flowers, Flowering Shrubs and Vines in Amesbury and Salisbury. *Proc. Essex Inst.* 3:271.
Wiegand, K.M.	1909	*Lathyrus palustris,* var. *pilosus* (Cham.) Ledeb. in Massachusetts. *Rhodora* 11:108.
	1911	The Genus *Amelanchier* in Eastern North America. *Rhodora* 14:117-161.
	1918	Some Species and Varieties of *Elymus* in Eastern North America. *Rhodora* 20:81-90.
	1920	*Eupatorium purpureum* and Its Allies. *Rhodora* 22:57-70.
	1920	Variations in *Lactuca canadensis. Rhodora* 22:9-11.
	1921	The Genus *Echinochloa* in North America. *Rhodora* 23:49-65.
	1922	*Carex laxiflora* and Its Relatives. *Rhodora* 24:189-201.
	1928	*Aster lateriflorus* and Some of Its Relatives. *Rhodora* 30:161-179.
Wilbur, Robert L.	1955	A Revision of the North American Genus *Sabatia (Gentianaceae). Rhodora* 57:1-33, 43-71, 79-104.
Williams, E.F.	1902	Preliminary Lists of New England Plants VIII. *Rhodora* 4:15-22.
	1919	George Goulding Kennedy. *Rhodora* 21:25-35.
Wilson, Mrs. Sophila A.	1928	Plants of Nahant. Pp. 211, 212 in *Some Annales of Nahant,* by Fred A. Wilson. Old Corner Book Store, Boston.

Stuart Kimball Harris

BIOGRAPHY OF STUART KIMBALL HARRIS

The subject of this brief biographical sketch was born in Haverhill, Massachusetts on January 30, 1906 and died on July 30, 1969 after an illness of several months. He is survived by his wife Calista Crane Harris, a son, a daughter and three grandchildren.

After graduating from high school, he went to Boston University where he received a degree of Bachelor of Business Administration in 1927. Fortunately he soon realized that the prospect of a business career was distasteful. He quickly made a new start, becoming a Bachelor of Science in 1930, and a Master of Arts two years later. Transferring to Harvard as a graduate student he was awarded an M.A. degree in 1933. In the same year he returned to Boston University as a member of the faculty of Liberal Arts and continued as such for thirty-six years, ranking as Professor of Biology from 1951 until his death. Concurrently he achieved a Harvard doctorate in botany in 1936 under the stimulating guidance of Professor M. L. Fernald. To meet successfully the stiff requirements demanded of a candidate for a Ph.D. by one institution of higher learning while teaching at a sister university in a neighboring city must be an unusual accomplishment. It connotes hard work effectively applied.

A Memorial to Professor Harris adopted by the faculty of his college at Boston University about two months after his death contains this paragraph:

> "In addition to his botanical courses he taught a variety of subjects including general biology, invertebrate zoology, comparative anatomy, ecology, and others. He brought to his classes a very thorough and orderly presentation, and was at his best before small groups of advanced students . . . Despite the wide variety of subjects that he taught, particularly in his earlier years at Boston University when the Department of Biology had only three or four teachers, he conscientiously kept abreast of contemporary developments and included them in his lectures."

Note well the adjectives *thorough, orderly* and the phrase conscientiously *kept abreast*. Imagine, if you will, the home-work involved in preparing lectures in five specialized subjects, outside his main field.

As his teaching load lightened, Harris was able to give increasing attention to extra-curricular activities. From a somewhat casual, though enthusiastic bird watcher, he eventually developed into a competent field ornithologist. He studied lichens intensively in preparation for a *Lichen Flora of Massachusetts*. This project was destined to be a thrust aside by an invitation to prepare a comprehensive *Flora of Essex County*, which he accepted on terms consistent with the high standards he habitually applied to his work. The procedures he adopted for this job well illustrate his painstaking work-habits.

As a preliminary step, he examined and indexed all of the herbarium specimens he could find (perhaps six to eight thousand) which had been collected in the county by him and his predecessors. These included the historically important material which had gravitated to the Peabody Museum of Salem over the years. The names of distinguished botanists appear repeatedly on the identifying labels. The earliest of these seems to have been the Ipswich botanist, William Oakes, who was particularly active from 1818 to 1832. John Robinson collected widely between 1865 and 1902 and Merritt L. Fernald contributed sheets dated from 1909 to 1932. Harris spotted on outline maps, one for each species or variety, the towns where they had been found, these showing him where and what to collect and what to avoid during subsequent field work. His intent was to show rather precise distributions of those elements of the flora not known to be of general occurrence. He also entered in a card index of species and varieties the collector's name, date and place of collection of each specimen, adding his own as he went along. It is this index which is the basis for the citations in this book. At the end of each year he sorted and bundled his new material and placed it in the custody of the New England Botanical Club where it is available for ready reference. At the time of his death, the text of his Flora was completed except for a Preface and parts of the Introduction. It represents many hours spent annually in the field from May through October over a period of years, and long evenings dealing with his specimens, keeping his records and finally typing the manuscript.

All this time-consuming labor, demanding accuracy in plant identification, citation and transcription by no means interfered with his professional duties nor excluded other activities. The work-load he carried in stride over the years was by ordinary standards well-nigh overwhelming. As years went by, he devoted more and more time to the affairs of the New England Botanical Club of which he became a member in 1930 and was its President at the time of his death thirty-nine years later. For many years, he was Recording Secretary, a post which in his view required regular attendance at Club and Council meetings. For an even longer period, he was an Associate Editor of its journal, *Rhodora*, an important scientific publication with a world-wide circulation.

Harris accepted membership on a small committee to arrange for the publication of *A Flora of Northern New Hampshire* by the late Arthur Stanley Pease. As was to be expected, when the committee passed the editorial buck to him, he took it with his usual good humor. The bulky manuscript required close scrutiny prior to retyping and final proof-reading. Accuracy was a prerequisite especially in transcribing the citations of many thousands of voucher specimens. I shudder at the drudgery of it all. He acted in a similar capacity for the Club in respect to the publication of *A Flora of Stoughton, Massachusetts* based on a catalog of plants collected over the years by the late Sydney F. Blake as a vacation hobby, and subsequently presented to his many friends and colleagues here and abroad by Mrs. Blake in his memory. Another valuable activity in behalf of the Club was service on a committee engaged in mapping and reporting on plant distribution as revealed by its several hundred thousand herbarium specimens. This work required critical judgement if misidentifications were to be prevented from distorting the distributional picture and thus leading to an occasional error in scientific literature. In this respect his competence was particularly helpful.

Although the Botanical Club and its need for shirt-sleeve workers was his first concern after fulfilling his obligations to Boston University, he did not neglect other opportunities to be useful. His interest in the Appalachian Mountain Club was unflagging throughout his adult life. As hut-man at the Lakes-of-of-the-Clouds, as Club Naturalist, as leader of annual spring flower tours, and as author of most of the text of an attractive little book entitled *Mountain Flowers of New England*, for which he drew sketches and which was published under the imprint of the Club, he thus devoted much time to furthering its objectives. Being a competent field ornithologist, he was elected to membership in the Nuttall Ornithological Club and served time as secretary in what is said to be the oldest scientific ornithological society in America. He was a member for many years and past President of the Essex County Ornithological Club. The Massachusetts Audubon Society's Ipswich River Sanctuary was not slow in drafting him as a member of the Development Committee. He was a member and past chairman of the Conservation Commission of the Town of Boxford where he lived.

Aside from the vocation of botany, an avocation as well, Harris gave expression to his artistic temperament by sketching, weaving and learning to be a silversmith. He liked to travel, going as far afield as the western states, and from the Inca country in Peru to Honduras and northwards to Alaska, the latter by the Alcan Highway in 1944 as cook for a presumably war-connected scientific expedition.

Among the seventy or more of Dr. Harris' published papers that are listed below is the long one of seventy-five pages on Navajo Indian Medical Ethnobotany of which he was the co-author. Like many of his writings it reveals the breadth of his scientific interests. The life of this modest, competent naturalist was a rewarding one to himself and family, his colleagues and his many friends.

Richard Jefferson Eaton

PUBLICATIONS OF STUART KIMBALL HARRIS

S.K. Harris. An Alpine Station for *Hieracium aurantiacum. Rhodora* 32:113, 1930.

S.K. Harris. Some Characteristic Plants of Mount Washington. *Bull. Boston Soc. Nat. Hist.* 55:5-11, 1930.

M.L. Fernald and S.K. Harris. Branching *Polygonatum pubescens. Rhodora* 35:404-406, 1933.

S.K. Harris. Notes on the Flora of Haverhill, Massachusetts. *Rhodora* 35:261-262, 1933.

S.K. Harris. Reports on the Flora of Massachusetts II. *Rhodora* 35:351-359, 1933.

C.A. Weatherby and S.K. Harris. A Second Station for *Lygodium palmatum* in New Hampshire. *Am. Fern. Journal* 23:124-126, 1933.

S.K. Harris. An Undescribed *Mutisia* from Equador. *Rhodora* 35:257-258, 1933.

M.L. Fernald and S.K. Harris. An Additional Note on the Branching Tendency in *Polygonatum. Rhodora* 36:59-60, 1934.

L.B. Smith and S.K. Harris. Three New Andean Species of *Lepanthes.* Bot. Mus. Leaflets, Harvard U. 2:33-36, 1934.

S.K. Harris. Lichen Flora of the United States Review. *Rhodora* 37:305, 1935.

S.K. Harris. Lichens, A Neglected Group. *Bull. Boston Soc. of Nat. Hist.* 74:3-8, 1935.

S.K. Harris. Heaths of the Presidential Range. *Appalachia N.S.* Vol. 6:24-34, 1940.

S.K. Harris. The Hut Doors Get "Blown" Open. *Appalachia N.S.* 6:268-269, 1940.

S.K. Harris. Ice in the Lakes of the Clouds. *Appalachia N.S.* 6:113-114, 1940.

S.K. Harris. California Shrubs (Review). *Rhodora* 3:212, 1941.

L.C. Wyman and S.K. Harris. Navajo Indian Medical Ethnobotany. *Univ. N. Mex. Bull. Anthropological Series* 5:1-76, 1941.

S.K. Harris. More Plants of the Presidential Range. *Appalachia* 7:320-328, 1941.

S.K. Harris. Further Flowers of the Presidential Range. *Appalachia* 8:54-63, 1942.

S.K. Harris. Wild Violets of North America Review. *Appalachia* 8:283, 1942.

S.K. Harris. Common Edible Mushrooms Review. *Appalachia* 9:430, 1943.

S.K. Harris. The Harry Hunter Monument and Other Notes. *Appalachia* 9:542.

S.K. Harris. Old Hutman's Vacation. *Appalachia* 9:496-505, 1945.

S.K. Harris. Plants of the Presidential Range IV. *Appalachia* 9:360-367, 1943.

S.K. Harris. Plants of the Presidential Range V. *Appalachia* 10:85-92, 1944.

S.K. Harris. Along the Alaska Highway with the Harvard Survey Party. *Appalachia* 11:380-382, 1945.

S.K. Harris. Plants of the Presidential Range VI. *Appalachia* 11:342-350, 1945.

S.K. Harris. Plants of the Presidential Range VII. *Appalachia* 12:66-76, 1946.

S.K. Harris. Plants of the Presidential Range VIII. *Appalachia* 13:370-376, 1947.

S.K. Harris. Plants of the Presidential Range IX. *Appalachia* 14:55-65, 1948.

S.K. Harris. Plants of the Presidential Range X. *Appalachia* 15:312-320, 1949.

S.K. Harris. Christmas Bird Count, Dummer Hill, New Hampshire. *Audubon Field Notes*, April, 1949.

S.K. Harris. Christmas Bird Count, Dummer Hill, New Hampshire. *Audubon Field Notes*, April, 1950.

L.C. Wyman and S.K. Harris. Ethnobotany of the Kayenta Navaho. *Univ. N. Mexico Publ. in Biol. #5*, 1951.

S.K. Harris. Christmas Bird Count, Dummer Hill, New Hampshire. *Audubon Field Notes* April, 1951.

S.K. Harris and R.C. Bean. *Wolffia columbiana* in Methuen, Massachusetts. *Rhodora* 53:272, 1951.

S.K. Harris. Field List of the Birds of Essex County, Massachusetts, 4th edition, 1952.

S.K. Harris. Christmas Bird Count, Dummer Hill, New Hampshire. *Audubon Field Notes*, April, 1952.

A.G. Humes and S.K. Harris. The Clam Hosts of *Najadicola ingens* (K.) Acarina in a Quebec Lake, 1952.

S.K. Harris. Christmas Bird Count, Dummer Hill, New Hampshire. *Audubon Field Notes*, April, 1953.

S.K. Harris. Flora of the Summit of Mount Washington. *Mt. Washington Observatory Bull.* #22.9, 1953.

S.K. Harris. The Mating Instinct. Review. *Appalachia* 17:130, 1954.

S.K. Harris. Christmas Bird Count, Dummer Hill, New Hampshire. *Audubon Field Notes*, April, 1954.

S.K. Harris. Christmas Bird Count, Dummer Hill, New Hampshire. *Audubon Field Notes,* April, 1955.

S.K. Harris. Christmas Bird Count, Dummer Hill, New Hampshire. *Audubon Field Notes*, April, 1956.

S.K. Harris. Christmas Bird Count, Dummer Hill, New Hampshire. *Audubon Field Notes*, April, 1957.

S.K. Harris. Christmas Bird Count, Dummer Hill, New Hampshire. *Audubon Field Notes*, April, 1958.

S.K. Harris. Bird Feeders, A Source of Adventive Weeds, *Bull. Mass. Audubon Society*. 42:248-250, 1958.

S.K. Harris. *Nelumbo lutea* in Essex County, Massachusetts. *Rhodora* 60:116, 1958.

S.K. Harris. *Cabomba caroliniana* in Essex County, Massachusetts. *Rhodora* 60:116, 1958.

S.K. Harris. *Rorippa sessiflora* in Essex County, Massachusetts. *Rhodora* 60:260, 1958.

S.K. Harris. A New Color Form of *Solidago sempervirens. Rhodora* 60:261-262, 1958.

S.K. Harris. *Elsholtzia ciliata* in Essex County, Massachusetts. *Rhodora* 61:63, 1959.

S.K. Harris. The Northernmost Station of *Magnolia virginiana*, its History and Present Status. *Rhodora* 61:65-70, 1959.

S.K. Harris. *Peucedanum palustre*, an Interesting Addition to the Flora of Essex County, Massachusetts. *Rhodora* 61:181, 1959.

S.K. Harris. Two Crucifers New to Essex County, Massachusetts. *Rhodora* 61:187, 1959.

S.K. Harris. Two Grasses New to Essex County, Massachusetts. *Rhodora* 61:247-248, 1959.

S.K. Harris. Plant Lists Are Where You Find Them: A List of Local Floras of Massachusetts Published Since 1898. *Rhodora* 62:20-29, 1960.

S.K. Harris. Three Grasses Apparently New to Massachusetts. *Rhodora* 62:262-263, 1960.

S.K. Harris. *Sesbania exaltata* (Raf.) Corey in Essex County, Massachusetts. *Rhodora* 62:234, 1960.

S.K. Harris. Trees, Shrubs and Woody Vines of Great Smoky Mountains National Park. Review. *Appalachia* 35:579, 1965.

S.K. Harris. *Silene pumilio*. *Appalachia* 35:779, 1965.

S.K. Harris. Two Saxifrages New to the Mt. Washington Area. *Appalachia* 37:152-153, 1968.

S.K. Harris. Amphibians and Reptiles of the Great Smoky Mountains. National Park Review. *Appalachia* 37:171, 1968.

S.K. Harris. Trees and Shrubs of Northern New England. Review. *Appalachia* 37:505, 1969.

S.K. Harris. Mountain Flowers of New England. Appalachian Mountain Club. 1964. Section on Mosses and Lichens by Jean H. Langenheim.

Arthur S. Pease. A Flora of New Hampshire. New England Botanical Club 1964. Typed, edited, indexed, with additions by S.K. Harris.

R.C. Bean, A.F. Hill and S.K. Harris. The sixteenth report of the committee on plant distribution. *Rhodora* 63:26-30, 1963.

R.C. Bean, A.F. Hill and S.K. Harris. The seventeenth report of the committee on plant distribution. *Rhodora* 67:306-318, 1967.

S.K. Harris. Notes on the Flora of Coos County, New Hampshire. *Rhodora* 67:195-197, 1965.

S.K. Harris. *Geocaulon lividum* in New England. *Rhodora* 67:407-409, 1965.

S.K. Harris. Notes on the Flora of Coos County, New Hampshire. *Rhodora* 69:29-34, 1967.

S.K. Harris. Notes on the Flora of Coos County, New Hampshire. *Rhodora* 71:166-174, 1969.

There is an obituary of S.K. Harris in *Appalachia,* 37:675. December 1969.

ABBREVIATIONS

AREAS: MNS is Marblehead Neck Sanctuary of the Massachusetts Audubon Society.

COLLECTORS: A & W is Atherton, Daniel and Wheat, Thomas; H is Harris, Stuart K.; Nat. is Native Plants; Int. is Introduced Plants.

INFORMATION

Entries in the Catalog enclosed thus [ ] were added by the editor, Dorothy Eastman Snyder.

P.A.S. (Peabody Academy of Science) is now P.M.S., (Peabody Museum of Salem).

Harris followed Gray, 8th edition for nomenclature with very few exceptions. However, he never capitalized species names, but followed the International Code in using lower case.

Numbers after area and collector H. (= Harris) in the catalogue are in the New England Botanical Club herbarium. Many early records are from the Peabody Museum of Salem Herbarium.

LIST OF TOWNS AND CITIES IN ORDER QUOTED

1. Salisbury
2. Amesbury
3. Merrimac
4. Haverhill
5. Methuen
6. Lawrence
7. Andover
8. North Andover
9. Boxford
10. Groveland
11. West Newbury
12. Newburyport
13. Newburyport, Plum Island
14. Newbury
15. Newbury, Plum Island
16. Georgetown
17. Rowley
18. Rowley, Plum Island
19. Ipswich
20. Ipswich, Plum Island
21. Topsfield
22. Middletown
23. Danvers
24. Wenham
25. Hamilton
26. Essex
27. Gloucester
28. Rockport
29. Manchester
30. Beverly
31. Salem
32. Peabody
33. Lynnfield
34. Saugus
35. Lynn
36. Swampscott
37. Marblehead
38. Nahant

EQUISETACEAE (Horsetail Family)

EQUISETUM L. Horsetail.

E. arvense L. Common Horsetail. A highly variable species with many minor described forms several of which are ignored in this enumeration. Open habitats; common. - Salisbury, Eaton 3 (1896); Amesbury, H. 18260 (1958); Haverhill, H. 2 (1931); Lawrence, Merrill (1852); North Andover, Morss (1885); Newburyport, Eaton 106 (1902); Georgetown, Horner (1875); Ipswich, Oakes; Ipswich, Plum Island, H., McGregor & Schweinfurth 12699 (1957); Topsfield, H. R. Towne; Middleton, H. 17691b (1958); Gloucester, Magnolia H. 19009 (1958); Rockport, Gates & Smith 844 (1928); Beverly Mackintosh (1924); Salem, Cole (1886); Saugus, Lund (1880).

E. arvense, forma *nemerosum* (A. Br.) Klinge - Amesbury, Eaton 5 (1897); Wenham, Robinson (1875); Beverly Farms, Mackintosh (1924).

E. arvense, forma *alpestre* (Wahlenb.) Luerss. - Amesbury, Eaton 10 (1900).

E. arvense, forma *campestre* (C. F. Schultz) Klinge - Salisbury, Eaton 12 (1896); Newburyport, Eaton (1901).

E. arvense, forma *pseudo-sylvaticum* (Milde) Luerss. - Salisbury, Eaton 6 (1897).

E. arvense, forma *diffusum* (A. A. Eaton) Clute - Amesbury, Eaton 9a (1900); Lawrence, Robinson (1875); Newburyport, Robinson (1875); Plum Island, St. John 848 (1913); Ipswich, St. John & White 558 (1913).

E. arvense, var. *boreale* (Bong.) Ledeb., forma *pseudo-alpestre* Vict. - Merrimac, H. 9253 (1953).

E. sylvaticum L., var. *pauciramosum* Milde, forma *multiramosum* Fern. Wood-Horsetail. Woodlands and thickets; common. - Amesbury, Eaton 35 (1902); Haverhill; H. 386 (1932); Andover, Robinson (1875); North Andover, Root & Zink (1945); West Boxford, Horner; Newburyport, Schweinfurth (1950); Ipswich, Oakes; Topsfield, Robinson (1875); Hamilton, Hubbard (1913); Essex Woods, Robinson (1876); Rockport, Smith 932 (1928); Salem, Robinson (1872); Swampscott, Robinson (1879); MNS, Snyder (1973).

x. *E. litorale* Kuhlewein Shores and swales; occasional. - Amesbury, Eaton (1896); Groveland, J. Russell (1859); Newburyport, Eaton (1897).

x. *E. litorale*, forma *arvensiforme* (A. A. Eaton) Vict. - Newburyport, Eaton 42 (1898).

x. *E litorale*, forma *humile* (Milde) Vict. - Amesbury, Eaton 40 (1896); Newburyport, Eaton (1898).

x. *E. litorale*, forma *gracile* (Milde) Vict. - Amesbury, Eaton (1899).

E. fluviatile L. Water-Horsetail. Wet shores and swales; frequent. - Amesbury, Eaton 49 (1899); Haverhill, H. 457 (1932); Methuen, H. 18384 (1958); Andover, H. 17904 (1958); Boxford, Robinson (1879); Groveland, H. 10606 (1955); West Newbury, White 520 (1913); Newburyport, Eaton (1903); Ipswich, Morong (1875); Topsfield, Mackintosh (1925); Danvers, Cole (1901); Wenham, Robinson (1875); Rockport, Smith 933 (1928).

E. fluviatile, forma *linnaeanum* (Döll) Broun - Newburyport, Eaton 49 (1897); Ipswich, Oakes; Topsfield, Robinson (1873); Danvers, Robinson (1879).

E. fluviatile, forma *minus* (A. Br.) Broun - Amesbury, Eaton 47 (1897).

E. hyemale L., var. *affine* (Engelm.) A. A. Eat. Scouring-rush. Dry or moist sandy shores, roadsides and open woods; frequent. - Amesbury, Eaton 73 (1902); Merrimac, Bean & H. 8367 (1953); Haverhill, Horner; Methuen, Robinson (1875); Andover, Pease 1559 (1902); North Andover, Morss; Newburyport, Horner; Topsfield, Mrs. Conley; Middleton, H. 1769la (1958); Danvers, Snyder (1950 on); Beverly, Robinson (1893).

E. hyemale, var. *affine*, forma *polystachyum* Prager - Amesbury, Eaton 74 (1902); Andover, Pease 94 (1902).

E. scirpoides Michx. Dwarf Scouring-rush. Mossy knolls or springy banks; rare. - Amesbury, Eaton (1895), 87 (1903).

LYCOPODIACEAE (Club-moss Family)

LYCOPODIUM L. Club-moss.

L. lucidulum Michx. Shining Club-moss. Cool woods; frequent. - Amesbury, Winslow 10175 (1910); Haverhill, H. 294 (1932); Methuen, Wheeler 174504 (1922); Andover, H. 17765 (1958); North Andover, H. 8299 (1953); Boxford, Hunnewell 8877 (1923); Georgetown, Horner; Ipswich, Oakes; Topsfield, Mrs. M. Peirce; Middleton, H. 18065 (1958); Wenham, Cole (1911); Gloucester, Magnolia, Farlow (1903); Manchester, Oakes; Beverly, Cole (1899); Peabody, Robinson (1874); Lynnfield, Young (1878).

L. inundatum L. Bog Club-moss. Moist sandy soil; frequent. - Andover, Pease 107 (1901); Ipswich, Conomo, Buxton (1926); Essex, Eaton (1896); West Gloucester, Robinson (1877); Rockport, Smith 716 (1927); Beverly, Robinson (1874); Peabody, H. 12587 (1956); Saugus, H. 13031 (1957); Marblehead, H. 12603 (1956).

L. inundatum, var. *bigelovii* Tuckerm. - Haverhill, Bradford, Horner; Newbury, Plum Island, Pease 2644 (1903); Ipswich, Prince (1902); Ipswich, Plum Island, Floyd (1897); Essex, Dodge (1895); Chebacco, Robinson (1874).

L. annotinum L. Interrupted Club-moss. Woods and clearings; occasional. - Wenham, Cole (1911); Essex, Eaton (1895); Beverly Farms, Mackintosh (1933).

L. annotinum, var. *acrifolium* Fern. - Wenham, Cole (1911); Essex, Dodge (1896); Chebacco, Robinson (1872); Manchester, Knowlton (1907).

L. clavatum L. Running Club-moss. Dry woods; common. - Amesbury, Bean & H. 7743 (1951); Merrimac, Bean & H. 7720 (1951); Haverhill, H. 185 (1931); Andover, Pease 103 (1902); North Andover, H. 9459 (1954); Boxford, Zink (1946); Georgetown, Horner; Ipswich, Oakes; Topsfield, Mrs. E. Peirce, Middleton, H. 8330 (1953); Wenham, Hunnewell (1916); Gloucester, Wheat (1971); Manchester, Robinson (1877); Lynnfield, Rodman (1887).

L. clavatum, var. *megastachyon* Fern. & Bissell - Boxford, Horner (1881), Zink (1943).

L. obscurum L. Flat-branch Ground-Pine. Woods and clearings; common. - Salisbury, H. 9474 (1954); Amesbury, Bean & H. 7741 (1951); Haverhill, H. 187 (1931); North Andover, H. 9460 (1954); Boxford, St. John & Pease 11944 (1917); Georgetown, Horner (1879); Rowley, H. 11960 (1955); Ipswich, Oakes; Essex, Winslow 10187 (1910); Chebacco, Robinson (1874).

L. obscurum, var. *dendroideum* (Michx.) D.C. Eat. Round-branch Ground-Pine. - Salisbury, H. 9475 (1954); Amesbury, Bean & H. 7742 (1951); Merrimac, Bean & H. 7710 (1951); Andover, Pease 1702 (1902); North Andover, H. 9461 (1954); Boxford, M.P. Perley (1883); Newbury, Smith 101 (1921); Ipswich, H. 18150 (1958); Topsfield, B.B. Towne; Middleton, Bean & H. 7773 (1951); Danvers, Bean & H. 9988 (1955); Wenham, Hunnewell 4624 (1916); Chebacco, Saunders (1876); Gloucester, Wheat (1971); Beverly, Robinson (1872); Saugus, H. 7796 (1951).

L. complanatum L., var. *flabelliforme* Fern. Christmas Green. Dry woods and clearings; common. - Amesbury, Bean & H. 7740 (1951); Merrimac, Bean & H. 7713 (1951); Haverhill, H. 188 (1931); Methuen, H. 9606 (1955); North Andover, H. 9458 (1954); Boxford, Robinson (1877); Newbury, Smith 102 (1921); Georgetown, Williams (1899); Rowley, H. 11961 (1955); Ipswich, H. 18164 (1958); Essex, Winslow 10184 (1910); Chebacco, Saunders (1876); Gloucester, Wheat (1971); Beverly, Robinson (1872).

L. tristachyum Pursh Ground-Pine. Dry woods; frequent; often confused with previous species. - Salisbury, H. 18413 (1958); Andover, Pease & Hopkins 24034 (1934); Boxford, H. 8206 (1952); Newbury, H. 8246 (1952); Georgetown, C. A. & A. W. Cheever (1916); Ipswich, H. 18413 (1958); Topsfield, Proctor; Wenham, Sears (1900); Hamilton-Essex Woods, Sears (1900); Gloucester, H. 12176 (1956); Beverly, Robinson (1872).

SELAGINELLACEAE (Spikemoss Family)

SELAGINELLA Beauv. Spikemoss.

S. apoda (L.) Fern. Meadow Spikemoss. Damp ground; not common. - Andover, Pease 120 (1902); Boxford, Horner (1876); Georgetown, Horner; Topsfield, Pease 7474 (1905); Middleton, H. 12639 (1957).

S. rupestris (L.) Spring Rock Spikemoss. Dry rocks; common. - Salisbury, Eaton (1896); Haverhill, H. 1280 (1934); Andover, Pease 118 (1902); Boxford, M. P. Perley (1872); Newbury, H. 8228 (1952); Rowley, H. 1592 (1934); Middleton, Bean & H. 7760 (1951); Danvers, Oakes; Gloucester, Fernald, Hunnewell & Long 8417 (1913); Salem, Forbes 99 (1903); Swampscott, H. 1450 (1934); Marblehead, H. 942 (1933).

ISOËTACEAE (Quillwort Family)

ISOËTES L. Quillwort.

I. muricata Dur. Shallow water; probably common but seldom collected. - Amesbury, H. 9172 (1953); Merrimac, H. 13729 (1957); Haverhill, H. 28395 (1965); Andover, Pease 2775 (1903); North Andover, H. 12917 (1957); Boxford, Sears (1888); Groveland, H. 9102 (1953); West Newbury, Bean & Pease (1940); Ipswich, H. 13240 (1957); Topsfield, H. 13227 (1957); Wenham, Robinson (1875); Hamilton, H. 12803 (1957); Peabody, H. 13314 (1957).

I. eatoni Dodge Ponds and steams; rare ? - Amesbury, Eaton (1897).

I. riparia Engelm. Shallow water; rare ? - Boxford, R. J. Eaton 5227 (1962), R. J. Eaton & Richardson 5299 (1962).

I. tuckermani A. Br. Pond margins and shores; occasional. - Amesbury, Eaton (1895, 1903); North Andover, Pease 2640 (1903).

OPHIOGLOSSACEAE (Adder's-Tongue Family)

BOTRYCHIUM Sw. Moonwort. Grape-Fern.

B. multifidum (Gmel.) Rupr., var. *intermedium* (D. C. Eat.) Farw. Leathery Grape-Fern. Moist and dry pastures; frequent. - Boxford, G. B. Parkhurst; Georgetown, Horner (1874); Topsfield, G. F. Dow; Danvers, Sears (1880); Essex Woods, Robinson (1873); Gloucester, Eaton (1903).

B. dissectum Spreng. Pastures, open thickets or woods; common. - Salisbury, Eaton (1899); Amesbury, Eaton (1899); Merrimac, Bean & H. 7719 (1951); Haverhill, H. 1490 (1934); West Boxford, Zink (1945); Groveland, H. 8832 (1953); Georgetown, Horner; Ipswich, H. 18440 (1958); Topsfield, Pease 7473 (1905); Danvers, Sears (1878); Essex, Knowlton (1907); Essex Woods, Robinson (1873); Gloucester, Witherby; Manchester, Merrill; Beverly, Lane (1880); Peabody, Russell (1856); Lynnfield, Robinson (1874); Swampscott, Robinson (1872).

B. dissectum, forma *obliquum* (Muhl.) Fern. - Salisbury, Eaton (1899); Amesbury, H. 18269 (1958); Haverhill, H. 1491 (1934); Andover, Pease 7 (1900); North Andover, Robinson (1884); Boxford, M. P. Perley (1885); Wenham, Robinson (1873); Essex, Knowlton (1907); Rockport, Forbes 5 (1903); Manchester, Merrill; Beverly, Lane (1880); Salem, Forbes 4 (1903); Swampscott, Robinson (1871).

B. dissectum, forma *elongatum* (Gilbert & Haberer) Weath. - Georgetown, Horner.

B. dissectum, forma *oneidense* (Gilbert) Clute - Amesbury, Eaton (1899); Essex Woods, Robinson (1873); Swampscott, Robinson (1872).

B. simplex E. Hitchc. Meadows and pastures; rare. - Groveland, Sears (1904); Salem, Sears (1880).

B. simplex, var. *tenebrosum* (A. A. Eat.) Clausen. Damp woods; rare. - Amesbury, Eaton (1891); Boxford, Hunnewell 8879 (1932); Groveland, Sears (1904); Byfield, Eaton (1898); TYPE.

B. matricariaefolium A. Br. Daisy-leaf Grape-Fern. Rich soil; occasional. - Salisbury, Eaton (1896); North Andover, Hunnewell 14094 (1936); Boxford, Sears & Robinson (1881); Groveland, Sears (1904); West Newbury, Conant (1879); Byfield, Eaton (1898); Middleton, Sears & Robinson (1881); Salem, Sears & Robinson (1881).

B. matricariaefolium, forma *gracile* (House) Weath. - Byfield, Eaton (1898); Saugus, Svenson (1917).

B. lanceolatum (Gmel.) Angstr., var. *angustisegmentum* Pease & Moore. Lance-leaved Grape-Fern. Rich soil; occasional. - Salisbury, Eaton (1895); Amesbury, Eaton (1899); Andover, Pease 5 (1902); Boxford, Forber (1910); Groveland, Sears (1904); West Newbury, Conant & Robinson (1879); Byfield, Eaton (1896); Georgetown, Horner (1876); Middleton, Robinson (1880); Gloucester, Magnolia, Sears (1880); Salem, Sears (1881); Saugus, Svenson (1917).

B. virginianum (L.) Sw. Rattlesnake Fern. Rich woods; frequent. - Haverhill, H. 1835 (1935); Boxford, Robinson (1877); Groveland, Horner; Georgetown, Horner (1874); Byfield, Eaton (1896); Rowley, Horner (1898); Middleton, H. 18066 (1958); Saugus, Svenson (1917).

OPHIOGLOSSUM L. Adder's Tongue.

O. vulgatum L., var. *pseudopodum* (Blake) Farw. Grassy meadows; probably more common than the number of collections indicate. - Andover, Pease 23 (1902); Boxford, Robinson (1876); Ipswich, Plum Island, E. Dodge (1898); Danvers, Sears (1880); Gloucester, Magnolia, V.P. Adams (1880); Rockport, Smith 312 (1923); Beverly, Davenport (1874); Salem, C.E. Faxon.

OSMUNDACEAE (Flowering Fern Family)

OSMUNDA L. Flowering Fern.

O. regalis L., var. *spectabilis* (Willd.) Gray Royal Fern. Damp ground; common. - Lawrence, H. 10076 (1955); West Boxford, Root & Zink; Georgetown, Robinson (1875); Ipswich, Oakes; Topsfield, H.R. Towne; Danvers, Frost; Wenham, Cole (1923); Chebacco, Robinson (1874); Gloucester, C.M. Taylor (1897); Rockport, Smith 725 (1927); Beverly Farms, Hubbard (1913); Salem, Cole (1910); Lynnfield, Bean (1919); Marblehead, Cole (1886).

O regalis, var. *spectabilis*, forma *anomala* (Farw.) Harris - Haverhill, Palfrey (1892); Lawrence, H. 10075 (1955); Boxford, H. 8445 (1953); Georgetown, Horner; Topsfield, Robinson (1874); Middleton, H. 8686 (1953); Danvers (1846); Wenham, Robinson (1871); Rockport, H. 10280 (1955); Manchester, H. 10766 (1955).

O. claytoniana L. Interrupted Fern. Damp ground; common. - Amesbury, H. 10588 (1955); Haverhill, H. 384 (1932); Lawrence, Merrill (1851); Andover, Robinson (1877); North Andover, Cole (1909); Boxford, M.E. Perley (1882); Georgetown, Horner; Ipswich, Oakes; Topsfield, H.R. Towne; Danvers, Robinson (1873); Manchester, Hubbard (1913); Salem, Cole (1886); [MNS, Snyder (1971)] .

O. cinnamomea L. Cinnamon Fern. Damp ground; common. - Amesbury, Eaton 51 (1899); Haverhill, H. 385 (1932); Lawrence, H. 10080 (1955); Andover, Robinson (1877); North Andover, Cole (1909); Boxford, M.P. Perley (1882); Georgetown, Horner; Ipswich, Oakes; Ipswich, Plum Island, H. 12100 (1956); Topsfield, H.R. Towne; Danvers, Frost; Hamilton, Robinson (1872); Rockport, Smith 961 (1928); West Manchester, Hubbard (1913); Salem, Robinson (1873); Lynn, Fuller (1897).

O. cinnamomea, forma *auriculata* (Hopkins) Kittredge - Essex County, Robinson (1877).

O. cinnamomea, forma *bipinnatifida* Clute - Beverly, Robinson (1877).

O. cinnamomea, forma *frondosa* (T, & G.) Britt. - Boxford, H. 8434 (1953); Georgetown, Horner; Ipswich, Oakes; Essex, Robinson (1874); Rockport, Smith 963 (1928); Peabody, Mackintosh (1898).

SCHIZAEACEAE (Curly-grass Family)

LYGODIUM Sw. Climbing Fern.

L. palmatum (Bernh.) Sw. Moist thickets in very acid soil. Probably now extinct in the County. - Saugus, Emery (1871), Sears (1885).

POLYPODIACEAE (Fern Family)

WOODSIA R. Br. Woodsia.

W. ilvensis (L.) R.Br. Rusty Woodsia. Rocky hills in exposed situations; not common. - Andover, Ballardvale, Zink (1944); Boxford, G.S. Morse; Danvers, Oakes; Salem, Robinson (1867); Marblehead, Snyder (1958).

W. obtusa (Spreng.) Torr. Blunt-lobed Woodsia. Rocky places in half shade; occasional. - Haverhill, H. 1831 (1935); West Boxford, Zink (1945); Georgetown, Horner; Byfield, Horner; Topsfield, Robinson (1873); Middleton, Robinson (1874); Danvers, Oakes; North Beverly, Robinson (1876); Salem, Robinson (1872).

CYSTOPTERIS Bernh. Bladder Fern.

C. fragilis (L.) Bernh. Fragile Fern. Damp rocks; much less common than the variety. - Boxford, Horner.

C. fragilis, var. *mackayii* Lawson - Merrimac, Bean & H. 8360 (1953); Haverhill, Bradford, Fernald 11271 (1914); Boxford, C.A. Cheever (1914); Newburyport, Smith 160 (1921); Ipswich, Oakes; Danvers, Sears (1901); Beverly, Prides Crossing, Hubbard (1913); Salem, Robinson (1873).

MATTEUCCIA Todaro. Ostrich Fern.

M. pensylvanica (Willd.) Raymond. Bottomland thickets; occasional. - Merrimac, Bean & H. 8365 (1953), Haverhill, Ward Hill, Root (1951); Methuen, Bean & H. 7749 (1951); Lawrence, Williams (1900); North Andover, Robinson (1879); Boxford, Robinson & Sears (1878); Groveland, M.E. Perley (1885); Gerogetown, Horner (1875).

ONOCLEA L. Sensitive Fern.

O. sensibilis L. Moist Places; common. - Haverhill, H. 382 (1932); Lawrence, H. 10081 (1955); North Andover, Root (1945); Boxford, M.E. Perley (1881); Georgetown, Horner; Rowley, Plum Island, H., McGregor & Schweinfurth 12724 (1957); Ipswich, Oakes; Ipswich, Plum Island, H., McGregor & Schweinfurth 12650 (1957); Topsfield, H.R. Towne; Danvers, Frost (1880); Essex, Bean (1938); Gloucester, Magnolia, Gilbert (1896); Rockport, Smith 2635 (1932); Manchester, Hubbard (1913); Salem, Robinson; [MNS, Snyder (1917)].

O. sensibilis, forma *hemiphyllodes* (Kiss & Kuemmerle) Weath. - Rockport, Witherby (1941).

O. sensibilis, forma *obtusilobata* (Schkuhr) Gilbert - Georgetown, Horner (1882); Hamilton, Robinson (1871); Gloucester, Magnolia, Clarke (1912); Salem, Robinson (1871).

DRYOPTERIS Adans. Shield-Fern.

D. thelypteris (L.) Gray, var. *pubescens* (Lawson) Nakai Marsh Fern. Moist places; common. - Haverhill, H. 563 (1932); Methuen, H. 8917 (1953); Georgetown, Horner; Ipswich, Oakes; Ipswich, Plum Island, H., McGregor & Schweinfurth 12703 (1957); Topsfield, Sears (1905); Danvers, Sears (1901); Wenham, Robinson (1871); Essex, H. 13351 (1957); Rockport, L.B. & R.C. Smith 2706 (1933); Beverly, Robinson (1871); Peabody, Mackintosh (1933); Lynnfield, Bean (1938).

D. simulata Davenp. Massachusetts Fern. Bogs; occasional. - Salisbury, Dodge & Davenport (1893); Haverhill, H. 500 (1932); North Andover, Pease 1591 (1900); Ipswich, Oakes; Essex, Knowlton (1907); Beverly, Robinson (1871); Peabody, Mackintosh (1924); [MNS, Snyder (1971)].

D. noveboracensis (L.) Gray. New York Fern. Damp places in woods; common. - Haverhill, H. 683 (1932); Methuen, H. 18378 (1958); Boxford, H. 18539 (1958); Georgetown, Horner; Ipswich, H. 18131 (1958); Topsfield, Sears (1905); Essex, Robinson (1873); Gloucester, C.M. Taylor (1897); Rockport, L.B. & R.C. Smith 2400 (1928); Beverly, Lowell (1846); Peabody, Mackintosh (1933).

D. disjuncta (Ledeb.) C.V. Mort. Oak-Fern. Cool woods; local. - Amesbury, H. 9524 (1954); Haverhill, Bradford, Fernald 11269 (1914); Georgetown, Horner; Middleton, Robinson (1885); Wenham, Lowell (1847); Essex; Beverly, Robinson (1873).

D. phegopteris (L.) Christens. Long Beech-Fern. Cool woods; local. - Merrimac, H. 18320 (1958); Haverhill, Bradford, Fernald 11270 (1914); Andover, Pease 78 (1900); West Boxford, Zink (1951); Georgetown, Horner; Danvers, Sears (1901); Essex, Robinson (1874); Rockport, L.B. & R.C. Smith 1091 (1928); Manchester, Hubbard (1913); Beverly, Robinson (1872).

D. hexagonoptera (Michx.) Christens. Broad Beech-Fern. Rich woods; rare. - Haverhill, H. 11882 (1951); North Andover, Bean (1947); Boxford, Root 11879 (1951); Georgetown, Horner (1875); Chebacco, Robinson & Sears (1876).

D. spinulosa (O.F. Muell.) Watt Spinulose Wood-Fern. Low woods; occasional. - Amesbury, H. 18232 (1958); Haverhill, H. 454 (1932); North Andover, H. 17982 (1958); Boxford, H. 17717 (1958); Georgetown, Horner; Ipswich, H. 13920 (1957); Topsfield, Sears (1905); Beverly, Robinson (1871).

D. spinulosa, var. *fructuosa* (Gilbert) Trudell - Andover, Knowlton (1903); Topsfield, Sears (1905); Danvers, Sears (1877); Wenham, Robinson (1871); Rockport, Smith 2719 (1933); Beverly, Prides Crossing, Robinson (1890); Swampscott, J.L.R. (1884).

D. spinulosa, var. *intermedia* (Muhl.) Underw. The most common variety. - Amesbury, Bean & H. 7739 (1951); Merrimac, Bean & H. 7718 (1951); Haverhill, H. 7788 (1951); Andover, H. 17758 (1958); Georgetown, Horner (1881); Danvers, Sears (1877); Wenham, Sears (1905); Essex, H. 11977 (1955); Manchester, Oakes; Saugus, Bean (1937); Marblehead, H. 12595 (1956).

D. spinulosa, var. *americana* (Fisch.) Fern. - Merrimac, H. 9255 (1953); Rowley, H. 11958 (1955).

x *D. boottii* (Tuckerm.) Underw. Boott's Wood-Fern. Swamps; occasional. - Amesbury, Winslow 10173 (1910); Andover, Pease 47 (1900); Georgetown, Horner; Topsfield, Sears (1905); Wenham, Robinson (1871); Chebacco, Robinson (1873); Beverly, Robinson (1883).

x *D. boottii* x *D. spinulosa* ? - Danvers, Sears (1905); det. C.A. Weatherby 1918.

D. cristata (L.) Gray. Crested Wood-Fern. Swamps; frequent. - Salisbury, Eaton 255 (1899); Amesbury, Winslow 10174 (1910); Haverhill, H. 455 (1932); Andover, H. 17934 (1958); North Andover, H. 17980 (1958); Boxford, Cole (1937); Georgetown, Horner; Ipswich, Oakes; Topsfield, Sears (1905); Middleton, H. 18107 (1958); Danvers, Robinson (1877); Wenham, Lowell (1857); Essex, Davenport (1879); Beverly, Robinson (1871); Lynnfield, Bean (1919).

D. cristata, var. *clintoniana* (D.C. Eat.) Underw. Clinton's Wood-Fern. - Amesbury, Eaton; Merrimac, Clarkson (1917); Danvers, Robinson (1877); Wenham, Sears (1905); Lynn, Fuller (1897).

D. cristata x *D. marginalis* - Merrimac, Dodge (1894); Boxford, Davenport (1895).

D. cristata, var. *clintoniana* x *D. spinulosa*, var. *intermedia* - Lynn, Fuller (1897).

D. marginalis (L.) Gray. Marginal Shield-Fern. Rocky places in woods; common. - Amesbury, Bean & H. 7737 (1951); Merrimac, Bean & H. 7721 (1951); Haverhill, H. 453 (1932); Boxford, M.E. Perley (1885); West Newbury, H. 9442 (1954); Newburyport, Dodge (1892); Georgetown, Horner (1897); Ipswich, Oakes; Middleton, Bean & H. 7771 (1951); Danvers, Oakes; Wenham, Sears (1905); Essex, H. 11978 (1955); Manchester, Hubbard (1913); Beverly, Robinson (1871); Salem, Robinson (1971); Saugus, H. 7794 (1951); Marblehead, Cole (1946).

D. marginalis, forma *elegans* (J. Robins.) F.W. Gray - Chebacco Woods, Robinson (1865); Swampscott, Robinson (1865) TYPE.

POLYSTICHUM Roth Shield-Fern.

P. acrostichoides (Michx.) Schott. Christmas Fern. Rocky woods; common. - Merrimac, Bean & H. 7715 (1951); Haverhill, H. 184 (1931); Georgetown, Horner; Ipswich, H. 18170 (1958); Topsfield, G.F. Dow; Danvers, Frost (1884); Chebacco Woods, Robinson (1865); Manchester, Oakes; Saugus, H. 7795 (1951); Marblehead, Cole (1930).

P. acrostichoides, forma *incisum* (Gray) Gilbert - Merrimac, Bean & H. 7716 (1951); Andover, Pease 33 (1900); Boxford, Alcott (1881); West Newbury, Conant (1879); Georgetown, Horner (1875); Salem, Sears (1884).

P. acrostichoides, forma *cristatum* Clute - Merrimac, Bean & H. 7717 (1951).

DENNSTAEDTIA Bernh.

D. punctilobula (Michx.) Moore. Hay-scented Fern. Shady places in moist woods; common. - Amesbury, H. 18229 (1958); Merrimac, H. 18363 (1958); Haverhill, H. 456 (1932); Lawrence, H. 10072 (1955); Georgetown, Horner (1881); Ipswich, H. 18173 (1958); Wenham, Sears (1905); Hamilton, Hubbard (1913); Essex Woods, Robinson (1875); Gloucester, Magnolia, J.M. Hunnewell (1899); Rockport, H. 10295 (1955); Manchester, Oakes; Beverly, Robinson (1872); Peabody, Mackintosh (1933). [MNS, Snyder (1971)].

ATHYRIUM Roth

A. thelypterioides (Michx.) Desv. Silvery Spleenwort. Rich woods; scarce. - Amesbury, Eaton (1897); Andover, Pease 9867 (1906); Ipswich, Oakes; Lynnfield, Russell; Swampscott, Robinson (1871).

A. filix-femina (L.) Roth, var. michauxii (Spreng.) Farw. Lady Fern. Low woods and thickets; occasional. - Merrimac, H. 18323 (1958); Georgetown, Horner (1895); Rowley, H. 11935 (1955); Ipswich, Plum Island, H., McGregor & Schweinfurth 12649 (1957); Middelton, Horner; Hamilton, H. 10776 (1955); Rockport, Rich (1897).

A. filix-femina, var. *michauxii,* forma *elatius* (Link) Clute - Rockport, Churchill (1887).

A. filix-femina, var. *michauxii,* forma *rubellum* (Gilbert) Farw. The most common form. - Salisbury, Eaton (1899); Amesbury, Eaton (1899); Haverhill, H. 682 (1932); Groveland, M.E. Perley (1885); Georgetown, Horner; Ipswich, Oakes; Middleton, Bean (1936); Danvers, A.F. Gray; Manchester, Hubbard (1913); Beverly, Robinson (1865).

A. filix-femina, var. *michauxii,* forma *confertum* (Butters) Fern. - Ipswich, Plum Island, H., McGregor & Schweinfurth 12648 (1957).

ASPLENIUM L. Spleenwort.

A. trichomanes L. Maidenhair-Spleenwort. Shaded rock crevices; occasional. - Haverhill, H. 1489 (1934); North Andover, Root (1960); Boxford, H. 8294 (1953); Georgetown, Horner; Danvers, Oakes; Hamilton, Eaton (1896); Salem, Cole (1905); Swampscott, Robinson (1872); Marblehead, Robinson (1872).

A. platyneuron (L.) Oakes. Ebony-Spleenwort. Wooded rocky banks; occasional. - Salisbury, Eaton (1898); Haverhill, H. 1488 (1934); Andover, Pease 53 (1900); Boxford, M.E. Perley (1881); West Newbury, H. 1268 (1934); Rowley, Oakes; Topsfield, Sears (1901); Middleton, Frost (1878); Danvers, Oakes; Essex Woods, Robinson (1874); Beverly, Lowell (1846); Salem, Cole (1905); Peabody, Robinson (1872); Swampscott, Cole (1889).

A. platyneuron, forma furcatum Clute - Beverly, Robinson (1872).

A. platyneuron, var. *incisum* (Howe) Robins. - Georgetown, Horner; Lynnfield, Mrs. J. Munroe (1889).

WOODWARDIA Sm. Chain-Fern.

W. Virginica (L.) Sm. Virginian Chain-Fern. Swamps and borders of woods; frequent. - Andover, Pease 89 (1900); Boxford, H. 13398 (1957); Georgetown, Horner; Danvers, Oakes; Wenham, H. 13758 (1957); Essex, Winslow 10181 (1910); Rockport, H. 19035 (1958); Beverly, Robinson (1876); Peabody, Mackintosh (1899); Saugus, Bean (1936); [Magnolia Snyder (1950s)]

W. areolata (L.) Moore. Netted Chain-Fern. Swamps; rare. - Salisbury, Eaton; Amesbury, Winslow 10177 (1910); Georgetown, Horner; Essex, Dodge (1896); Gloucester, Magnolia, Robinson & Sears (1877); Manchester, Sears (1885); Lynn, Sears (1901); Swampscott, C.C. Kingman (1898).

ADIANTUM L. Maidenhair.

A. pedatum L. Maidenhair-Fern. Rich hardwoods; occasional. - Haverhill, H. 1288 (1934); Andover, Knowlton (1903); North Andover, Bean (1947); Boxford, L.B. Twitchell (1883); Georgetown, Horner; Danvers, Oakes; Salem, Sears (1885); Peabody, Collins; Swampscott, Cole (1888).

PTERIDIUM Gleditsch Bracken.

P. aquilinum (L.) Kuhn, var. *latiusculum* (Desv.) Underw. Pasture-Brake. Dry sterile woods and clearings; common. - Salisbury, H. 18985 (1958); North Andover, Root (1957); Boxford, H. 18536 (1958); Georgetown, Horner; Ipswich, Lowell; Danvers, Robinson (1868); Rockport, Robinson (1875); Manchester, Robinson (1876); Peabody, Mackintosh (1933); [MNS, Snyder (1950s on)].

POLYPODIUM L. Polypody.

P. virginianum L. Rock-Polypody. Shaded rocks; common. - Haverhill, H. 348 (1932); Andover, Pease 89 (1900); Boxford, M.E. Perley (1881); Georgetown, Horner (1881); Rowley; Ipswich; Topsfield, H.R. Towne; Essex Woods, Robinson (1873); Gloucester, Magnolia, J.M. Hunnewell (1898); Rockport, Smith & R.C. Gates (1928); Manchester, Hubbard (1913); Salem, Cole (1886); Salem, Baker's Island, Robinson (1892); Saugus, H. 7789 (1951), [MNS, 5 Dec. (1972)].

MARSILEACEAE (Marsilea Family)

MARSILEA L.

M. quadrifolia L. Introduced from Europe. - Boxford, Eaton (1900); Salem, Cole (1899), A.P. Morse (1912). The Boxford station was found prior to 1900 and is still flourishing. The Salem station was known to Robinson some years before 1880, there appear to have been no collections made since 1912.

TAXACEAE (Yew Family)

TAXUS L. Yew.

T. canadensis Marsh. Ground Hemlock. Rich woods; local. - Salisbury, H. 9476 (1954); Amesbury, Knowlton (1906); Danvers, Sears (1872); Gloucester, Magnolia, H. 19063 (1959).

PINACEAE (Pine Family)

TSUGA (Endl.) Carr. Hemlock.

T. canadensis (L.) Carr. Rocky woods; frequent. - Salisbury, H. 9477 (1954); Amesbury, H. 9473 (1954); Haverhill, H. 279 (1932); Andover, Pease 1892 (1903); North Andover, H. 9453 (1954); Boxford, Robinson (1890); West Newbury, H. 9447 (1954); Middleton, Sears (1885); Wenham, Robinson (1876); Rockport, L.C. & R.C. Smith 845 (1928); Manchester, Hubbard (1913); Saugus, Bean & H. 9795 (1955).

PICEA Dietr. Spruce.

P. rubens Sarg. Red Spruce. Low ground; rare. - Methuen, H. 7663 (1951); Andover, H. 17920 (1958); Wenham, Robinson (1884); Rockport, Sears (1897).

P. mariana (Mill.) BSP. Black Spruce. Swamps; local. - Methuen, H. 7662 (1951); Andover, Pease 407 (1902); Boxford, Duncan & H. 7658 (1951); Ipswich, W. Faxon (1896); Danvers, Sears (1888); Essex, Sears (1897); Manchester, Hubbard (1913).

LARIX Mill. Larch.

L. laricina (DuRoi) K. Koch. American Larch. Swamps; local. - Methuen, H. 7664 (1951); Andover, H. 9464 (1954); Boxford, H. 8296 (1953); Ipswich; Topsfield, A.M. Dow; Danvers, Frost (1881); Wenham, Sears (1883).

L. decidua Mill. European Larch. Introduced from Europe. Commonly planted and rarely spreading from cultivation. - North Andover, Root (1957); Newbury, Bean (1946).

PINUS L. Pine

P. strobus L. White Pine. Common throughout the County. - Salisbury, H. 9480 (1954); Amesbury, H. 9438 (1954); Haverhill, H. 284 (1932); Methuen, S. Havens (1884); Andover, Pease 3735 (1904); North Andover, H. 9457 (1954); Boxford, M.E. Perley (1881); Groveland, H. 9452 (1954); West Newbury, H. 9445 (1954); Newbury, White 235 (1913); Georgetown, H. 9472 (1954); Rowley, H. 9432 (1954); Topsfield, H.R. Towne; Danvers, Frost (1881); Wenham, Robinson (1875); Manchester, Hubbard (1913); Beverly, Sears (1885); Peabody, Sears (1876); Marblehead H. 12606 (1956).

P. resinosa Ait. Red Pine. Frequently planted; apparently native in a few localities. - Andover, Pease 3642 (1904); Boxford, Robinson (1876); Peabody, Mackintosh (1933).

P. sylvestris L. Scotch Pine. Native of Europe. Frequently cultivated and rarely spreading by seed. - Boxford, Kennedy (1906); Newburyport, B.L. Robinson (1903); Danvers, Sears (1885); Wenham, Robinson (1884); Salem, Cole (1921); Lynnfield, Rodman (1887); Lynn, Buck (1898).

P. rigida Mill. Pitch Pine. Sandy soil; common. - Amesbury, H. 9437 (1954); Merrimac, H. 9441 (1954); Haverhill, H. 285 (1932); Andover, Pease 124 (1902); North Andover, H. 9455 (1954); Boxford, Robinson (1878); Newbury, White 140 (1913); Newbury, Plum Island, H. McGregor & Schweinfurth 12735 (1957); Georgetown, H. 9469 (1954); Rowley, Plum Island, H., McGregor & Schweinfurth 12734 (1957); Ipswich, H. 13935 (1957); Danvers, Frost (1881); Wenham, Mrs. Kimball (1885); Rockport, H. 19024 (1958); Manchester, Deane (1887); Beverly, Robinson (1875); Salem, A.P. Morse (1912); Marblehead, H. 12605 (1956).

THUJA L. Arbor Vitae.

T. occidentalis L. Native to the North and West of the County. Widely planted and rarely spreading by seed. - Andover, Pease 131 (1901).

CHAMAECYPARIS Spach White Cedar

C. thyoides (L.) BSP. Swamps; chiefly found in the eastern portion of the County. - Amesbury, Bean & H. 7725 (1951); Andover, Pease 121 (1900); North Andover, H. 8298 (1953); Boxford, Hopkins & Pease 23166 (1932); Topsfield, Pease 23202 (1932); Middleton, Hopkins & Pease 24049 (1934); Wenham, Robinson (1875); Essex, Duncan & H. 7652 (1951); Manchester, Duncan & H. 7651 (1951); Peabody, H. 7659 (1951); Lynnfield, H. 7801 (1952).

JUNIPERUS L. Juniper.

J. communis L., var. *depressa* Pursh. Ground Juniper. Pastures and sterile hillsides; common. - Salisbury, H. 9478 (1954); Amesbury, H. 9433 (1954); Merrimac, H. 9439 (1954); Haverhill, H. 189 (1931); Andover, Pease 122 (1901); North Andover, H. 9456 (1954); Boxford, Pease 23154

(1932); Groveland, H. 9451 (1954); West Newbury, Pease 23151 (1932); Newbury, H. 18471 (1958); Georgetown, Pease 23153 (1932); Rowley, Pease 23155 (1932); Ipswich, Oakes; Topsfield, Robinson (1875); Danvers, Frost (1881); Wenham, Robinson (1891); Essex, Hubbard (1913); Marblehead, Cole (1924).

J. communis, var. *saxatilis* Pallas - Salem, Forbes 132 (1903).

J. horizontalis moench Creeping Savin. Rocks or sandy banks; rare. - Newbury, Eaton (1885).

J. virginiana L. Red Cedar. Dry hills; common. - Salisbury, H. 9479 (1954); Amesbury, H. 9434 (1954); Merrimac, H. 9440 (1954); Haverhill, H. 287 (1932); Andover, Pease 1855 (1903); North Andover, H. 9454 (1954); Boxford, M.E. Perley (1884); Groveland, H. 9450 (1954); West Newbury, H. 9444 (1954); Newbury, H. 18472 (1958); Georgetown, H. 9470 (1954); Rowley, Pease 23145 (1932); Rowley, Plum Island, H., McGregor & Schweinfurth 12710 (1957); Ipswich, Oakes; Ipswich, Plum Island, H. 12104 (1956); Middleton, Robinson (1877); Wenham, Robinson (1888); Hamilton, Hubbard (1913); Beverly, Lowell (1847); Salem, Forbes (1903); Lynn, Buck (1898); Swampscott, A.P. Morse (1925); [MNS, Snyder (1950s on)].

TYPHACEAE (Cat-tail Family)

TYPHA L. Cat-tail.

T. latifolia L. Common Cat-tail. Marshes and shallow water; common. - Haverhill, H. 501 (1932); Andover, Pease 135 (1902); North Andover, H. 12918 (1957); Boxford, M.E. Perley (1882); West Newbury, H. 9114 (1953); Newburyport, Robinson (1875); Ipswich, H. 14042 (1957); Ipswich, Plum Island, H., McGregor & Schweinfurth 12660 (1957); Manchester, H. 10721 (1955); Beverly, Hubbard (1913); Salem, Cole (1909); Saugus, H. 10882 (1955); Lynn, H. 13170 (1957); Marblehead, Cole (1930).

T. latifolia, forma *ambigua* (Sonder) Kronf. - Haverhill, H. 502 (1932).

T. angustifolia L. Narrow-leaved Cat-tail. Marshes and shallow water; more common in brackish or saline conditions. - Salisbury, H. 10476 (1955); Haverhill, Bradbord, H. 13151 (1957); Newbury, Williams (1899); Rowley, H. 11948 (1955); Rowley, Plum Island, H. 12457 (1956); Ipswich, Fassett 435 (1922); Ipswich, Plum Island, H., McGregor & Schweinfurth 12657 (1957); Middleton, Pease 28056 (1940); Rockport, L.B. & R.C. Smith 2702 (1933); Salem, Sears (1880); Marblehead, Cole (1930).

SPARGANIACEAE (Bur-reed Family)

SPARGANIUM L. Bur-reed.

S. eurycarpum Engelm. Shallow water; occasional. - Amesbury, H. 13014 (1957); Haverhill, H. 785 (1932); West Newbury, N.E.B.C. (1951); Newburyport; H. 866 (1933); Middleton, Bean (1950); Beverly, F.H. Peabody (1894).

S. androcladum (Engelm.) Morong. Shores and shallow waters; common. - Merrimac, H. 13697 (1957); North Andover, Root (1945); Boxford, M.E. Perley (1882); Groveland, H. 9088 (1953); Danvers, Sears (1885); Wenham, Sears (1886); Hamilton, H. 12811 (1957); Chebacco, Robinson (1875).

S. americanum Nutt. Shores and shallow water; common. - Merrimac; H. 9218 (1953); Haverhill, H. 569 (1932); Andover, Pease 132 (1900); Boxford, H. 8191 (1952); Groveland, M.E. Perley (1885); Georgetown, H. 12982 (1957); Ipswich, H. 18122 (1958); Topsfield, H.R. Towne; Middleton, H. 12904 (1957); Danvers, Cushman 2827 (1908); Rockport, Smith 2714 (1933); Manchester, Hubbard (1913); Beverly, Lowell (1847); Peabody, Mackintosh (1933); Lynn, Forbes (1911).

S. chlorocarpum Rydb. Shores and shallow water; frequent. - Haverhill, H. 3 (1931); North Andover, H. 9499 (1954); Boxford, H. 8195 (1952); Groveland, H. 9076 (1953); West Newbury, N.E.B.C. (1951); Ipswich, H. 18123 (1958); Middleton, Bean (1950); Danvers, H. 11615 (1955); Manchester, H. 10749 (1955).

S. angustifolium Michx. Shores and deep or shallow water; occasional. - Boxford, H. 13419 (1957); Groveland, H. 9094 (1953); Rowley, H. 13521 (1957).

S. multipedunculatum (Morong) Rydb. Lakes and ponds; rare. - Beverly, Robinson (1875).

S. fluctuans (morong) Robins. Cold ponds; occasional. - Amesbury, H. 13027 (1957); Boxford, H. 13196 (1957); West Newbury, Robinson (1879); Georgetown, H. 12963 (1957); Rowley, H. 13518 (1957); Ipswich, H. 13423 (1957); Topsfield, H. 13288 (1957).

S. minimum (Hartm.) Fries. Shallow pools; occasional. - Boxford, H. 8965 (1953); Groveland, H. 12887 (1957); Georgetown, H. 12959 (1957).

ZOSTERACEAE (Pondweed Family)

ZOSTERA L. Eelgrass.

Z. marina L., var. *stenophylla* Aschers & Graebn. Shallow sea-water; probably common but seldom collected. - Ipswich, St. John 17642 (1926); Gloucester, Sears (1880); Beverly, Lowell (1847); Salem, Russell (1853); Lynn, Torrey (1914); Marblehead, Sears (1887); Nahant, C.P. (1853).

POTAMOGETON L. Pondweed.

P. robbinsii Oakes. Muddy waters; common. - Haverhill, Lunt (1913); Methuen, H. 12857 (1957); Andover, Pease 2458 (1903); North Andover, H. 12946 (1957); Boxford, H. 13415 (1957); Groveland, H. 12876 (1957); West Newbury, H. 9143 (1953); Georgetown, H. 12977 (1957); Ipswich, H. 13480 (1957); Topsfield, H. 8886 (1953); Wenham, Oakes; Hamilton,Fernald & Svenson 732 (1928); North Beverly, Deane (1886); Lynnfield, H. 14122 (1957); Lynn, Young (1880).

P. zosteriformis Fern. Quiet waters; occasional. - Boxford, H. 13554 (1957); Groveland, H. 9104 (1953); Georgetown, H. 12987 (1957); Ipswich, H. 13265 (1957); Topsfield, H. 13234 (1957); Lynnfield, Young (1879); Saugus, Svenson 3003 (1928).

P. foliosus Raf. Fresh or brackish water; rare. - Boxford, Robinson (1876); Chebacco, Robinson (1875).

P. foliosus, var. *macellus* Fern. - Georgetown, H. 12954 (1957).

P. pusillus L. Basic or alkaline waters; rare. - Andover, Pease 2837 (1903); North Andover, H. 9503 (1954); Boxford, H. 8963 (1953); Essex, H. 13880 (1957); Saugus, Svenson 3000 (1928).

P. gemmiparus Robbins. Quiet waters; rare. - west Gloucester, (1892).

P. obtusifolius Mert. & Koch. Cold streams and ponds; occasional. - Parker River, Robinson (1879); Wenham, Robinson (1875); West Peabody, H. 9340 (1953).

P. berchtoldi Fieber. Shallow water; occasional. - North Andover, Pease 2637 (1903); Boxford, H. 8970 (1953); Peabody, H. 13323 (1957); Lynn, Young (1879).

P. berchtoldi, var. *acuminatus* Fieber - Danvers, Russell, (1857); Marblehead, Russell (1859).

P. berchtoldi, var. *tenuissimus* (Mert. & Koch) Fern. - Merrimac, H. 13727 (1957); Andover, Pease 2607 (1903); Boxford, H. 13416 (1957); Topsfield, H. 13598 (1957); Wenham, Robinson (1876); Essex, Robinson (1875); North Beverly, Deane 89 (1886).

P. berchtoldi, var. *lacunatus* (Hagstr.) Fern. - Boxford, H. 13266 (1957).

P. vaseyi Robbins. Muddy water; rare. - Amesbury, H. 13008 (1957); Wenham, Robinson (1878).

P. spirillus Tuckerm. Quiet waters; common. - Amesbury, H. 13009 (1957); Merrimac, H. 13731 (1957); Haverhill, H. 616 (1932); Andover, Pease 137 (1901); North Andover, H. 12947 (1957); Boxford, Robinson (1876); Georgetown, H. 12949 (1957); Ipswich, Morong; Topsfield, H. 8885 (1953); Hamilton, H. 12828 (1957); Chebacco, Robinson.

P. capillaceus Poir. Quiet waters; frequent. - Amesbury, H. 13023 (1957); Andover, Pease 4251 (1904); Boxford, H. 8196 (1952); Danvers, Lowell (1848); Hamilton, Morong; Peabody, Russell (1854); Lynn, H. 13182 (1957).

P. epihydrus Raf., var. *nuttallii* (C. & S) Fern. Quiet waters; common. - Amesbury, H. 13022 (1957); Haverhill, H. 615 (1932); Methuen, Root (1949); Andover, Pease 136 (1901); North Andover, H. 13077 (1957); Boxford, Robinson (1875); Rowley, H. 13523 (1957); Ipswich, H. 13267 (1957); Topsfield, B.B. Towne; Middleton, H. 12910 (1957); Danvers, H. 9030 (1953); Hamilton, H. 12809 (1957); Essex, H. 13379 (1957); Gloucester, H. 13822 (1957); Manchester, H. 10733 (1955); Peabody, H. 13313 (1957); Lynnfield, Young (1879); Saugus, Robinson (1874).

P. amplifolius Tuckerm. Lake and streams; common. - Haverhill, Bradford, H. 13166 (1957); Boxford, H. 8972 (1953); Groveland, H. 12875 (1957); Georgetown, H. 12950 (1957); Rowley, H. 13522 (1957); Ipswich, H. 13260 (1957); Topsfield, H. 13229 (1957); Middleton, H. 12911 (1957); Wenham, Robinson (1875); Lynn, H. 13402 (1957).

P. pulcher Tuckerm. Muddy acid waters; common. - Andover, Pease 2011 (1903); North Andover, H. 13074 (1957); Boxford, H. 8728 (1953); West Newbury, Robinson (1879); Ipswich H. 13468 (1957); Topsfield, H. 8889 (1953); Middleton, H. 18111 (1958); Danvers, H. 9029 (1953); Wenham, C.E. Faxon; Hamilton, H. 9416 (1953); Chebacco, Robinson (1875); Beverly, Robinson (1875).

P. nodosus Poir. Ponds and streams; rare. - Methuen, Boott (1865); Ipswich River, Robinson (1876).

P. gramineus L. Ponds and streams; occasional. Topsfield, A.F. Towne; Wenham, Robinson (1876).

P. gramineus, var. *maximus* Morong - Boxford, H. 13668 (1957); Ipswich, H. 13449 (1957); Wenham, H. 13098 (1957).

P. gramineus, var. *myriophyllus* Robbins - Merrimac, H. 13700 (1957); Boxford, H. 13806 (1957); Georgetown, Sears (1876); Rowley, H. 13502 (1957); North Beverly, Deane (1886).

P. gramineus x *P. illinoensis* - Wenham, E. & C.E. Faxon (1880).

P. gramineus x *P. perfoliatus,* var. *bupleuroides* - Wenham Pond, E. Faxon (1880).

P. illinoensis Morong. Calcareous ponds; rare. - Wenham, Wenham Pond, Robbins. (1867); Pleasant Pond, Robinson (1875).

P. natans L. Lakes and quiet streams; common. - Amesbury, H. 13025 (1957); Haverhill, H. 190 (1931); Methuen, Root (1949); Andover, H. 12788 (1957); North Andover, H. 9510 (1954); Boxford, H. 8961 (1953); Groveland, H. 12882 (1957); Georgetown, H. 12960 (1957); Ipswich, H. 13259 (1957); Topsfield, H. 13618 (1957); Wenham, C.E. Faxon (1880); Peabody, Mackintosh (1933).

P. oakesianus Robbins. Acid pools; rare. - Beverly, Robinson (1875); Lynn, E. Faxon (1880).

P. praelongus Wulfen. Cold lakes and streams; occasional. - Haverhill, Bradford, H. 539 (1932); Wenham, Wenham Lake, Robinson (1875); Pleasant Pond, C.E. Faxon (1880).

P. perfoliatus L., var. *bupleuroides* (Fern.) Farw. Calcareous or brackish waters; frequent. - Merrimac, H. 13703 (1957); Haverhill, Bradford, H. 540 (1932); Andover, Boott (1868); West Newbury, H. 9144 (1953); Newbury, Plum Island, H. 11290 (1955); Georgetown, H. 12951 (1957); Ipswich, H. 13264 (1957); Wenham, Robinson (1875); Beverly, Fernald & Svenson 751 (1928); Lynnfield, H. 14118 (1957); Lynn, H. 13186 (1957).

RUPPIA L. Ditch-grass.

R. maritima L., var. *longipes* Hagstr. Saline pools and ditches; frequent. - Ipswich, Oakes; Danvers, Sears (1879); Rockport, L.B. & R.C. Smith 2703 (1933); Salem, Sears (1887).

R. maritima, var. *rostrata* Agardh - Newbury, Bean (1913); Ipswich, Sears (1886); Gloucester, Seymour 4000 (1933); Salem, Sears (1887).

R. maritima, var. *subcapitata* Fern. & Wieg. - Beverly, Lowell (1849); Salem, Winter Island, Russell.

ZANNICHELLIA L. Horned Pondweed.

Z. palustris L., var. *major* (Boenn.) W.D.J. Koch. Fresh, brackish or alkaline waters; rare. - Salem, Robinson (1879); Saugus, Svenson 3001 (1928).

NAJADACEAE (Naiad Family)

NAJAS L. Naiad.

N. flexilis (Willd.) Rostk. & Schmidt. Streams and ponds; common. - Amesbury, H. 13024 (1957); Merrimac, H. 13687 (1957); Haverihill, H. 503 (1932); Andover, Pease 3068 (1903); North Andover, H. 12937 (1957); Boxford, H. 13223 (1957); Rowley, H. 13515 (1957); Ipswich, Morong (1875); Topsfield, H. 13601 (1957); Middleton, Phillips (1910); Peabody, H. 13317 (1957); Saugus, Robinson (1874).

N. gracillima (A.Br.) Magnus. Muddy, peaty or sandy ponds; rare. - North Andover, Pease 2636 (1903); Newburyport, Robinson (1886).

JUNCAGINACEAE (Arrow-grass Family)

TRIGLOCHIN L. Arrow-grass.

T. maritima L. Salt marsh; common. - Salisbury, Alcott (1882); Newbury, H. 10398 (1955); Newbury, Plum Island, H. 868 (1933); Ipswich, Oakes; Ipswich, Plum Island, H. 12112 (1956); Danvers, Robinson (1874); Essex, H. 9364 (1953); Gloucester, Sanford (1954); Rockport, Cushman 3368 (1908); Manchester, Hubbard (1913); Salem, Cole (1889); Lynn, F. Boott (1816).

SCHEUCHZERIA L.

S. palustris L., var. *americana* Fern. Peaty bogs; rare. - Peabody, Cedar Pond, Sears (1886); Essex County, Oakes, Pickering. Robinson states " 'Wenham swamp, 1824' (memo. Dr. Chas. Pickering)."

ALISMATACEAE (Water-plantain Family)

ALISMA L. Water-plantain.

A. triviale Pursh. Shallow water; rare. - Rockport, Smith 2722 (1933); Marblehead, Kennedy (1887).

A. subcordatum Raf. Shallow water; common. - Haverhill, H. 504 (1932); Methuen, H. 8920 (1953); North Andover, Morss (1885); Boxford, Williams (1899); Groveland, M.E. Perley (1885); West Newbury, H. 9138 (1953); Georgetown, Horner; Rowley, Williams (1906); Topsfield, H.R. Towne; Danvers, Sears (1884); Beverly, Lowell (1846); Salem, Phippen (1857); Saugus, H. 10875 (1955); Swampscott, H. 10665 (1955); Marblehead, Cole (1927).

LOPHOTOCARPUS Th. Durand

L. spongiosus (Engelm.) J.G. Sm., Tidal mud of estuaries; rare. - Newburyport, Eaton; Rowley, Fassett 1070 (1923); Saugus, Fassett & Peattie (1921).

SAGITTARIA L. Arrowhead.

S. subulata (L.) Buchenau, var. *gracillima* (S. Wats.) J.G. Sm. Deep waters of streams; rare. - Methuen, Root (1949).

S. rigida Pursh. Calcareous or brackish mud or water; occasional. - Haverhill, Root (1949); Lawrence, Robinson (1879); Pease 146 (1902); West Newbury, H. 11113 (1955).

S. rigida, forma *elliptica* (Engelm.) Fern. - South Lawrence, Pease 145 (1902); West Newbury, N.E.B.C. (1951); Newburyport, Robinson (1903).

S. graminea Michx. Wet sand or mud or in shallow water; common. - Amesbury, Eaton (1898); Merrimac, H. 13698 (1957); Haverhill, H. 670 (1932); Methuen, H. 8954 (1953); Lawrence, Sears (1885); North Andover, H. 12934 (1957); Boxford, H. 13556 (1957); Groveland, H. 9101 (1953); Newburyport, Moulton (1886); Wenham, Robinson (1877); Essex, H. 13367 (1957).

S. engelmannia J.G. Sm. Wet sand and peat; rare. - Gloucester, Magnolia, J.M. Hunnewell (1901).

S. latifolia Willd. In water or wet places; common. - Merrimac, H. 13709 (1957); Haverhill, H. 4 (1931); Boxford, Williams (1899); Georgetown, Horner; Gloucester, Magnolia, J.M. Hunnewell (1899); Manchester, H. 10710 (1955); Beverly, Robinson (1875).

S. latifolia, forma *hastata* (Pursh) Robins. - Amesbury, Bean (1913); Merrimac, H. 13686 (1957); Haverhill, H. 542 (1932); North Andover, Morss (1885); Boxford, M.E. Perley (1885); Groveland, M.E. Perley (1885); Newburyport, Williams (1898); Ipswich, Forbes (1913); Hamilton, Phippen (1857); Peabody, Mackintosh (1925).

S. latifolia, forma *gracilis* (Pursh) Robins. - Salisbury, Eaton; Amesbury, H. 9155 (1953); Merrimac, H. 13674 (1957); North Andover, Morss (1885); Georgetown, Horner; Rowley, H. 13498 (1957); Wenham, Robinson (1875); West Gloucester, Robinson (1877).

S. latifolia, var. *obtusa* (Muhl.) Wieg. - Rowley, H. 13509 (1957); Topsfield, H.R. Towne; Hamilton, Hunnewll 4620 (1916); Gloucester, Magnolia, Fitch (1883); Salem, Cole (1888); Peabody, Mackintosh (1933).

HYDROCHARITACEAE (Frog's-bit Family)

ELODEA Michx. Waterweed.

E. canadensis Michx. Quiet waters; frequent. - Amesbury, H. 9187 (1953); Merrimac, H. 13702 (1957); Andover, J. Blake 963 (1883); North Andover, Sears (1885); West Newbury, N.E.B.C. (1951); Topsfield, H. 8887 (1953); Middleton, H. 12892 (1957); Wenham, H. 13120 (1957); Peabody, H. 9338 (1953); Lynnfield, H. 14115 (1957).

E. nuttallii (Planch.) St. John. Shallow waters; frequent. - Methuen, H. 12832 (1957); North Andover, Pease 2638 (1903); Boxford, H. 13749 (1957); Georgetown, H. 12976 (1957); Ipswich, H. 13266 (1957); Danvers, H. 9032 (1953); Essex, H. 13371 (1957); Peabody, H. 9291 (1953); Lynn, H. 13188 (1957).

VALLISNERIA L. Eelgrass.

V. americana Michx. Water-celery. Quiet waters; frequent. - Merrimac, H. 13704 (1957); Andover, H. 12786 (1957); North Andover, Pease 2639 (1903); Boxford, H. 13214 (1957); Ipswich, H. 13257 (1957); Topsfield, H. 13232 (1957); Middleton, H. 12894 (1957); Wenham, Robinson (1875); Hamilton, H. 12808 (1957); Essex, H. 13352 (1957); Beverly, Deane (1886); [Salem, Snyder (1950s)] ; Lynn, Young (1879).

GRAMINEAE (Grass Family)

BROMUS L. Brome-grass.

B. latiglumis (Shear) Hitchc. Rich thickets and woods; occasional. - Merrimac, Eaton (1900); Haverhill, H. 7784 (1952); North Andover, H. 11020 (1955); West Newbury, H. 11155 (1955); Ipswich, Lowell; Beverly.

B. purgans L. Rich woods; rare. - Boxford, H. 18540 (1958).

B. purgans, forma *laevivaginatus* Wieg. - Boxford, Williams (1906).

B. ciliatus L., var. *intonsus* Fern. Thickets and rocky slopes; occasional. - Andover, Pease 2330 (1903); North Andover, H. 11016 (1955); West Newbury, Robinson (1879); Ipswich, Oakes; Wenham, Robinson (1875); Lynnfield, Bean (1916).

B. dudleyi Fern. rare. - South Lawrence Grant (1879).

B. kalmii Gray. Rich thickets and woods; rare. - Beverly, Lowell (1846).

B. inermis Leyss. Awnless Brome. Introduced from Europe. Roadsides and fields; occasional. - North Andover, H. 18041 (1958); Topsfield, H. 18306 (1958).

B. marginatus Nees. Introduced from the West. Roadsides; rare. - Rowley, H. & Wetherall 9559 (1954).

B. secalinus L. Cheat or Chess. Naturalized from Europe. Waste places; frequent. - Andover, Hubbard 637 (1913); Boxford, H. 8632 (1953); Newburyport, H. & Wetherall 9565 (1954); Newburyport, Plum Island, Dodge (1890); Rowley, H. 9541 (1954); Ipswich, Oakes; Salem, Robinson (1879); Swampscott, Churchill (1891).

B. racemosus L. Adventive from Europe. Waste places; rare. - Danvers, Sears (1879); West Gloucester, Robinson (1877).

B. commutatus Schrad. Naturalized from Europe. Waste places; occasional. - Essex County, Oakes; Rowley, H. 10423 (1955); Ipswich, H. & Wetherall 9558 (1954); Danvers, H. 7964 (1952); Beverly, Robinson (1877); Salem, Morong (1879).

B. mollis L. Soft Chess. Naturalized from Europe. Waste places; occasional. - Essex County, Oakes; Boxford, H. 8146 (1952); West Newbury, H. 11109 (1955); Ipswich, Plum Island, H. 12097 (1956); Salem, H. 9711 (1955); Swampscott, Conant (1890).

B. mollis, forma *leiostychys* (Hartm.) Fern. Naturalized from Europe. - Rowley, H. 9544 (1954); Salem, Robinson (1879).

B. rigidus Roth. Adventive from Europe. Waste places; rare. - Boxford, H. 9635 (1954).

B. sterilis L. Naturalized from Europe. Waste places; rare. - Haverhill, Ward Hill, H. 8313 (1953); Boxford, M.E. Perley (1882); Essex Co., Conant.

B. tectorum L. Naturalized from Euorpe. Waste places; common in the early season. - Salisbury, H. 10494 (1955); Haverhill, Bradford, H. 10457 (1955); Methuen, H. 7986 (1952); North Andover,

H. 17999 (1958); Boxford, M.E. Perley (1882); Newburyport, Plum Island, H. 12002 (1956); Newbury, Plum Island, H. 12033 (1956); Rowley, H. 9540 (1954); Ipswich, H. 10232 (1955); Topsfield, H. 18299 (1958); Danvers, Sears (1884); Gloucester, H. 10317 (1955); Beverly, Conant (1881); Salem, H. 9706 (1955); Lynnfield, H. 8335 (1953); Saugus, H. 18630 (1958); Lynn, Porter (1950); Marblehead, Bean & H. 9727 (1955); Nahant, S.S. Clark 36 (1916).

FESTUCA L. Fescue-grass.

F. ovina L. Sheep's-Fescue. Naturalized from Europe. Dry soil; frequent. - Lawrence, H. 9882 (1955); West Newbury, H. 7896 (1952); Ipswich, Deane (1926); Gloucester, H. 10149 (1955); Rockport, Smith 693 (1927); Manchester, Hubbard 560 (1913); Beverly Farms, Hubbard 582 (1913); Lynnfield, H. 8333 (1953); Nahant, S.S. Clark (1916).

F. ovina, forma *hispidula* (Hack.) Holmb. Naturalized from Europe. - Amesbury, H. 10545a (1955); Boxford, H. 7866 (1953); Manchester, Hubbard 560a (1913).

F. ovina, var. *duriuscula* (L.) W.D.J. Koch. Naturalized from Euorpe. - Salisbury, H. 10503 (1955); Topsfield, H. 8062 (1952); Gloucester, Magnolia, Clarke (1912); Lynnfield, H. 8331 (1953).

F. capillata Lam. Naturalized from Europe. Dry open soil; frequent. - Amesbury, H. 10543 (1955); Merrimac, Bean & H. 8592 (1953); Andover, Pease 3610 (1904); Boxford, Fernald (1909); Groveland, Bean & H. 8517 (1953); West Newbury, H. 7895 (1952); Newburyport, H. 10370 (1955); Rowley, H. 7943 (1952); Danvers, Sears (1904); Nahant, S.S. Clark (1916).

F. rubra L. Sandy soils; common. - Salisbury, White 53 (1913); Amesbury, H. 10524 (1955); Merrimac, Bean & H. 8598 (1953); Haverhill, H. 371 (1932); Andover, Pease 3661 (1904); Newbury, Kents Island, Eaton (1896); Newbury, Plum Island, White 118 (1913); Rowley, Plum Island, H. 12047 (1956); Ipswich, Plum Island, H. 12120 (1956); Middleton, Bean & H. 7772 (1951); Gloucester, H. 12138 (1956); Rockport, C.E. Faxon & Robinson (1877); Beverly, Sears; Salem, Robinson (1879); Salem, Little Mirsery Island, H. 13295 (1957); Lynnfield, H. 8332 (1953); Saugus, Bean & H. 9775 (1955); Marblehead, H. 9698 (1955).

F. rubra, forma *squarrosa* (Fries) Holmb. - Gloucester, H. 12163 (1956).

F. rubra, forma *megastachys* (Gaudin) Holmb. - Amesbury, Eaton (1896).

F. rubra, var. *commutata* Gaudin - Amesbury, Eaton (1896); Newburyport, H. 10466 (1955).

F. rubra, var. *juncea* (Hack.) Richter - Newburyport, H. 10372 (1955).

F. elatior L. Meadow-Fescue. Naturalized from Europe. Meadows and roadsides; frequent. - Merrimac, H. 10592 (1955); Andover, Pease 2129 (1903); North Andover, H. 18000 (1958); Rowley, Plum Island, H. 12045 (1956); Middleton, H. 8699 (1953); Danvers, H. 7965 (1952); Wenham, Robinson (1875); Chebacco, Robinson (1879); Gloucester, Magnolia, Clarke (1912); Rockport, Cushman (1908); Manchester, Hubbard 574 (1913); Beverly, Lowell (1846); Salem, H. 9712 (1955).

F. elatior, forma *aristata* Holmb. Naturalized from Euorpe. - Boxford, H. 17694 (1958).

F. obtusa Biehler. Moist woods; not common. - Methuen, H. 7994 (1952); Andover, Pease 3644 (1904); Boxford, E. & C.E. Faxon (1878); Groveland, H. 8813 (1953); Newburyport, Plum Island,

Eaton; Rowley, Plum Island, H., McGregor & Schweinfurth 12715 (1957); Danvers, Oakes.

VULPIA K.C. Gmel.

V. octoflora (Walt.) Rydb., var. *tenella* (Willd.) Fern. Dry open soil; occasional. - Amesbury, Huntington (1902); Merrimac, Conant (1878); Andover, Ballardvale, Pease 3711 (1904); West Boxford, Fernald (1909); Ipswich, Oakes; Ipswich, Plum Island, Oakes; Middleton, Manning (1884); Rockport, C.E. Faxon & Robinson (1877); Peabody, Fuller (1897).

PUCCINELLIA Parl. Goose-grass.

P. maritima (Huds.) Parl. Near the coast; frequent. - Newburyport, Dodge (1890); Newbury, Conant (1878); Newbury, Plum Island, Pease 13672 (1912); Ipswich Beach, Hubbard 642 (1913); Essex H. 10254 (1955); Rockport, Smith 615 (1927); Manchester, Hubbard 577 (1913); Beverly, Lowell (1846); Lynn, B.M. Porter.

P. fasciculata (Torr.) Bickn. Seashore; uncommon. - Plum Island, Oakes; Rowley, B.M. Porter (1938).

P. distans (L.) Parl. Naturalized from Europe. Waste places near coast; occasional. - Essex County, Oakes; Plum Island, Robinson; Essex, B.M. Porter (1949); Salem, Robinson (1879); West Lynn, Hubbard 565 (1913).

GLYCERIA R.Br. Manna-grass.

G. borealis (Nash) Batchelder. Float-grass. Wet places or shallow water; frequent. - Amesbury, Huntington & Kennedy (1900); Merrimac, H. 3711 (1957); Methuen, H. 12836 (1957); Andover, Porter (1941); North Andover, H. 12935 (1957); Boxford, Sears (1878); Middleton, H. 8701 (1953); Gloucester, Magnolia, Clarke (1912).

G. septentrionalis Hitchc. Floating Manna-grass. Swamps; occasional. - Amesbury, H. 10519 (1955); Methuen, H. 18409 (1958); Georgetown, Porter (1935); Lynnfield, Young (1879).

G. acutiflora Torr. Pond margins; frequent. - Andover, Pease 2017 (1903); Boxford, M.E. Perley (1882); Groveland, Bean & H. 8476 (1953); West Newbury, Conant (1878); Newbury, Ogden, Schubert, Smith & Weatherby (1938); Ipswich, Oakes; Wenham, Rich (1890); Gloucester, Mangnolia, Clarke (1912); Rockport, Robinson (1877); Manchester, Hubbard 586 (1913); Beverly, Robinson (1879).

G. melicaria (Michx.) F.T. Hubbard. Wet woods; occasional. - East Haverhill, Sears (1901); North Andover, H. 17988 (1958); Essex County, Oakes.

G. obtusa (Muhl.) Trin. Swamps and wet places; frequent. - Haverhill, H. 5698 (1933); Methuen, H. 8950 (1953); Boxford, Pease 23170 (1932); Georgetown, Horner; Ipswich, Oakes; Wenham, Hunnewell (1914); Essex Woods, Lowell (1846); Gloucester, Annisquam, Church 701 (1940); Beverly, Lowell (1847); Salem, Robinson (1879).

G. canadensis (Michx.) Trin. Rattlesnake-grass. Damp thickets and shores; common. - Haverhill, H. 8 (1931); Methuen, H. 18408 (1958); Lawrence, H. 12291 (1956); Andover, Pease 3184 (1903); North Andover, H. 13069 (1957); Boxford, H.E. Parkhurst (1882); Newburyport, White 201 (1913); Ipswich, H. 18125 (1953); Middleton, H. 8709 (1953); Danvers, Lowell (1846); Wenham, Robinson (1879); Hamilton, Hubbard 647 (1913); Essex, Bean (1919); Gloucester, Annisquam,

Church 829 (1940); Rockport, Cushman (1908); Manchester, H. 10771 (1955); Beverly, Robinson (1879); Salem, Cole (1911); Peabody, Mackintosh (1933); Saugus, H. 10879 (1955); Marblehead, Cole (1886).

G. laxa Scribn. Swamps; occasional. - Amesbury, H. 18248 (1958); North Andover, H. 17966 (1958); Ipswich, H. 14018 (1957); Middleton, H. 18076 (1958).

G. striata (Lam.) Hitchc. Fowl-meadow Grass. Moist ground; common. - Salisbury, H. 12226 (1956); Amesbury, H. 18270 (1958); Merrimac, H. 18355 (1958); Haverhill, Bradford, Cushman (1908); Lawrence, H. 9869 (1955); Andover, Pease 178 (1901); North Andover, H. 17961 (1958); Boxford, M.E. Perley (1882); Groveland, Bean & H. 8375 (1953); West Newbury, Conant (1878); Georgetown, Horner; Byfield, Porter (1941); Ipswich, Church 91 (1936); Middleton, Porter (1941); Danvers, Lowell (1846); Wenham, Robinson (1875); West Gloucester, Church 702 (1940); Rockport, Smith 728 (1927); Manchester, Church 824 (1940); Beverly, Conant (1880); Salem, Cole (1886).

G. striata, var. *stricta* (Scribn.) Fern. - West Manchester, Hubbard 575 (1913).

G. grandis S. Wats. Reed-meadow Grass. Wet meadows and banks of streams; frequent. - Methuen, H. 9856 (1953); North Andover, H. 12923 (1957); West Newbury; Conant (1878); Newburyport, White 198 (1913); Georgetown, Horner; Wenham, Bean & H. 9999 (1955); Gloucester, Magnolia, Clarke (1912); Rockport, Rouleau & Smith 4318 (1945); Peabody, Fuller (1893); Lynnfield, Bean (1916); Lynn, Sears (1878); Swampscott, Porter (1938).

G. pallida (Torr.) Trin. Pools and pond margins; occasional. - Andover, Pease 8041 (1905); North Andover, H. 9494 (1954); Boxford, C.E. Faxon & Robinson (1877); Newburyport, Williams (1898); Topsfield, Wenham, Essex, Church 805 (1940); Gloucester, Magnolia, Clarke (1912); Manchester, Hubbard 617 (1913); Beverly, Lowell (1849); Peabody, H. 12749 (1957).

G. fernaldii (Hitchc.) St. John. Wet places; rare. - Ipswich, H. 18138 (1958).

POA L. Meadow-grass.

P. annua L. Annual Bluegrass. Naturalized from Europe. Cultivated and waste ground; common. - Haverhill, H. 296 (1932); Andover, Pease 3528 (1904); Boxford, H. 8853; Georgetown, Horner; Ipswich, Plum Island, H., McGregor & Schweinfurth 12678 (1957); Danvers, H. 7963 (1952); Gloucester, H. 11999 (1955); Rockport, Anderson, Shafer & Smith (1931); Manchester, Hubbard (1913); Salem, Robinson (1877); Lynnfield, H. 14116 (1957); Lynn, Porter (1939); Marblehead, Bean & H. 9750 (1955).

P. chapmaniana Scribn. Adventive from South; rare. - Haverhill, Ward Hill, H. 369 (1932).

P. compressa L. Canada Bluegrass. Naturalized from Europe. Dry ground; common. - Amesbury, H. 18262 (1958); Haverhill, H. 7 (1931); Lawrence, H. 10099 (1955); Andover, Pease 202 (1901); North Andover, H. 9489 (1954); West Boxford, Horner; Groveland, Bean & H. 8497 (1953); Newbury, Plum Island, H. 12020 (1956); Georgetown, Horner; Rowley, Bean & H. 8425 (1953); Rowley, Plum Island, H. 12041 (1956); Ipswich, Oakes; Ipswich, Plum Island, H. 12125 (1956); Topsfield, H. 8058 (1952); Danvers, Sears (1878); Wenham, Conant (1881); Hamilton, Hubbard 585 (1913); Essex, Porter (1940); Gloucester, Magnolia, Clarke (1912); Rockport, Robinsor (1875); Beverly, H. 10199 (1955); Salem, Robinson (1877); Lynnfield, Lowell (1849); Saugus, Bean & H. 9832 (1955); Marblehead, H. 9691 (1955).

P. pratensis L. Kentucky Bluegrass. Common both in a cultivated and a wild state. - Merrimac, Bean & H. 8604 (1953); East Haverhill, Conant (1878); Lawrence, H. 9847 (1955); Andover, Pease 201 (1901); North Andover, H. 18033 (1958); Boxford, M.E. Perley (1882); Groveland, Bean & H. 8539 (1953); West Newbury, Conant (1879); Newbury, Plum Island, Knowlton (1913); Rowley, Bean & H. 8424 (1953); Ipswich, Oakes; Ipswich, Plum Island, H. 12130 (1956); Danvers, Sears (1877); Rockport, H. 10277 (1955); Manchester, Hubbard 572 (1913); Lynnfield, H. 8334 (1953).

P. angustifolia L. Dry open places; occasional. - Andover, Pease 200 (1901); Newbury, Plum Island, H. 12029 (1956); Rowley, Plum Island, H. 12057 (1956); Ipswich, Plum Island, H. 12080 (1956); Marblehead, H. 9693 (1956).

P. trivialis L. Rough-stalked Meadow-grass. Damp places; rare. - Amesbury, H. 18210 (1958); Haverhill, Conant (1878); Groveland, Conant (1878); Ipswich, Oakes; Beverly Farms, Hubbard 588 (1913).

P. nemoralis L. Thickets and open woods; occasional. - Boxford, H. 8020 (1952); Topsfield, H. 8063 (1952); Wenham, Robinson (1875); West Manchester, Hubbard 571 (1913).

P. palustris L. Fowl-meadow Grass. Meadows; frequent. - Amesbury, Eaton (1897); Methuen, H. 7989 (1952); Lawrence, H. 12288 (1956); North Andover, H. 11053 (1955); Boxford, H. 8021 (1952); West Newbury, Conant (1879); Georgetown, Horner; Gloucester, Magnolia, Clarke (1912); Beverly, Lowell (1849); Salem, Robinson (1878); Swampscott, H. 10663 (1955).

BRIZA L. Quaking Grass.

B. media L. Introduced from Europe. Roadsides and fields; frequent. - Haverhill, H. 467 (1932); Andover, Pease (1904); North Andover, H. 7908 (1952); Boxford, M.E. Perley (1882); Groveland, Bean & H. 8482 (1952); Topsfield, H. 8055 (1952); Middleton, Pease 28069 (1940); Danvers, Oakes; Rockport, Gates & Smith 1065 (1928); Beverly, Lowell (1849); Salem, O. Russell (1859); Peabody, Robinson (1877).

DACTYLIS L. Orchard-grass.

D. glomerata L. Naturalized from Europe. Fields and roadsides; common. - Salisbury, Bean & H. 8391 (1953); Amesbury, H. 10493 (1955); Merrimac, Bean & H. 8600 (1953); Haverhill, H. 368 (1932); Boxford, H. 7869 (1952); Groveland, H. 7884 (1952); West Newbury, H. 10334 (1955); Georgetown, M.E. Perley (1884); Ipswich, H. 13990 (1957); Topsfield, H. 8070 (1952); Wenham, Bean & H. 10022 (1955); Hamilton, H. 10792 (1955); Essex, H. 10245 (1955); Gloucester, Robinson & Sears (1877); Rockport, Gates & Smith 1085 (1928); Manchester, Hubbard (1913); Beverly, H. 10214 (1955); Salem, H. 9709 (1955); Lynn, H. 10922 (1955); Swampscott, H. 10679 (1955); Marblehead, Bean & H. 9765 (1955); Nahant, S.S. Clark 34 (1916).

D. glomerata, var. *ciliata* Peterm. - Andover, Pease 169 (1901); Middleton, M.E. Perley (1884).

ERAGROSTIS Beauv. Love-grass.

E. megastachya (Koel.) Link. Stink-grass. Naturalized from Europe. Roadsides and waste places; common. - Haverhill, H. 639 (1932); Methuen, H. 11735 (1955); Lawrence, Pease 1607 (1902); North Andover, H. 18734 (1958); Groveland, Horner; Newburyport, H. 20744 (1959); Ipswich, H. 13397 (1957); Topsfield, H. 11378 (1955); Danvers, H. 11634 (1955); Hamilton, H. 10838 (1955); Essex, H. 11302 (1955); Gloucester, H. 12363 (1956); Salem, Morse (1925); Peabody, H. 11544 (1955); Saugus, H. 11850 (1955); Marblehead, H. 11471 (1955).

E. pectinacea (Michx.) Nees. Ditches and roadsides; frequent. - Amesbury, Eaton (1897); Lawrence, Pease 2669 (1903); Andover, J. Blake (1885); Groveland, H. 28375 (1965); Newbury, Eaton (1895); Georgetown, Horner; Rowley, H. 18519 (1958); Topsfield, H. 11364 (1955); Hamilton, H. 10831 (1955); Peabody, Mackintosh (1924); Saugus, H. 18666 (1958).

E. pilosa (L.) Beauv. Naturalized from Europe. Waste places; uncommon. - Salisbury, Morse (1924); Gloucester, Fernald, Hunnewell & Long 8771 (1913).

E. multicaulis Steud. Adventive from eastern Asia; rare. - Boxford, H. 30756 (1966); Salem, Conant (1879).

E. frankii C.A. Mey. Damp places; rare. - Danvers, H. 18580 (1958); Salem, H. 10623 (1955).

E. capillaris (L.) Nees. Lace-grass. Dry soil; uncommon. - Andover, J. Blake (1883); Danvers, H. 11622 (1955); Essex, Hunnewell 189; Saugus, H. 18695 (1958).

E. spectabilis (Pursh) Steud. Tumble-grass. Dry places; rare. - Topsfield, Sears (1879).

E. spectabilis, var. *sparsihirsuta* Farw. Dry places; frequent. - Amesbury, Eaton (1895); Haverhill, H. 601 (1932); Andover, Pease 1632 (1901); North Andover, H. 11034 (1955); Boxford, H. 30745 (1966); Groveland, H. 8808 (1953); West Newbury, Porter (1941); Georgetown, Horner (1875).

TRIODIA R.Br.

T. flava (L.) Smyth, *forma cuprea* (Jacq.) Fosberg. Andover, A.P. Morse (1929).

TRIPSALIS Beauv.

T. purpurea (Walt.) Chapm. Sand-grass. In sand; rare. - Salisbury, Eaton (1896); Ipswich, Robinson (1875); Nahant.

CYNOSURUS L. Dog's-tail.

C. cristatus L. Crested Dog's-tail. Adventive from Europe; rare. - Salem, Robinson (1879).

DISTICHLIS Raf. Alkali-grass.

D. spicata (L.) Greene. Salt marsh; common. - Salisbury, White 354 (1913); Newbury, Bean & Stockman (1954); Newbury, Plum Island, White 320 (1913); Rowley, Horner; Ipswich, St. John 22 (1911); Danvers, Sears (1880); Essex, H. 9391 (1953); Lynn, Robinson (1875); Marblehead, H. 11905 (1955).

PHRAGMITES Trin. Reed.

P. communis Trin., var. *berlandieri* (Fourn.) Fern. Fresh and saline marshes; frequent. - Salisbury, Eaton (1898); Andover, H. 19060 (1959); Rowley, Plum Island, H. 12544 (1956); Ipswich, H. 14020 (1957); Topsfield, W.S. Horner; Danvers, Sears (1875); Essex, H. 18996 (1958); West Gloucester, H. 18997 (1958); Rockport, Lufkin (1890); [Salem, Snyder (1950s on)] ; Saugus, Emilio & Stubbs (1926); Lynn, H. 11835 (1955); Marblehead, H. 18931 (1958); Nahant, Mackintosh (1931).

AGROPYRON Gaertn.

A. trachycaulum (Link) Malte, var. *novae-angliae* (Scribn.) Fern. Thickets; rare. - Newbury, Eaton (1897); Middleton, Pease 28070 (1940).

A. pungens (Pers.) R & S. Sandy coast; rare. - Plum Island, J. Perkins (1942); Ipswich Beach, Sears (1904).

A. repens (L.) Beauv. Witch-grass. A common and aggressive weed, probably native in northern Maine but introduced from Europe in Essex County; most frequently found in one of its confusing varieties and forms. - Newbury, H. 10397 (1955); Danvers, Oakes; Rockport, Forbes (1904); Beverly, Lowell (1853); Swampscott, H. 10647 (1955).

A. repens, forma *aristatum* (Schum.) Holmb. - Salisbury, White 54 (1913); Amesbury, H. 10559 (1955); Merrimac, Bean & H. 8603 (1953); Haverhill, H. 468 (1932); Methuen, H. 9625 (1954); Boxford, M.P. Perley (1882); Groveland, Bean & H. 8534 (1953); Ipswich, Plum Island, H. 12092 (1956); Danvers, H. 7960 (1952); Hamilton, H. 10839 (1955); Essex, H. 10252 (1955); Rockport, H. 10264 (1955); Lynn, Porter (1936); Swampscott, H. 10669 (1955).

A. repens, forma *trichorrhachis* Rohlena. - Swampscott, H. 10660 (1955).

A. repens, forma *pilosum* (Scribn.) Fern. - Topsfield, Sears (1877); Wenham, Robinson (1875); Marblehead, Sears (1883).

A. repens, var. *subulatum* (Schreb.) Reichenb. - South Lawrence, Pease 4893 (1904); North Andover, Morss (1884); Plum Island, St. John 691 (1913); West Manchester, Hubbard 644 (1913).

A. repens, var. *subulatum,* forma *vaillantianum* (Wulf. & Schreb.) Fern. - Methuen, H. 9624 (1954); Newbury, Bean (1913); Plum Island, Knowlton (1913); Ipswich, Oakes; Middleton, H. 8698 (1953); Hamilton, H. 12502 (1956); West Gloucester, Robinson & Sears (1877); Peabody, H. 8761 (1953).

A. repens, var. *subulatum*, forma *heberhachis* Fern. - Andover, Pease 148 (1901).

A. repens, var. *subulatum*, forma *setiferum* Fern. - Groveland, Bean & H. 8532 (1953); West Newbury, H. 10960 (1955); West Manchester, Hubbard 643 (1913).

LOLIUM L. Darnel.

L. perenne L. Perennial Rye-grass. Naturalized from Europe. Fields and roadsides; occasional. - North Andover, H. 18017 (1958); Boxford, H. 8898 (1953); Ipswich, Plum Island, H. 12098 (1956); Topsfield, H. 7982 (1952); Manchester, Hubbard 559 (1913); Beverly, Lowell (1846); Salem, Robinson (1879).

L. multiflorum Lam. Italian Rye-grass. Naturalized from Europe. Fields and roadsides; occasional. - North Andover, H. 10848 (1955); Newburyport, H. 11285 (1955); Salem, H. 9710 (1955); Saugus, H. 18691 (1958).

L. multiflorum, forma *submuticum* (Mutel) Hayek - Rowley, H. 18514 (1958); Saugus, Porter (1941); Lynn, Porter (1937).

L. multiflorum, var. *diminutum* Mutel - Boxford, H. 8435 (1953); West Newbury, Conant (1877); Rowley, H. 9548 (1954).

L. temulatum L. Bearded Darnel. Adventive from Europe. Fields and waste places; rare. - Rowley, Oakes; Ipswich, Oakes (1830); Topsfield, H. 9587 (1954).

HORDEUM L. Barley.

H. jubatum L. Open ground; frequent near the shore. - Salisbury, White 47 (1913); North Andover, H. 18042 (1958); Rowley, W.B. Horner; Ipswich, Oakes; Rockport, Smith 745 (1927); Beverly, Lowell (1846); Saugus, Seymour 5187 (1938); Lynn, Porter (1938); Marblehead Neck, Frost.

H. vulgare L. Barley. Adventive from Europe. Waste places; not common. - Boxford, H. 8101 (1952); Rowley, H. 18495 (1958); Essex, Robinson & Sears (1877); Saugus, H. 18673 (1958).

TRITICUM L. Wheat.

T. aestivum L. Introduced from Eurasia. Escaped from cultivation, waste places; occasional. - Boxford, H. 8607 (1953); Topsfield, H. 8742 (1953); Danvers, Sears (1879); Beverly, Montserrat, Hubbard 681 (1913); Salem, Robinson (1877).

SECALE L. Rye.

S. cereale L. Introduced from Eurasia. Escaped from cultivation, waste places; occasional. - Newbury, Plum Island, H. 12030 (1956); Rowley, H. 9546 (1954); Rowley, Plum Island, H. 12070 (1956); Ipswich, Plum Island, H. 12089 (1956); Topsfield, H. 18298 (1958); Wenham, Bean & H. 10008 (1955); Salem, Robinson (1877).

ELYMUS L. Wild Rye.

E. arenarius L., var. *villosus* Mey. Strand-wheat. Beaches; rare. - Gloucester, Jenks (1881).

E. virginicus L. Terrell Grass. Rich thickets and shores; common. - Salisbury, White 270 (1913); Amesbury, H. 18966 (1958); Merrimac, H. 9281 (1953); Haverhill, H. 602 (1932); Methuen, Bean & H. 7750 (1951); Lawrence, Jenks (1884); Andover, Pease 3208 (1903); North Andover, Bean (1916); Groveland, H. 8810 (1953); West Newbury, Bean & Pease (1940); Newbury, Kent's Island, Eaton (1897); Rowley, H. 8290 (1952); Ipswich, H. 13445 (1957); Gloucester, Dame (1881); Rockport, Pigeon Cove, Churchill (1887); Beverly, Lowell (1846); Salem, Robinson (1877).

E. virginicus, forma *hirsutiglumis* (Scribn.) Fern. - Amesbury, Eaton 535 (1898); Georgetown, Porter (1941).

E. virginicus, var. *halophilus* (Bickn.) Wieg. - Amesbury, Bean (1950); West Newbury, NEBC. (1951); Rowley, Porter (1941); Gloucester, Fernald, Hunnewell & Long 8823 (1913).

E. riparius Wieg. Rich thickets and borders of streams; occasional. - Amesbury, Eaton (1896); Merrimac, H. 9275 (1953); Lawrence, H. 11251 (1955); Andover, Pease 2657 (1903); North Andover, H. 10014 (1955); Groveland, H. 11138 (1955); Ipswich, Oakes.

E. villosus Muhl. Thickets and shores; occasional. - Haverhill, Ward Hill, Bean (1916); Groveland, H. 8809 (1953); West Newbury, Robinson (1879); Ipswich, Oakes; Beverly, Lowell (1848).

E. villosus, forma *arkansanus* (Scribn. & Ball) Fern. - Plum Island, Grape Swamp, Eaton (1896).

E. canadensis L. Dry rocky or sandy soil; occasional. - Merrimac, Bean & H. 7734 (1951); Haverhill, H. 7785 (1951); Methuen, Bean & H. 7751 (1951); West Newbury, Porter (1941).

E. canadensis, forma *glaucifolius* (Muhl.) Fern. - Andover, Pease 1615 (1901); West Newbury, Robinson (1879).

HYSTRIX Moench Bottle-brush Grass.

H. patula Moench. Rich woods; occasional. - Haverhill, H. 1836 (1935); Georgetown, Horner (1885); Middleton, Bean (1940); Essex, Eaton (1895).

H. patula, var. *bigeloviana* (Fern.) Deam - West Boxford, Horner; Middleton, Pease 28028 (1940); Wenham or Hamilton, Robinson (1877).

ARRHENATHERUM Beauv. Oat-grass.

A. elatius (L.) Mert. & Koch. Tall Oat-grass. Naturalized from Europe. Fields and roadsides; occasional. - Haverhill, E. & C.E. Faxon; Andover, H. 17734 (1958); Boxford, H. 8350 (1953); Groveland, E.S. Parker (1878); Danvers, Sears (1878); Wenham, Conant (1881); Gloucester, Magnolia, Clarke (1912); Beverly Farms, Hubbard 580 (1913); Peabody, Mackintosh (1925); Marblehead, Bean & H. 9722 (1955).

HOLCUS L.

H. lanatus L. Velvet-grass. Naturalized from Europe. Sterile fields and waste places; common. - Salisbury, H. 10473 (1955); Andover, Pease 3783 (1904); Boxford, M.E. Perley (1882); Groveland, H. 871 (1933); West Newbury, Conant (1879); Georgetown, Horner; Rowley, Bean (1940); Ipswich, Oakes; Danvers, Sears (1879); Hamilton, H. 10793 (1955); Essex, H. 10258 (1955); Rockport, Smith 727 (1927); Manchester, Lane (1877); Beverly Farms, Hubbard 579 (1913); Peabody, Mackintosh (1933); Swampscott, H. 10664 (1955).

SPHENOPHOLIS Scribn.

S. intermedia Rydb. Damp places; rare. - Salisbury Plain, Eaton (1896); West Newbury, Conant (1879); Danvers, Sears (1878).

TRISETUM L.

T. spicatum (L.) Richter, var. *molle* (Michx.) Beal. Ledges and cool shores; rare. - Haverhill, Bradford, Fernald 11284 (1914); Andover, Pease 2654 (1903).

T. pensylvanicum (L.) Beauv. Swamp-oats. Damp meadows; rare. - Andover, Pease 2811 (1903).

AVENA L. Oat.

A. sativa L. Introduced from Europe. Waste places; occasional. - Haverhill, H. 544 (1932); Boxford, H. 8100 (1952); Rowley, H. 18506 (1958); Danvers, H. 18584 (1958); Gloucester, Robinson & Sears (1877); Peabody, H. 27068 (1964); Saugus, H. 18708 (1958); Lynn, Robinson (1875).

A. hybrida Koch. Introduced from Europe. Waste places; occasional. - Newburyport, Williams (1898).

DESCHAMPSIA Beauv. Hairgrass.

D. flexuosa (L.) Trin. Common Hairgrass. Dry soil; common. - Salisbury, D. & E.W. White 16 (1913); Andover, Knowlton (1903); Boxford, Horner; Groveland, Conant (1879); Georgetown, Horner (1884); Rowley, Horner; Rowley, Plum Island, H. 12069 (1956); Hamilton-Essex, A.B. Seymour (1908); Essex Woods, Oakes; Gloucester, Swan (1881); Rockport, C.E. Faxon & Robinson (1877); West Manchester, Hubbard 557 (1913); Beverly, Lowell (1846).

DANTHONIA DC. Wild Oat-grass.

D. spicata (L.) Beauv. Poverty-grass. Dry soil; common. - Salisbury, H. 10481 (1955); Amesbury, H. 18267 (1958); Merrimac, H. 18335 (1958); Methuen, H. 9611 (1954); Lawrence, H. 9886 (1955); Andover, H. 10065 (1955); Boxford, H. 8018 (1952); Groveland, Bean & H. 8489 (1953); Rowley, H. 7935 (1952); Ipswich, H. 9570 (1954); Ipswich, Plum Island, H., McGregor & Schweinfurth 12689 (1957); Topsfield, H. 8065 (1952); Danvers, Robinson (1875); Gloucester, H. 12169 (1956); Rockport, Smith 1092 (1928); Beverly Farms, Hubbard 562 (1913); Salem, Baker's Island, Robinson (1890); Lynnfield, Bean (1916); Saugus, Bean & H. 9796 (1955).

D. alleni Aust. Arid soil; uncommon. - Amesbury, Eaton (1897); Boxford, H. 8035 (1952); Groveland, Bean & H. 8540 (1953).

D. compressa Aust. Woodlands and clearings; uncommon. - Haverhill, H. 508 (1932); Georgetown, Horner; Ipswich, H. 18174 (1958); Peabody, Sears (1884); Lynn, Cushman (1908).

SPOROBOLUS R.Br. Drop-seed.

S. cryptandrus (Torr.) Gray. Sand Drop-seed. Coastal sand; occasional. - Salisbury, Eaton (1896); Ipswich, Oakes; Nahant, Parsons.

S. vaginiflorus (Torr.) Wood. Poverty-grass. Dry sterile soil; occasional. - Lawrence, H. 11769 (1955); Andover, Pease 206 (1901); West Newbury, Conant (1877); Byfield, Porter (1941); Danvers, Lowell (1848); Wenham, Robinson (1877).

S. vaginiflorus, var. *inaequalis* Fern. - Beverly, H. 11678 (1955); Saugus, H. 18648 (1958).

S. neglectus Nash. Dry sterile soil; rare. - Ipswich, H. 13993 (1957).

S. asper (Michx.) Kunth. Dry open soil; occasional. - Newbury, Plum Island, H. 13809 (1957); Ipswich, Oakes; Danvers, Sears (1878).

CALAMAGROSTIS Adans. Reed-Bentgrass.

C. pickeringii Gray. Acid peats; rare. - West Andover, Hubbard 634 (1913).

C. pickeringii, var. *debilis* (Kearney) Fern. & Wieg. - West Andover, Robinson (1878); North Andover, Conant (1880). No collections since 1904.

C. canadensis (Michx.) Nutt. Blue-joint. Meadows and bogs; common. - Salisbury, H. 12216 (1956); Amesbury, H. 10547 (1955); Merrimac, H. 9274 (1953); Haverhill, H. 507 (1932);

Methuen, Bean & H. 7752 (1951); Lawrence, H. 10096 (1955); North Andover, H. 13064 (1957); Boxford, H. 8043 (1952); Groveland, H. 10602 (1955); West Newbury, Conant (1879); Georgetown, Horner; Rowley, H. 8276 (1952); Topsfield, H. 8082 (1952); Middleton, H. 18101 (1958); Danvers, Sears (1884); Wenham, Robinson (1875); Rockport, Cushman (1908); Beverly, Lowell (1845); Peabody, Mackintosh (1933); Lynnfield, Bean (1919); Saugus, H. 10881 (1955); Lynn, H. 13172 (1957); Swampscott, H. 10666 (1955).

C. canadensis var. *macouniana* (Vasey) Stebbins - Boxford, Williams (1899).

C. cinnoides (Muhl.) Bart. Damp sandy or peaty soils; not common. - Salisbury, Eaton (1895); Ipswich, Oakes; Danvers, Sears (1879); Essex Woods, Russell (1849); Manchester, Tuckerman; Beverly, Lowell (1845).

C. epigejos (L.) Roth., var. georgica (K. Koch) Ledeb. Naturalized from southern Russia. Sandy woods and openings; rare. - Gloucester, Weatherby 5974 (1930).

AMMOPHILA Host. Sand-reed.

A. breviligulata Fern. Beachgrass. Coastal dunes; common. - Newburyport, Plum Island, Cole (1886); Newbury, Plum Island, Pease 3103 (1903); Rowley, Plum Island, H. 12569 (1956); Ipswich, Prince (1902); Ipswich, Plum Island, H. 12449 (1956); Gloucester, Williams (1898); Beverly, Lowell (1846); Lynn, Robinson (1875).

AGROSTIS L. Bentgrass.

A. alba L. Redtop. Naturalized from Europe. Fields and roadsides, common. - Salisbury, White 56 (1913); Amesbury, H. 10545 (1955); Andover, Pease 3806 (1904); Newburyport, Plum Island, H. 13051 (1957); Newbury, Plum Island, H. 12018 (1956); Georgetown, Horner; Rowley, H. 12254 (1956); Topsfield, H. 12184 (1956); Middleton, H. 8700 (1953); Hamilton, Hubbard 646 (1913); Essex, H. 10262 (1955); Gloucester, H. 12141 (1956); Rockport, H. 10311 (1955); Manchester, H. 10719 (1955); Saugus, H. 10867 (1955); Swampscott, H. 10677 (1955); Marblehead, H. 9773 (1956).

A. alba, forma *aristigera* Fern. - Ipswich, H. 18119 (1958); Beverly, Lowell (1846).

A. alba, var. *palustris* (Huds.) Pers. Shores and damp places; common. - Salisbury, H. 10506 (1955); Amesbury, H. 10552 (1955); Merrimac, Bean & H. 8593 (1953); Haverhill, H. 12631 (1932); Boxford, M.P. Perley (1882); Newbury, Plum Island, White 107 (1913); Ipswich Beach, Hubbard 640 (1913); Middleton, H. 18075 (1958); Danvers, Sears (1878); Essex, H. 9379 (1953); Gloucester, Robinson (1877); Rockport, Robinson (1877); Salem, Robinson (1878); East Saugus, Rich (1901).

A. tenuis Sibth. Rhode Island Bent. Naturalized from Europe. Fields and roadsides; occasional. - Haverhill, Conant (1878); Methuen, W. Boott (1866); Andover, Pease 3067 (1903); North Andover, Morss (1884); West Manchester, Hubbard 645 (1913); Beverly, Lowell (1845); Salem, Manning (1882).

A. tenuis, forma *aristata* (Sincl.) Wieg. - Merrimac, H. 18336 (1958); Boxford, M.P. Perley (1882); West Newbury, Conant (1878); Ipswich, Oakes; Rockport, Smith 1140 (1930).

A. scabra Willd. Hairgrass. Sterile soil; common. - Haverhill, H. 12632 (1932); Methuen, H. 7959 (1952); Andover, Pease 153 (1901); North Andover, H. 14053 (1957); Boxford, J. Parkhurst

(1882); Groveland, H. 10608 (1955); Georgetown, Horner; Rowley, H. 7931 (1952); Ipswich Beach, Hubbard 641 (1913); Danvers, Lowell (1846); Wenham, H. 13775 (1957); Hamilton-Essex, A.B. Seymour (1908); Gloucester, Robinson (1877); Rockport, Smith 1004 (1928); Manchester, Beverly, Hubbard 627 (1913).

A. perennans (Walt.) Tuckerm. Upland Bent. Dry open soil; occasional. - Haverhill, H. 12633 (1932); Andover, H. 17760 (1958); Ipswich, H. 9571 (1954); Danvers, Sears (1878); Beverly, Mackintosh (1924); Lynn, Rich (1889).

A. perennans, var. *aestivalis* Vasey Thingrass. Thickets and shores; occasional. - Boxford, H. 13648 (1957); West Newbury, Porter (1941); Lynn, Rich (1892).

A. canina L. Brown Bent. Peaty shores; occasional. - Andover, Pease 3069 (1903); Ipswich, Fernald (1913); Beverly, POrter (1941); Essex County, Oakes.

A. spica-venti L. Adventive from Europe; rare. - Haverhill, Ward Hill, H. 367 (1932).

CINNA L. Wood Reed Grass.

C. arundinacea L. Moist woods; frequent. - Merrimac, H. 9259 (1953); Haverhill, Bradford, Lunt (1913); Lawrence, H. 11249 (1955); North Andover, Pease 2928 (1905); Groveland, Horner (1882); West Newbury, H. 11121 (1955); Ipswich, Lowell; Danvers, Sears (1880); Wenham, Sears (1878).

C. latifolia (Trev.) Griseb. Woods and thickets; occasional. - Haverhill, Bradford, Fernald 11285 (1914); Essex Woods, Lowell (1848); Essex County, Pickering.

PHLEUM L. Timothy.

P. pratense L. Naturalized from Europe. Fields and roadsides; common. - Salisbury, White 85 (1913); Amesbury, H. 10518 (1955); Merrimac, Bean & H. 8594 (1953); Haverhill, H. 466 (1932); Andover, Pease 197 (1901); North Andover, Morss (1884); Boxford, M.E. Perley (1884); West Newbury, Conant (1878); Newburyport, H. 10380 (1955); Newbury, Plum Island, H. 12034 (1956); Rowley, Plum Island, H. 10240 (1956); Ipswich, Oakes; Ipswich, Plum Island, H. 12084 (1956); Wenham, Robinson (1875); Hamilton, H. 10803 (1955); Essex, H. 10260 (1955); Rockport, Smith 1066 (1928); Manchester, Hubbard (1913); Beverly, H. 10202 (1955); Salem, Robinson (1878); Swampscott, H. 10670 (1955).

P. pratense, var. *nodosum* (L.) Huds. Naturalized from Europe. Fields and roadsides; rare in collections but probably frequently overlooked. - Beverly, F.H. Peabody (1892).

P. arenarium L. Native of Europe. Waif under bird-feeders. - Boxford, H. 8617b (1953).

P. subulatum (Savi) Aschers & Graebn. Native of Europe. Waif under bird-feeders. - Boxford, H. 8617a, 8991 (1953); Topsfield, H. 9642 (1954).

ALOPECURUS L. Foxtail.

A. pratensis L. Meadow-Foxtail. Naturalized from Europe. Meadows and roadsides; common. - Amesbury, Eaton (1895); Haverhill, H. 389 (1932); Andover, Alcott (1863); Boxford, H. 7850 (1952); Groveland, H. 7883 (1952); West Newbury, Conant (1879); Newburyport, H. & Wetherall

9564 (1954); Ipswich, Oakes; Middleton, H. 7879 (1952); Essex, H. 10261 (1955); Rockport, Smith 939 (1928); West Manchester, Hubbard (1913); Salem, Robinson (1879); Peabody, Mackintosh (1925); Marblehead, H. 9677 (1955).

A. geniculatus L. Marsh-Foxtail. Naturalized from Europe. Ditches and wet clearings; frequent. - Haverhill, Ward Hill, H. 351 (1932); Andover, Alcott (1866); West Newbury, Conant (1879); Newburyport, H. 10465 (1955); Ipswich, Oakes; Danvers, Robinson (1879); Gloucester, Rand & B.L. Robinson (1896); Rockport, Smith 946 (1928); Beverly.

A. carolinensis Walt. Adventive from the South. - Haverhill, Ward Hill, H. 9888 (1955).

A. aequalis Sobol. Shallow water and shores; occasional. - Haverhill, Ward Hill, Bean (1935); Andover, Pease 154 (1901); North Andover, H. 14077 (1957); Boxford, H. 9716 (1955); West Newbury, Alcott (1878); Ipswich, H. 10229 (1955); Danvers, Sears (1878).

MUHLENBERGIA Schreb.

M. schreberi J.F. Gmel. Drop-seed. Thickets and roadsides; occasional. - Haverhill, H. 599 (1932); Boxford, H. 30739 (1966); Newburyport, H. 20737 (1959); Ipswich, H. 11598 (1955); Gloucester, H. 12320 (1956); Salem, Robinson (1904); Saugus, H. 18717 (1958); Lynn, Porter (1941).

M. sobolifera (Muhl.) Trin. Dry rocky woods; rare. - Boxford, H. 18542 (1958); Groveland, Sears; Beverly, Sears (1878).

M. sylvatica Torr., forma *attenuata* (Scribn.) Palmer & Steyerm. Damp woods; rare. - North Andover, Pease 1537 (1901).

M. mexicana (L.) Trin. Shores and damp clearings; rare. - Andover, Pease 1774 (1902); Georgetown, Horner; Peabody, H. 18592 (1958).

M. mexicana, forma *ambigua* (Torr.) Fern. - Amesbury, Eaton 269 (1899); Andover, Pease 2618 (1903).

M. frondosa (Poir.) Fern. Damp open woods and waste places; common. - Salisbury, H. 18984 (1958); Andover, Sears (1878); North Andover, H. 8998 (1953); Boxford, H. 8175 (1952); Groveland, H. 9061 (1953); West Newbury, Conant (1877); Newburyport, H. 20732 (1959); Danvers, Robinson (1877); Hamilton, Hunnewell 4633 (1916); Essex, H. 18847 (1958); West Gloucester, Clarke (1912); Beverly, Sears (1878); Salem, Sears (1905); Peabody, H. 11521 (1955); Saugus, H. 18619 (1958); Marblehead, H. 11423 (1955).

M. frondosa, forma *commutata* (Scribn.) Fern. - Amesbury, Eaton (1898); Lawrence, H. 11235 (1955); Georgetown, Horner; Beverly, Sprague.

M. glomerata (Willd.) Trin. Meadows and wet shores; occasional. - Andover, Sears (1878); Boxford, Robinson (1877); West Newbury, Conant (1878); Topsfield, Sears (1901); Danvers; Essex County, Pickering.

M. uniflora (Muhl.) Fern. Bogs and damp places; occasional. - Merrimac, H. 9225 (1953); Boxford, H. 9287 (1953); West Newbury, Conant; Newbury, H. 8227 (1952); Danvers, Sears (1878); Rockport, Anderson & Smith 2416 (1930); Peabody, H. 11542a (1955); Lynn, Young (1879).

BRACHYELYTRUM Beauv.

B. erectum (Schreb.) Beauv., var. *septentrionale* Babel Woods and thickets; occasional. - Salisbury, Eaton (1895); Amesbury, H. 18219 (1958); Haverhill, H. 465 (1932); Andover, Robinson (1879); West Newbury, Conant (1879); Beverly Farms, Hubbard 589 (1913).

ORYZOPSIS Michx. Mountain-Rice.

O. asperifolia Michx. Woods and thickets; occasional. - Haverhill, H. 350 (1932); Andover, Lowell Junction, Hopkins & Pease 24039 (1934); Manchester, Oakes.

O. racemosa (Sm.) Ricker. Rich woods; rare. - Georgetown, Robinson (1875); Essex, Eaton (1896); Chebacco, Robinson (1875).

O. pungens (Torr.) Hitchc. Rocky soil; frequent. - Haverhill, H. 349 (1932); Andover, Alcott (1863); Boxford, Robinson (1875); Middleton, Sears (1878); Manchester, Oakes; Beverly Farms, Hubbard (1913).

STIPA L. Feathergrass.

S. avenacea L. Black Oat-grass. Dry woods; rare. - Manchester, Hubbard 618 (1913); Peabody, Fuller (1897); Lynn, Cushman (1908).

ARISTIDA L. Needlegrass.

A. tuberculosa Nutt. Dry coastal sand; frequent on Plum Island. - Newbury, Plum Island, H. 11291 (1955); Ipswich, Castle Neck, H. 13949 (1957); Ipswich, Plum Island, St. John & White 551 (1913); Plum Island, Boott (1856), Robinson (1876), Conant (1878).

A. oligantha Michx. Adventive from the South or West. - Newburyport, H. 7692 (1951).

A. dichotoma Michx. Poverty-grass. Dry sterile soil; frequent. - Salisbury, Eaton (1896); Merrimac, H. 9203 (1953); Boxford, H. 8171 (1952); Groveland, H. 11172 (1955); Ipswich, Wade 18796 (1958); Danvers, Lowell (1848); Wenham, Robinson (1877); Beverly, Conant; Peabody, Conant (1878); Marblehead, H. 11469 (1955).

A. longispica Poir., var. *geniculata* (Raf.) Fern. Sandy soil; occasional. - Salisbury, Eaton (1896); Plum Island, Eaton (1896); Ipswich, St. John 20 (1911); Danvers, Sears (1878).

A. purpurascens Poir. Sandy soil; occasional. - Andover, Pease 158 (1902); Peabody, Conant (1878); Lynnfield, Fernald (1917).

SPARTINA Schreb. Cord-grass.

S. pectinata Link. Fresh-water Cord-grass. Damp soil; common. - Salisbury, H. 10487 (1955); Amesbury, H. 18963 (1958); Merrimac, J.Y.B. & Horner (1884); Boxford, Alcott (1883); Groveland, H. 10991 (1955); West Newbury, H. 9149 (1953); Newburyport, Eaton; Georgetown, Williams (1906); Rowley, Horner (1883); Ipswich, H. 13988 (1957); Ipswich, Plum Island, H. 12423 (1956); Topsfield, Robinson (1876); Danvers, Sears (1878); Essex, H. 9385 (1953); Gloucester, Magnolia, Jenks 82; Rockport, Robinson (1874); Manchester, Lane (1877); Beverly, Lowell (1846); Salem, Cole (1886); Marblehead, Sanford (1944).

S. pectinata, var. *suttiei* (Farw.) Fern. - Salisbury, Eaton & Fernald (1902).

S. pectinata X *alterniflora* ? - Swampscott, Sanford (1944).

S. alterniflora Loisel. Salt-water Cord-grass. Saline shores and marshes; occasional. - Gloucester, Annisquam, Ripley 16405 (1916); Peabody, Mackintosh (1933); Marblehead, H. 11915 (1955).

S. alterniflora, var. *pilosa* (Merr.) Fern. More frequent. - Newbury, Bean & Stockman (1954); Rowley, H. 8272 (1952); Ipswich, H. 13891 (1957); Plum Island, Robinson (1876); Gloucester, Smith (1945); Lynn, Robinson (1875).

S. alterniflora, var. *glabra* (Muhl.) Fern. Most common. - Salisbury, White 292 (1913); Rowley, H. 8254 (1952); Ipswich, Townsend (1910); Plum Island, Horner; Essex, H. 9389 (1953); Gloucester, Cushman (1908); Manchester, Thoreau; Beverly, Lowell (1846); Saugus, H. 11861 (1955).

S. patens (Ait.) Muhl. Salt-meadow Grass. Salt marsh; common. - Newbury, Bean & Stockman (1954); Rowley, Pingree Island, Horner; Ipswich, Townsend (1910); Ipswich, Plum Island, St. John & White 582 (1913); Essex, H. 9368 (1953); Gloucester, Cushman (1908); Rockport, Williams (1898); Beverly, Lowell (1846); Salem, Cole (1930); Saugus, H. 18665 (1958); Lynn, Robinson (1875); Marblehead, H. 11910 (1955).

S. patens, var. *morogyna* (M.A. Curtis) Fern. - Plum Island, Eaton (1896); Gloucester, Annisquam, Church 308 (1938).

X *S. caespitosa* A.A. Eat. Borders of salt marshes; rare. - Salisbury, Eaton (1898); Newbury, Eaton (1897); Plum Island, Eaton; Gloucester, Annisquam, Church 78 (1936).

CYNODON Richard Bermuda Grass.

C. dactylon (L.) Pers. Native of Europe. Adventive in waste places. - Lawrence, H. 18573 (1958).

BOUTELOUA Lag. Grama-grass.

B. gracilis (HBK.) Lag. Adventive from the West. About woolen mills. - Danvers, Conant & Robinson (1880).

ELEUSINE Gaertn. Yard-grass.

E. indica (L.) Gaertn. Wiregrass. Naturalized from the Old World. Yards and waste places; uncommon in the County but now an aggressive weed in Back Bay, Boston. - Methuen, H. 11730 (1955); Salem, Robinson (1879); H. 11487 (1955); Saugus, H. 18607 (1958).

PHALARIS L. Canary-grass.

P. canariensis L. Birdseed-grass. Adventive from Europe. Waste places; frequent on dumps. - Haverhill, Bradford, H. 637 (1932); Lawrence, H. 18749 (1958); Boxford, H. 8105 (1952); Newburyport, Robinson (1879); Rowley, H. 18500 (1958); Ipswich, H. 13856 (1957); Topsfield, H. 8744 (1953); Danvers, H. 18581 (1958); Gloucester, Magnolia, H. 12366 (1956); Salem, Robinson (1877); Peabody, H. 18598 (1958); Saugus, H. 18617 (1958); Marblehead Neck, H. 870 (1933).

P. arundinacea L. Reed Canary-grass. Shores and meadows; common. - Andover, H. 10059 (1955); North Andover, Pease 1717 (1901); Boxford, H. 8008 (1952); Groveland, H. 12864 (1957); West Newbury, Conant (1877); Rowley, Smith 631 (1927); Ipswich, H. 13441 (1957); Topsfield, Porter (1936); Hamilton, Hubbard (1913); Lynnfield, Young (1879); Marblehead, Bean & H. 9754 (1955).

P. arundinacea, forma *variegata* (Parnell) Druce Ribbon-grass. Escape from cultivation; occasional. - Newburyport, White 176 (1913); Beverly, Robinson (1879); Salem, Robinson (1875).

ANTHOXANTHUM L. Sweet Vernal Grass.

A. odoratum L. Naturalized from Europe. Fields and waste places; common. - Salisbury, H. 10483 (1955); Merrimac, Bean & H. 8602 (1953); Haverhill, H. 6 (1931); Andover, Pease 157 (1901); North Andover, Robinson (1885); Boxford, Alcott (1882); West Newbury, H. 10335 (1955); Newburyport, Eaton (1896); Georgetown, Horner; Rowley, H. 7932 (1952); Ipswich, Ware (1907); Middleton, H. 31914 (1968); Danvers, Essex, H. 10248 (1955); Rockport, Smith 744 (1927); West Manchester, Hubbard (1913); Beverly, H. 10186 (1955); Salem, Cole (1886); Marblehead, Bean & H. 9745 (1955); Nahant, S.S. Clark 10 (1916).

HIEROCHLOE R.Br. Holy Grass.

H. odorata (L.) Beauv. Sweet Grass. Meadows and shores; most frequent near the coast. - Salisbury, D. & E.W. White 34 (1913); Andover, Pease 182 (1902); West Newbury, Porter (1941); Newburyport, Forbes 294 (1903); Georgetown, Horner (1875); Plum Island, Mackintosh (1925); Rockport, Smith 935 (1928); Beverly Farms, Hubbard (1913); Salem, Sears (1896); Peabody, Mackintosh (1925); Marblehead, H. 9673 (1955).

LEERSIA Sw. Cutgrass.

L. virginica Willd., var. *ovata* (Poir.) Fern. Wet thickets and swales; frequent. - Merrimac, H. 9285 (1953); Haverhill, Ward Hill, H. 543 (1932); South Lawrence, Pease 1579 (1902); Andover, Robinson (1879); Groveland, H. 8817 (1953); West Newbury, Pease 28539 (1940); Georgetown, Horner; Ipswich, H. 18177 (1958); Danvers, Sears (1880); Manchester, H. 11344 (1955).

L. oryzoides (L.) Sw. Rice-cutgrass. Swamps and shores; frequent. - Andover, Pease 186 (1901); North Andover, H. 13066 (1957); Boxford, H. 8788 (1953); Newburyport, Eaton & Fernald (1902); Georgetown, Horner; Danvers, H. 9042 (1953); Wenham, Robinson (1875); Peabody, Mackintosh (1933).

L. oryzoides, forma *glabra* A.A. Eat. Estuaries. - West Newbury, H. 11135 (1955); Newburyport, Eaton & Fernald (1902).

L. oryzoides, forma *inclusa* (Wiesb.) Dörfler - Amesbury, H. 9159 (1953); Merrimac, H. 9243 (1953); Boxford, H. 8187 (1952); Groveland, H. 9105 (1953); Newburyport, Eaton (1902); Danvers, H. 9035 (1953); Peabody, H. 9333 (1953); Lynnfield, H. 14106 (1957).

ZIZANIA L. Wild Rice.

Z. aquatica L. Quite waters; local except along Merrimac River where common. - Amesbury, Bean (1949); Haverhill, Horner & H.E. Parkhurst (1883); Lawrence, Pease 207 (1901); Andover, Pease 30431 (1943); West Newbury, Pease 28538 (1940); Newbury, White 214 (1913); Newbury,

Plum Island, Hopkins & Pease 24085 (1934); Ipswich; Topsfield, H. 13617 (1957); Wenham, Cole (1911).

Z. aquatica, var. *brevis* Fassett - Amesbury, Fassett 2332 (1924); Rowley, Fassett 2335 (1924).

DIGITARIA Heist. Crab-grass.

D. filiformis (L.) Koel. Slender Crab-grass. Sterile soil; rare. - Salisbury, Eaton (1898); Haverhill, H. 28389 (1965); Andover, Pease 1548 (1902).

D. ischaemum (Schreb.) Muhl. Naturalized from Europe. Waste ground, occasional. - Haverhill, H. 597 (1932); Lawrence, H. 11228 (1955); Boxford, H. 26201 (1963); Groveland, H. 11169 (1955); West Newbury, H. 11145 (1955); Rowley, Plum Island, H. 12759 (1956); Ipswich, Robinson (1877); Rockport, Forbes 241 (1903); Salem, Robinson (1892); Lynnfield, H. 14110b (1957); Saugus, H. 18685 (1958).

D. sanguinalis (L.) Scop. Naturalized from Old World. A very common and troublesome weed. - Salisbury, White 381 (1913); Methuen, H. 11733 (1955); North Andover, H. 11029 (1955); Boxford, M.E. Perley (1885); Groveland, H. 28379 (1965); West Newbury, H. 11154 (1955); Newbury, Plum Island, H. 12533 (1956); Georgetown, Horner; Rowley, H. 18518 (1958); Ipswich, H. 13876 (1957); Topsfield, H. 11375 (1955); Hamilton, H. 10830 (1955); Gloucester, H. 13828 (1957); Manchester, H. 10713 (1955); Beverly, Lowell (1845); Peabody, H. 8759 (1953); Lynnfield, Rich (1892); Saugus, H. 8205 (1952).

D. sanguinalis, var. *ciliaris* (Retz.) Parl. - Methuen, H. 11753 (1955).

PASPALUM L.

P. ciliatifolium Michx., var. *muhlenbergii* (Nash) Fern. Open ground or thin woods; occasional. - Amesbury, Rich (1906); Andover, Pease 2883 (1903); Boxford, M.P. Perley (1882); Georgetown, Horner; Danvers, Oakes; Essex, Fernald, Hunnewell & Long 8507 (1913); Lynnfield, Rich (1892).

PANICUM L. Panic-grass.

P. dichotomiflorum Michx. Waste places; frequent. - Lawrence, H. 11770 (1955); Boxford, H. 28356 (1965); Topsfield, H. 11390 (1955); Danvers, Lowell (1846); Essex, Fernald, Hunnewell & Long 8521 (1913); Manchester, H. 11339 (1955); Beverly, H. 11702 (1955); Lynnfield, H. 14130 (1957); Lynn, Rich (1899).

P. dichotomiflorum, var. *geniculatum* (Wood) Fern. - South Lawrence, Pease 2801 (1903); Newburyport, White 466 (1913); Danvers, H. 9044 (1953); Essex, H. 11305 (1955); Salem, H. 11517 (1955); Saugus, H. 11860 (1955); Lynn, H. 27062 (1964).

P. capillare L. Old-witch Grass. Open soil; common. - Merrimac, H. 9235 (1953); Haverhill, H. 598 (1932); Lawrence, H. 11268 (1955); Andover, H. 14103 (1957); North Andover, H. 11035 (1955); Boxford, H. 8158 (1952); Groveland, H. 13969 (1957); West Newbury, H. 9142 (1953); Georgetown, H. 13584 (1957); Ipswich, H. 14014 (1957); Topsfield, H. 11392 (1955); Danvers, H. 20772 (1959); Hamilton, H. 12488 (1956); Essex, H. 11307 (1955); Gloucester, H. 13830 (1957); Lynnfield, H. 14132 (1957); Marblehead, H. 11424 (1955); Nahant, Mackintosh (1931).

P. capillare, var. *occidentale* Rydb. - Lawrence, Pease 1538 (1901); Andover, Pease 187 (1901); Georgetown, Horner; Peabody, Mackintosh (1933).

P. philadelphicum Bernh. Open soil; rare. - Andover, Sears (1885); Boxford, Sears (1882); H. 8835 (1953).

P. tuckermani Fern. Sandy shores and open soil; rare. - Amesbury, H. 9177 (1953); Andover, Pease 2605 (1903); Groveland, Pease 2056 (1903), H. 9060 (1953).

P. miliaceum L. Millet. Adventive from Old World. Waste places; common. - Haverhill, Bradford, H. 720 (1932); Methuen, H. 11710 (1955); Lawrence, H. 11787 (1955); Andover, H. 12299 (1956); Boxford, H. 8770 (1953); Newbury, H. 20708 (1959); Rowley, H. 18496 (1958); Ipswich, H. 13852 (1957); Topsfield, H. 8743 (1953); Danvers, H. 11621 (1955); Essex, H. 11311 (1955); Gloucester, Magnolia, H. 12361 (1956); Beverly, Conant (1879); Salem, H. 11511 (1955); Peabody, H. 11527 (1955); Saugus, H. 18676 (1958); Marblehead, H. 11450 (1955).

P. virgatum L. Switchgrass. Dry soils; rare. - Lawrence, Pease 1582 (1902); North Andover, H. 11032 (1955).

P. virgatum, var. *spissum* Linder. Sandy fresh to brackish shores; common. - Salisbury-Amesbury, White 248 (1913); Andover, Pease 2953 (1903); Boxford, Robinson (1877); Groveland, Horner; Newbury, Oak Island, White 230 (1913); Rowley, H. 8273 (1952); Ipswich, H. 13882 (1957); Ipswich, Plum Island, H. 12413 (1956); Gloucester, C.A. & U.F. Weatherby (1930); Rockport, Forbes 270 (1902); Manchester, Clarke (1912); Beverly, Lowell (1846); Peabody, Buxton (1925); Saugus, Rich (1889).

P. agrostoides Spreng. Sandy shores and meadows; common. - Haverhill, Lunt (1913); Andover, Pease 2642 (1903); Boxford, H. 8164 (1952); Georgetown, Horner; Rowley, H. 13458 (1957); Danvers, Sears (1877); Essex, Winslow (1910); Saugus, Robinson (1847).

P. depauperatum Muhl. Dry open soil; occasional. - Haverhill, Ward Hill, H. 388 (1932); Lawrence, H. 9871 (1955); Rowley, H. 7927 (1952); Ipswich, H. 9568 (1954); Topsfield, H. 8073 (1952); Peabody, Robinson (1877); Lynnfield, Bean (1916).

P. depauperatum, var. *psilophyllum* Fern. Common. - Merrimac, Bean & H. 8576 (1953); Haverhill, H. 458 (1932); Andover, Pease 1527 (1902); Boxford, H. 8030 (1952); Groveland, Bean & H. 8542 (1953); Newbury, Plum Island, White 129 (1913); Georgetown, H. 19309 (1959); Rowley, H. 7934 (1952); Ipswich, Oakes; Hamilton, Hubbard 583 (1913); Gloucester, H. 12140 (1956); Manchester, Hubbard 561 (1912); Beverly, Lowell (1846); Lynnfield, Young (1879); Saugus, Bean & H. 9786 (1955); Lynn, Cushman 2975 (1908).

P. linearifolium Scribn. Rare. - Lynn, Mt. Spickett, Cushman 2982 (1908).

P. linearifolium, var. *werneri* (Scribn.) Fern. Common. - Haverhill, H. 12630 (1932); North Andover, H. 17957 (1958); Boxford, Hunnewell 8869 (1923); Groveland, Bean & H. 8487 (1953); Topsfield, H. 10137 (1955); Hamilton, Hubbard 584 (1913); Gloucester, Magnolia, Clarke (1912); Manchester, Hubbard 614 (1913); Beverly, Hubbard 625a (1913); Lynn, H. 10912 (1955).

P. boreale Nash Meadows and fields; occasional. - Salisbury, White 67 (1913); Haverhill, Bradford, Cushman 2956 (1908); Andover, Pease 3199 (1903); Boxford, H.E. Parkhurst (1882); Ipswich, H. 18414 (1958).

P. dichotomum L. Dry woods and thickets; occasional. - Amesbury, H. 18253 (1958); Andover, Pease 2170 (1903); North Andover, H. 17947 (1958); Boxford, H. 10219 (1955); Georgetown,

Horner; Essex, Hubbard 563 (1913); West Manchester, Hubbard 139 (1913); Beverly Farms, Hubbard 592 (1913).

P. lucidum Ashe. Swamps and wet woods; rare. - Manchester, Hubbard 608 (1913); West Manchester, Hubbard 193 (1911).

P. spretum Schultes Wet peats and sands; frequent. - Andover, Hubbard 629 (1913); Boxford, H. 8724 (1953); Georgetown, Horner; Ipswich, Oakes; Topsfield, Sears (1877); Danvers, Lowell (1846); Manchester, Hubbard 173 (1911); Beverly, Lowell (1846); Lynn, Rich (1889).

P. meridionale Ashe. Dry open woods; occasional. - Haverhill, H. 460 (1932); Ipswich, H. 9569 (1954); Manchester, Hubbard 483 (1912); Beverly Farms, Hubbard 595 (1913).

P. lanuginosum Ell., var. *fasciculatum* (Torr.) Fern. Sterile soil; common. - Salisbury, H. 10489 (1955); Haverhill, Bradford, Fernald 11263 (1914); Methuen, H. 9601 (1954); Lawrence, H. 9852 (1955); North Andover, H. 8994 (1953); Boxford, H. 9005 (1953); Groveland, Bean & H. 8536 (1953); Newbury, Plum Island, Fernald 8593 (1913); Rowley, Plum Island, H. 12067 (1956); Ipswich, Oakes; Ipswich, Plum Island, H. 12083 (1956); Wenham, Robinson (1875); West Gloucester, Hubbard 205 (1911); Rockport, H. 10308 (1955); Manchester, Hubbard 488 (1913).

P. lanuginosum, var. *implicatum* (Scribn.) Fern. Common. - Salisbury, D. & E.W. White 30 (1913); Amesbury, Eaton; Merrimac, Bean & H. 8584 (1953); Haverhill, H. 459 (1932); Methuen, H. 8955 (1953); Andover, Pease (1902); Boxford, Pease 13648 (1912); Groveland, H. 8822 (1953); Newbury, Bean & H. 8414 (1953): Rowley H. 8120 (1952); Ipswich, H. 13904 (1957); Ipswich, Plum Island, H. 12434 (1956); Topsfield, H. 8077 (1952); Middleton, Bean (1950); Danvers, H. 7974 (1952); Hamilton, Hubbard 648 (1913); Gloucester, H. 12166 (1956); Rockport, Smith 4321 (1945); Manchester, Hubbard 478 (1912); Beverly, Lowell (1846); Salem, Cole.

P. lanuginosum, var. *lindheimeri* (Nash) Fern. Frequent. - Salisbury, H. 10648 (1955); Merrimac, H. 9204 (1953); North Andover, H. 9899 (1955); Ipswich, H. 13953 (1957); Gloucester, H. 10148 (1955); Manchester, Hubbard 482 (1912); Lynn, H. 10913 (1955).

P. lanuginosum, var. *septentrionale* Fern. Occasional. - Methuen, H. 18397 (1958); Andover, H. 10069 (1955); Boxford, H. 8900 (1953); Middleton, H. 18088 (1958); West Gloucester, Hubbard 205 (1911); Manchester, Hubbard 655 (1913).

P. subvillosum Ashe. Dry sandy soil; occasional. - Salisbury, White 86 (1913); Ipswich, Oakes; West Gloucester, Hubbard 485 (1912); Manchester, Hubbard 587 (1913).

P. columbianum Scribn. Dry open ground; frequent. - Haverhill, H. 461 (1932); Andover, Pease 2180 (1903); Newburyport, White 497 (1913); West Gloucester, Hubbard 469 (1912); Manchester, Hubbard 650 (1913); Beverly Farms, Hubbard 474 (1912).

P. columbianum, var. *oricola* (Hitchc. & Chase) Fern. Dry coastal sand; rare. - Ipswich Beach, Hubbard 473 (1912); Ipswich, Plum Island, St. John & White 554 (1913).

P. sphaerocarpon Ell. Dry open soil; occasional. - Methuen, H. 9615 (1954); Andover, Pease 2176 (1903); Groveland, Bean & H. 8535 (1953); Newbury, South Byfield, Sears (1901).

P. oligosanthes Schultes, var. *scribnerianum* (Nash) Fern. Dry thin woods and openings; common. - Salisbury, D. & E.W. White 24 (1913); Haverhill, Ward Hill, Cushman 2943 (1908); Methuen, H. 7984 (1952); Andover, Pease 188 (1901); North Andover, Bean (1916); Boxford, H.

8646 (1953); Groveland, Bean & H. 8488 (1953); Ipswich, Oakes; Topsfield, H. 8080 (1952); Middleton, Bean (1950); Danvers, Sears (1878); Hamilton, H. 10825 (1955); Rockport, H. 10309 (1955); Manchester, Hubbard 573 (1913); Beverly, Robinson (1884); Swampscott, H. 10656 (1955); Marblehead, H. 9679 (1955).

P. xanthophysum Gray. Dry open soil; rare. - Andover, Lowell Junction, Pease 2453 (1903).

P. commutatum Schultes, var. *ashei* (Parson) Fern. Thin woods; rare. - Manchester, Hubbard 609 (1913); Lynn, Mt. Spickett, Cushman 2980 (1908).

P. clandestinum L. Moist or dry thickets or shores; common. - Salisbury, H. 12218 (1956); Amesbury, Bean (1950); Merrimac, Bean & H. 8568 (1953); Haverhill, Ward Hill, Bean (1916); South Lawrence, Pease 2365 (1903); North Andover, H. 9898 (1955); Boxford, Pease 13669 (1912); Groveland, Bean & H. 8537 (1953); West Newbury, Conant (1879); Georgetown, Horner; Ipswich, Danvers, Sears (1878); Hamilton, Robinson (1876); Rockport, H. 19032 (1958); Saugus, H. 10910 (1955).

P. latifolium L. Dry open rich woods; occasional. - Andover, Pease 1526 (1902); North Andover, H. 17958 (1958); Boxford, M.E. Perley (1884); Danvers, Lowell (1847); Gloucester, Magnolia, Clarke (1912); Manchester, Hubbard 492 (1912); Beverly Farms, Hubbard 591 (1913).

ECHINOCHLOA Beauv.

E. frumentacea (Roxb.) Link. Japanese Millet. Introduced from east Asia. Waste places; not common. - Methuen, H. 11729 (1955); Saugus, H. 10901 (1955).

E. crusgalli (L.) Beauv. Barnyard-grass. Naturalized from Old World. Waste places; common. - Haverhill, H. 5 (1931); North Andover, H. 18737 (1958); Boxford, H. 30775 (1966); Newburyport, White 402 (1913); Georgetown, H. 13579 (1957); Topsfield, H. 8847 (1953); Danvers, Cushman 4838 (1908); West Gloucester, Robinson (1877); Beverly, Montserrat, Hubbard 679 (1913); Salem, H. 10637 (1955); Peabody, Mackintosh (1933); Lynnfield, H. 14137 (1957); Saugus, H. 10905 (1955).

E. crusgalli, forma *longiseta* (Trin.) Farw. Naturalized from Old World. Waste places; common. - Lawrence, H. 11245 (1955); Boxford, M.E. Perley (1885); West Newbury, H. 11123 (1955); Newburyport, White 467 (1913); Byfield, Porter (1941); Georgetown, Horner; Ipswich, Williams (1899); Danvers, H. 18575 (1958); Wenham, Fernald, Hunnewell & Long 8673 (1913); Manchester, H. 11336 (1955); Salem, Cole (1886); Lynnfield, Rich (1892).

E. pungens (Poir.) Rydb. Low ground; frequent. - Salisbury, White 360 (1913); Groveland, H. 10977 (1955); West Newbury, H. 9119 (1953); Rowley, Plum Island, H. 12454 (1956); Danvers, Sears (1902); Gloucester, Fernald, Hunnewell & Long 8672 (1913); Peabody, H. 11525 (1955).

SETARIA Beauv. Bristly Foxtail.

S. glauca (L.) Beauv. Foxtail. Naturalized from Eurasia. Waste places; common. - Haverhill, H. 505 (1932); Boxford, H. 8771 (1953); Groveland, H. 10978 (1956); West Newbury, H. 11147 (1955); Newburyport, White 401 (1913); Georgetown, Horner; Rowley, Plum Island, H. 12475 (1956); Topsfield, H. 8849 (1953); Hamilton, H. 12504 (1956); Gloucester, H. 12314 (1956); Manchester, H. 11355 (1955); Beverly, Montserrat, Hubbard 680 (1913); Peabody, Mackintosh (1933); Saugus, H. 10900 (1955); Lynn, H. 27053 (1964).

S. verticillata (L.) Beauv. Naturalized from Eurasia. Waste places; occasional. - Newburyport, Raymond Dodge; Salem, Cole (1934).

S. faberii Herrm. Naturalized from eastern Asia. Disturbed soil; rare. - Lynn, H. 27059 (1964).

S. viridis (L.) Beauv. Green Foxtail. Naturalized from Eurasia. Waste places; common. - Methuen, H. 11746 (1955); Andover, Pease 204 (1901); Boxford, H. 8157 (1952); Newburyport, H. 10933 (1955); Rowley, H. 8115 (1952); Rowley, Plum Island, H. 12474 (1956); Ipswich, Plum Island, H. 12388 (1956); Danvers, Kennedy (1907); Gloucester, H. 12329 (1956); Beverly, Montserrat, Hubbard 677 (1913); Salem, Lowell (1847); Saugus, H. 18713 (1958); Marblehead, H. 11451 (1955).

S. italica (L.) Beauv. Hungarian Millet. Introduced from Eurasia. Waste places; frequent. - Haverhill, H. 18815 (1958); Lawrence, H. 18764 (1958); Rowley, H. 18497 (1958); Chebacco, Robinson (1875); Danvers, Sears (1878); Peabody, H. 27067 (1964); Saugus, H. 18764 (1958); Lynn, H. 18817 (1958).

CENCHRUS L. Burgrass.

G. longispinus (Hack.) Fern. Beach sand and waste places; occasional. - Haverhill, Bradford, Fernald 11282 (1914), H. 11202 (1955); Lawrence, Sears (1885), H. 11791 (1955); Saugus, H. 18621 (1958).

MISCANTHUS Anderss.

M. sacchariflorus (Maxim.) Hack. Introduced from Asia. Escape from cultivation. - Beverly, H. 18888 (1958).

ANDROPOGON L. Beardgrass.

A. scoparius Michx., var. *frequens* F.T. Hubbard. Broom-beardgrass. Dry sterlis soil; common. - Merrimac, Bean & H. 7709 (1951); Haverhill, H. 595 (1932); Lawrence, H. 11227 (1955); North Andover, H. 11049 (1955); Boxford, M.E. Perley (1883); Georgetown, Robinson (1875); Rowley, H. 8271 (1952); Topsfield, H. 11380 (1955); Hamilton, H. 10844 (1955); Rockport, Forbes 216 (1902); Beverly, Prides Crossing, Kidder (1885).

A. scoparius, var. *septentrionalis* Fern. & Grisc. Common. - Salisbury, White 340 (1913); Merrimac, H. 9252 (1953); Rowley, Plum Island, H. 12577 (1956); Danvers, Sears (1883); West Gloucester, Robinson (1877); Beverly, Lowell (1846); Peabody, Mackintosh (1933); Marblehead, H. 11430 (1955).

A. scoparius, var. *littoralis* (Nash) Hitchc. Upper portion sea beaches; rare. - Rowley, H. 8282 (1952).

A. gerardi Vitman Swales, shores and dry open ground; common. - Amesbury, Bean (1950); Haverhill, Ward Hill, Bean (1916); Methuen, Bean & H. 7753 (1951); Andover, Pease 156 (1901); North Andover, H. 11048 (1955); Boxford, H. 13530 (1957); Newbury, H. 20701 (1959); Ipswich, Hunnewell 2992 (1914); Danvers, Robinson (1875); Essex, Buxton (1925); Beverly, Lowell (1846); Salem, Forbes 219 (1903); Peabody, Mackintosh (1925).

SORGUM Adans. Sorghum.

S. halapense (L.) Pers. Johnson-grass. Introduced from Eurasia. Waste places; rare. - Byfield, Porter (1941).

S. vulgare Pers. Sorghum. Introduced from Eurasia. Waste places; rare. - Lawrence, H. 18768 (1958); Boxford, H. 12512 (1956); Peabody, H. 27077 (1964); Saugus, H. 18675 (1958).

SORGHASTRUM Nash.

S. nutans (L.) Nash. Indian Grass. Borders of woods; occasional. - Andover, Pease 2614 (1903); Georgetown, W.S. Horner; Ipswich, Oakes; Middleton, Porter (1938); Danvers, Lowell (1847); Lynnfield, Moore & Smith 1150 (1930).

ZEA L.

Z. mays L. Maize. Casual on dumps. - Haverhill, H. 18814 (1958); Lawrence, H. 18738 (1958); North Andover (1884); Boxford, H. 9028 (1953); Danvers, H. 20762 (1959); Essex, H. 18866 (1958); Lynn, Robinson (1875); Marblehead, H. 11457 (1955).

CYPERACEAE (Sedge Family)

CYPERUS L. Galingale.

C. diandrus Torr. Wet sandy or gravelly soils; frequent. - Merrimac, H. 13672 (1957); Lawrence, Robinson (1877); Andover, Pease (1900); Boxford, H. 14043 (1957); Groveland, H. 28373 (1965); Georgetown, Horner (1885); Rowley, H. 13492 (1957); Danvers, H. 9046 (1953); Wenham, Hubbard 95 (1911); West Gloucester, Robinson (1877); Rockport, Fernald, Hunnewell & Long 8827 (1913); Salem, F.F. Forbes 544 (1903).

C. rivularis Kunth. Wet sandy or gravelly soils; occasional. - Andover, Williams (1899); Wenham, Hubbard 95a (1911); Hamilton, Hubbard 181 (1911); Rockport, Fernald, Hunnewell & Long 8830 (1913); Salem, Forbes 545 (1903).

C. filicinus Vahl Brackish to fresh sands; occasional. - Newbury, Plum Island, Pease 2896 (1903); Rowley and Byfield, Sears (1904); Ipswich Beach, F.T. & G.G. Hubbard 160 (1911); Ipswich, Plum Island, H. 12410 (1956); West Gloucester, Robinson & Sears (1877); Rockport, Fernald, Hunnewell & Long 8835 (1913); Salem, Cole (1887); Nahant, Parsons (1885).

C. erythrorhizos Muhl. Alluvial soil; occasional. - Lawrence, Robinson & Sears (1877); Newburyport, Eaton (1896); Ipswich, Plum Island, H. 12411 (1956).

C. odoratus L. Low ground; rare. - Essex County, Oakes.

C. esculentus L. Yellow Nut-grass. Damp sandy soil; occasional. It has not become a serious weed as Robinson feared. - Amesbury, Eaton (1897); Haverhill, Bradford, Fernald 11294 (1914); Lawrence, Sears (1885); Boxford, H. 9004 (1953); West Newbury, Conant (1878); Danvers, H. 20758 (1959); Essex, H. 13337 (1957); Peabody, H. 27063 (1964).

C. dentatus, Torr. Sandy or gravelly shores; common. - Amesbury, Robinson (1879); Haverhill, Lunt 176 (1911); Andover, H. 14093 (1957); North Andover, Sears (1885); Boxford, A.W. & C.A. Cheever (1913); Groveland, H. 9097 (1953); West Newbury, White 507 (1913); Newbury, Plum Island, Pease 2894 (1903); Georgetown, Sears (1876); Ipswich Beach, F.T. & G.G. Hubbard 159 (1911); Danvers, Cushman 4832 (1908); Peabody, H. 9321 (1953); Lynnfield, H. 14136 (1957).

C. dentatus, forma *ctenostachys* Fern. - Plum Island, Robinson (1878).

C. strigosus L. Meadows, swales and shores; common. - Amesbury, H. 9166 (1953); Haverhill, H. 13 (1931); Methuen, H. 11726 (1955); Lawrence, Jenks (1884); Andover, Pease 1756 (1902); North Andover, Bean (1936); Boxford, H. 8197 (1952); Groveland, H. 11186 (1955); West Newbury, H. 10965 (1955); Georgetown, Horner; Rowley, Plum Island, H. 12461 (1956); Ipswich, H. 13254 (1957); Topsfield, H. 8851 (1953); Danvers, Cushman 4834 (1908); Wenham, Robinson (1877); Essex, H. 13338 (1957); West Manchester, Hubbard 171 (1911); Beverly, Lowell (1846); Salem, H. 11503 (1955); Peabody, H. 9334 (1953); Saugus, H. 18658 (1958).

C. grayii Torr. Dry coastal sands; occasional. - Salisbury, White 288 (1913); Newbury, Plum Island, Pease 2837 (1903); Ipswich, Fernald (1911); Ipswich, Plum Island, St. John & White 552 (1913); Gloucester, Magnolia, Clarke (1911).

C. filiculmis Vahl. Sandy ground; rare. - West Gloucester, Hubbard 200 (1911).

C. filiculmis, var. *macilentus* Fern. Common. - Salisbury, White 287 (1913); Amesbury, H. 9169 (1953); Merrimac, H. 9254 (1953); Haverhill, H. 14 (1931); Methuen, H. 11731 (1955); Andover, Boott (1868); Boxford, H. 8778 (1953); Groveland, H. 11167 (1955); Newburyport, Plus Island, H. 13043 (1957); Newbury, Plum Island, H. 11292 (1955); Rowley, Plum Island, H. 12464 (1956); Ipswich, Oakes; Ipswich, Plum Island, H. 12437 (1956); Wenham, Hubbard 101 (1911); Hamilton, H. 10836 (1955); Gloucester, Magnolia, Clarke (1911); Beverly, Lowell (1845); Salem, Cole (1888); Peabody, Mackintosh (1933); Saugus, H. 10885 (1955); Swampscott, H. 10657 (1955); Marblehead, H. 11472 (1955); Nahant, Kennedy (1894).

DULICHIUM Pers.

D. arundinaceum (L.) Britt. Three-way Sedge. Wet places; common. - Amesbury, H. 9193 (1953); Merrimac, H. 9215 (1953); Haverhill, H. 15 (1931); Lawrence, H. 12292 (1956); Andover, Pease 3091 (1903); North Andover, Porter (1939); Boxford, H. 8201 (1952); Groveland, H. 9090 (1953); Georgetown, Horner (1881); Rowley, H. 13511 (1957); Middleton, H. 8704 (1953); Wenham, Robinson (1875); Hamilton, H. 9414 (1953); West Gloucester, Hubbard 108 (1911); Rockport, Gates & Smith 1090 (1928); Manchester, H. 10763 (1955); Beverly, Lowell (1846); Peabody, H. 9308 (1953).

ELEOCHARIS R.Br. Spike-rush.

E. robbinsii Oakes. Shallow ponds; occasional. - Boxford, H. 13394 (1957); Ipswich, H. 13261 (1957); Essex, H. 13376 (1957); West Gloucester, Hubbard 123 (1911); Lynn, Young (1880).

E. parvula (R. & S.) Link. Wet saline or brackish shores; occasional. - Salisbury-Amesbury, White 250 (1913); Newburyport, White 442 (1913); Plum Island, St. John 870 (1913); West Gloucester, Robinson & Sears (1877); Beverly, Lowell (1849); Salem, Sears; Saugus, Svenson 3005 (1928); Nahant, Boott (1854).

E. rostellata Torr. Saline or brackish marshes; rare. - Newburyport, Dodge (1890); Beverly Farms, Hubbard 91 (1911).

E. acicularis (L.) R. & S. Damp shores; common. - Salisbury, White 366 (1913); Amesbury, Bean (1950); Merrimac, H. 13732 (1957); Haverhill, H. 570 (1932); Methuen, H. 12851 (1957); Lawrence, Robinson (1879); Boxford, H. 8784 (1953); Groveland, H. 9106 (1953); Newbury, Kent's Island, Sears (1904); Ipswich, H. 13469 (1957); Topsfield, H. 13629 (1957); Danvers, Lowell (1849); Beverly, Robinson (1875); Peabody, H. 13316 (1957); Lynnfield, H. 14150 (1957); Swampscott, Robinson (?) (1977).

E. acicularis, forma *longicaulis* (Desmaz.) Hegi - Boxford, H. 13790 (1957); Saugus, Fernald & Svenson 823 (1928).

E. acicularis, var. *submersa* (Hj.Nilss.) Svenson - Wenham Lake, Phillips (1910).

E. olivacea Torr. Wet sands and peat; occasional. - Salisbury, White 376 (1913); Merrimac, H. 13719 (1957); Methuen, H. 8915 (1953); Andover, Pease 2604 (1903); Ipswich, H. 13241 (1957); Wenham, H. 13756 (1957); Rockport, Fernald, Hunnewell & Long 8874 (1913); Manchester, Hubbard 132 (1911); Saugus, Young (1879); Lynn, E. & C.E. Faxon (1881).

E. engelmanni Steud. Wet sand; rare. - Saugus, Fernald & Svenson 818 (1928); Lynn, Young (1879).

E. obtusa (Willd.) Schultes. Muddy or wet places; common. - Salisbury, White 369 (1913); Amesbury, H. 9151 (1953); Merrimac, H. 9212 (1953); Haverhill, H. 16 (1931); Andover, Pease 2643 (1903); North Andover, Porter (1939); Boxford, Sears (1880); Groveland, H. 9096 (1953); Georgetown, Horner; Rowley, H. 8130 (1952); Topsfield, H. 8844 (1953); Middleton, H. 9426 (1953); Danvers, Lowell (1846); Wenham, Robinson (1876); Essex, Fernald, Hunnewell & Long 8881 (1913); Gloucester, Lanesville, Robinson (1877); Manchester, Hubbard 75 (1911); Beverly, Mackintosh (1933); Peabody, H. 9300 (1953); Saugus, Fernald & Svenson 819 (1928).

E. obtusa, var. *jejuna* Fern. - North Andover, H. 14084 (1957).

E. palustris (L.) R. & S., var. major Sonder Shallow water; frequent. - Methuen, H. 18400 (1958); North Andover, H. 12930 (1957); Boxford, H. 8874 (1953); Georgetown, H. 12966 (1957); Rowley, H. 13484 (1957); Middleton, H. 12899 (1957); Wenham, Robinson (1875); Hamilton, H. 9411 (1953); Gloucester, H. 13825 (1957); Salem, Robinson (1879); Peabody, Mackintosh (1933); Lynn, H. 13174 (1957).

E. smallii Britt. Wet sandy swamps and shores; occasional. - Amesbury, Eaton (1900); Andover, Pease 3198 (1903); Boxford, H. 8727 (1953); Groveland, H. 9085 (1953); Ipswich, Oakes; Wenham, Hunnewell 4626 (1916); Beverly, Montserrat, Hubbard 54 (1911); Peabody, H. 9309 (1953); Lynnfield, Young (1879).

E. halophila Fern. Saline or brackish shores; rare. - Salisbury, White 20 (1913); Newbury, Williams (1898); Rockport, Forbes 586+ (1902).

E. tenuis (Willd.) Schultes. Damp sands and gravels; frequent. - Amesbury, Eaton; Haverhill, Bradford, Cushman 2957 (1908); North Andover, Porter (1938); Boxford, M.E. Perley (1882); Georgetown, Horner; Rowley, H. 8129 (1952); Wenham, Robinson (1875); Rockport, Cushman 3373 (1908); West Manchester, Hubbard 32 (1911).

E. elliptica Kunth. Damp sands and gravels; occasional. - Amesbury, Eaton (1898); Andover, Pease 2130 (1903); West Newbury, Conant (1877); Georgetown, Horner (1883); Manchester, Hubbard 42 (1911).

E. tuberculosa (Michx.) R. & S. forma retrorsa Svenson. Wet sandy shores; occasional. - Danvers, Lowell (1849); Essex, Hubbard 90 (1911); Gloucester, Swan (1881); Manchester, Oakes.

BULBOSTYLIS (Kunth) C.B. Clarke

B. capillaris (L.) C.B. Clarke. Dry open soil; frequent. - Amesbury, H. 9168 (1953); Merrimac, H. 9232 (1953); Haverhill, H. 18 (1931); Boxford, H. 8172 (1952); Groveland, H. 11173 (1955); Middleton, Porter (1937); Danvers, Lowell (1849); Peabody, H. 9302 (1953).

B. capillaris, var. *crebra* Fern. Common. - Andover, Pease 4271 (1904); North Andover, H. 11051 (1955); Boxford, H. 8652 (1953); Newburyport, Plum Island, Dodge (1890); Ipswich, Oakes; Topsfield, H. 18288 (1958); Danvers, H. 11614 (1955); Wenham, Robinson (1877); Chebacco, Robinson (1875); Essex, H. 13341 (1957); West Gloucester, Hubbard 107 (1911); Manchester, Hubbard 192 (1911).

FIMBRISTYLIS Vahl

F. autumnalis (L.) R. & S. Sandy shores; frequent. - Haverhill, H. 571 (1932); Methuen, H. 8947 (1953); Andover, Pease 2649 (1903); North Andover, H. 14065 (1957); Boxford, Robinson (1880); Groveland, H. 13968 (1957); Ipswich, Oakes; Danvers, Sears (1878); Wenham, Hubbard 100 (1911); East Gloucester, Swan (1881).

F. autumnalis, var. *mucronulata* (Michx.) Fern. - Boxford, Sears.

SCIRPUS L. Bulrush.

S. verecundus Fern. Dry woods; occasional. - Amesbury, Huntington (1899); West Boxford, Fernald (1909); Georgetown, Horner (1884); Ipswich, Morong (1875); Danvers, Sears (1878); Lynnfield.

S. hudsonianus (Michx.) Fern. Boggy meadows; local. - Amesbury, Huntington (1909); Boxford, E. & C.E. Faxon (1878); South Georgetown, W.S. Horner (1885); Danvers, Sears (1883); Wenham, Sears (1905); Hamilton, Oakes; Beverly, Lowell (1853); Lynnfield, Hitchings (1890).

S. smithii Gray. Sandy shores; rare. - Amesbury, H. 9178 (1953); Andover, Blake (1883); North Andover, Blake (1883).

S. smithii, forma *setosus* Fern. - Lawrence, Conant (1879); South Groveland, Pease 1536 (1901).

S. purshianus Fern. Muddy or sandy shores; occasional. - Lawrence, Robinson & Sears (1877); South Groveland, H. 13966 (1957); South Georgetown, Horner; Danvers, Lowell (1846).

S. subterminalis Torr. Swaying Rush. Shallow water; frequent. - Andover, Pease 4878 (1904); Boxford, H. 8969 (1953); Groveland, H. 12866 (1957); Georgetown, H. 12964 (1957); Rowley, H. 13500 (1957); Ipswich, H. 13247 (1957); Middleton, H. 12893 (1957); Wenham, H. 13770 (1957); Hamilton, Robinson; Essex, H. 13342 (1957); Lynnfield, Young (1879).

S. torreyi Olney. Shore Merrimac River; rare. - Lawrence, Robinson (1879); Jenks (1884).

S. americanus Pers. Chair-maker's Rush. Fresh or saline shores and marshes; common. - Amesbury, Bean (1950); Haverhill, Bradford, H. 572 (1932); Andover, Pease 264 (1900); Ipswich Beach, Hubbard 763 (1911); Ipswich, Plum Island, H. 12387 (1956); Danvers, H. 9041 (1953); Wenham, Hubbard 102 (1911); Hamilton, Hubbard 88 (1911); Chebacco, Robinson (1876); Essex, H. 9366 (1953); West Gloucester, Robinson & Sears (1877); Manchester, Lane (1877); Beverly, F.H. Peabody (1894); Swampscott, Swan (1886); Nahant, Parsons.

S. olneyi Gray. Saline marshes; rare. - Newbury, Kents Island, Eaton (1897).

S. validus Vahl, var. *creber* Fern. Great Bulrush. Brackish or fresh shallow water and marshes; common. - Merrimac, H. 9216 (1953); Haverhill, H. 20 (1931); North Andover, Morss (1885); Boxford, M.P. Perley (1883); Groveland, H. 9087 (1953); West Newbury, H. 9116 (1953); Newbury, White 208 (1913); Byfield, Robinson (1875); Georgetown, Horner (1883); Rowley, H. 13508 (1957); Ipswich, H. 13263 (1957); Topsfield, H. 13604 (1957); Danvers, Sears (1878); Wenham, H. 13101 (1957); East Gloucester, Swan (1881); Nahant, Deane (1884).

S. acutus Muhl. Hard-stem Bulrush. Pond margins; rare. - Boxford, H. 13660 (1957); Topsfield, Williams (1903); Wenham, Lowell (1847); Essex County, Oakes.

S. fluviatilis (Torr.) Gray. River-bulrush. Borders of streams; rare. - Amesbury, Fassett 2334 (1924); Ipswich, St. John 11891 (1916); Ipswich River Trip, Robinson (1876).

S. robustus Pursh. Saline marshes; rare. - Newbury, Williams (1893).

S. maritimus L., var. *fernaldi* (Bickn.) Beetle. Salt-marsh Bulrush. Saltmarsh and brackish to fresh tidal shores; common. - Newburyport, White 177 (1913); Newbury, White 209 (1913); Rowley, H. 8253 (1952); Rowley, Plum Island, H. 12545 (1956); Ipswich, Sears (1886); Danvers Port, Sears (1878); Rockport, Forbes 627 (1902); Salem, Robinson (1879).

S. maritimus, var. *fernaldi,* forma *agonus* Fern. - Manchester, Thoreau; West Manchester, Hubbard 73 (1911).

S. paludosus Nels., var. *atlanticus* Fern. Bayonet-grass. Saline marshes and shores; common. - Salisbury, White 10 (1913); Newburyport, Bean (1940); Newbury, Bean & Stockman (1954); Newbury, Plum Island, Pease 3098 (1903); Rowley, Plum Island, H. 12463 (1956); Ipswich, Plum Island, H. 12401 (1956); Danvers, Robinson (1879); Essex, H. 9365 (1953); West Gloucester, Hubbard 198 (1911); Beverly Farms, Hubbard 84 (1911); Salem, Robinson (1875); Peabody, Mackintosh (1933).

S. rubrotinctus Fern. Marshes; occasional. - Haverhill, H. 12613 (1932); Lawrence, Sears (1879); Andover, Pease 1725 (1901); Newburyport, White 200 (1913); Georgetown, M.E. Perley (1884); Rockport, Smith 1006 (1928); Beverly, Lowell (1853); Marblehead, Sears (1885).

S. expansus Fern. Low thickets; rare. - Lawrence, Robinson (1879); Andover, Young (1882).

S. atrovirens Willd. Meadows and low thickets; rare. - Boxford, Porter (1938); Manchester, Hubbard 93 (1911).

S. atrovirens, var. *georgianus* (Harper) Fern. Occasional. - Merrimac Valley, Robinson (1879); Salisbury, H. 12219 (1956); Merrimac, H. 9240 (1953); Methuen, H. 8957 (1953); Andover, Pease 268 (1901); Georgetown, Horner; Ipswich, H. 18118 (1958); Topsfield, Sears (1879); Ipswich River, Sears (1878); Saugus, H. 10858 (1955).

S. cyperinus (L.) Kunth. Wet meadows and swamps; common. - Salisbury, White 371 (1913); Amesbury, H. 9157 (1953); Merrimac, H. 9237 (1953); Methuen, H. 8942 (1953); Lawrence, H. 11229 (1955); Andover, H. 19061 (1958); Boxford, H. 8192 (1952); Groveland, H. 9086 (1953); West Newbury, H. 9121 (1953); Newbury, H. 8239 (1952); Rowley, Plum Island, H. 12453 (1956); Ipswich, H. 13928 (1957); Ipswich, Plum Island, H. 12421 (1956); Manchester, Hubbard 92 (1911); Salem, Robinson (1879); Lynn, H. 13171 (1957); Marblehead, H. 12599 (1956).

S. cyperinus, var. *pelius* Fern. Occasional. - Haverhill, H. 640 (1932); Newbury, Williams (1899); Georgetown, Horner; Topsfield, H.R. Towne; West Gloucester, Hubbard 109 (1911).

S. cyperinus, var. *pelius*, forma *condensatus* (Fern.) S.F. Blake - Hamilton, Hunnewell 4631 (1916).

S. pedicellatus Fern. Rich shores; rare. - Amesbury, Eaton (1897); Haverhill, H. 546 (1932); Merrimac Valley, Robinson (1879); Wenham, Knowlton (1908).

S. atrocinctus Fern. Meadows and swamps; rare. - Boxford, M.P. Perley (1882); Danvers, Sears (1878); Lynnfield, Boott (1854).

S. atrocinctus, forma *brachypoda* (Fern.) S.F. Blake - Salisbury, Eaton (1898); Amesbury, Eaton (1897); Boxford, H. 8140 (1952).

ERIOPHORUM L. Cotton-grass.

E. spissum Fern. Hare's-tail. Acid bogs; rare. - Andover, Pease 1535 (1902).

E. gracile W.D.J. Koch. Wet peat; occasional. - Methuen, H. 18374 (1958); Andover, H. 17900 (1958); Boxford, Forbes (1909); Georgetown, Horner; Ipswich, Oakes; Topsfield, Oakes; Danvers, Lowell (1846).

E. tenellum Nutt. Peaty soil; occasional. - Merrimac, H. 9223 (1953); Haverhill, H. 1292 (1934); Newbury, White 225 (1913); Hamilton, Hubbard 85 (1911); Beverly, Montserrat, Hubbard (1913); Lynnfield, Boott (1854).

E. viridi-carinatum (Engelm.) Fern. Peaty meadows; frequent. - Haverhill, Bradford, Cushman (1908); Andover, Pease 253 (1901); Boxford, E. & C.E. Faxon (1878); West Newbury, Conant (1878); Newburyport, Conant (1878); Georgetown, Horner (1880); Ipswich, Oakes; Topsfield, Eva Morissey & Barbara Woodbury; Danvers, Sears (1878); Hamilton, Robinson (1879); Beverly, Lowell (1846).

E. virginicum L. Tawny Cotton-grass. Bogs and peaty meadows; frequent. - Haverhill, H. 721 (1932); Methuen, H. 8951 (1953); Andover, Blake (1883); Boxford, H. 8872 (1953); Newbury, White 246 (1913); Georgetown, Horner; Danvers, Sears (1886); Wenham, Conant (1878); Chebacco, Robinson (1875); West Gloucester, Hubbard 104 (1911); Rockport, Smith 2715 (1933); Manchester, Lane (1877); Beverly, Lowell (1845); Peabody, Sears (1886); Lynnfield, Boott (1852); Marblehead Neck, Kennedy (1887).

E. virginicum, forma *album* (Gray) Wieg. - Haverhill, H. 722 (1932); Andover, Pease 255 (1900); Danvers, Sears (1885).

HEMICARPHA Nees

H. micrantha (Vahl) Pax. Known in County only from sandy shores of Lake Cochichewick. - North Andover, Blake (1883), Pease 24101 (1934), H. 14058 (1957).

RHYNCHOSPORA Vahl Beak-rush.

R. capitellata (Michx.) Vahl. Damp shores; common. - Amesbury, Eaton; Merrimac, H. 9227 (1953); Haverhill, H. 509 (1932); Methuen, H. 8944 (1953); Andover, Pease 260 (1901); North Andover, Porter (1937); Boxford, M.E. Perley (1882); Topsfield, R.J. Eaton (1931); Danvers, Lowell (1849); Rockport, Smith 1093 (1928); Manchester, Oakes; Lynnfield, H. 9356 (1953).

R. alba (L.) Vahl. Bogs and wet sands; common. - Methuen, H. 12839 (1957); Andover, Pease 259 (1901); Boxford, H. 13382 (1957); Georgetown, Horner; Danvers, Sears (1885); Wenham, Robinson (1875); Essex, Hubbard 89 (1911); Beverly, Lowell (1846); Salem, Robinson (1875); Lynnfield, Bean (1934); Marblehead Neck, Kennedy (1887); Essex County, Manasseh Cutler.

R. fusca (L.) Ait. Bogs; occasional. - Boxford, C.E. Faxon (1878), H. 8738; Georgetown, H. 12968 (1957); Manchester, Oakes; Beverly, Lowell (1874).

CLADIUM P.Br. Twig-rush.

C. mariscoides (Muhl.) Torr. Marshes and pond margins; frequent. - Salisbury, White 26 (1913); Boxford, H. 8295 (1952); Georgetown, H. 9468 (1954); Rowley, H. 11945 (1955); Ipswich, Oakes; Middleton, Sears (1876); Wenham, Sears (1902); Hamilton, H. 12822 (1957); Essex, Lowell (1848); East Gloucester, Swan (1881); Manchester, Oakes; Lynn, Fuller (1896).

CAREX L. Sedge.

C. foena Willd. Dry open soils; occasional. - Salisbury, Eaton (1896); Andover, Pease 3653 (1904); Boxford, Pease & St. John 11947 (1917); Ipswich, Oakes; Salem, Robinson (1879); Lynnfield, Forbes (1912).

C. retroflexa Muhl. Dry rocky woods; rare. - Boxford, C.E. Faxon (1878).

C. convoluta Mackenz. Dry woods and thickets; occasional. - Merrimac, Bean & H. 8371 (1953); Haverhill, H. 710 (1932); Lawrence, H. 9856 (1955); Boxford, H. 7951 (1952); Georgetown, Horner; Essex County, Lowell, Sears (1878).

C. radiata (Wahlenb.) Dew. Open woods and thickets; rare. - Andover, Pease 3648 (1904).

C. cephalophora Muhl. Dry woods and openings; common. - Haverhill, Ward Hill, H. 787 (1932); North Andover, H. 9894 (1955); Boxford, Robinson (1878); Georgetown, Horner; Rowley, H. 7926 (1952); Ipswich, Oakes; Topsfield, Bean & H. 9963 (1955); Danvers, Lowell (1848); Wenham, Bean & H. 10034 (1955); West Manchester, Hubbard 9 (1911); Beverly, Montserrat, Hubbard 59 (1911); Marblehead, H. 9680a (1955).

C. spicata Huds. Naturalized from Europe. Dry fields and roadsides; rare. - West Newbury, Conant (1878); Gloucester, H. 12173 (1956); Salem, Robinson (1879); Morong (1879).

C. muhlenbergii Schkuhr. Dry woods and fields; occasional. - Haverhill, H. 711 (1932); Andover, Pease 3613 (1904); Boxford, Robinson (1878); Topsfield, H. 8074 (1952); Beverly, Robinson (1879); Essex County, Oakes.

C. sparganioides Muhl. Rich woods; rare. - Essex County, Oakes.

C. vulpinoidea Michx. Low ground; common. - Salisbury, H. 12212 (1956); Methuen, H. 12838 (1957); West Newbury, Conant (1879); Newburyport, White (1913); Newbury, Bean & H. 8412 (1953); Ipswich, Oakes; Danvers, Bean & H. 9991 (1955); Essex, H. 10240 (1955); Manchester, Hubbard 62 (1911); Beverly, H. 10613 (1955).

C. annectens Bickn. Dry or moist sterlie soils; occasional. - Andover, Pease 3086 (1903); Newbury, Schweinfurth (1950); Ipswich, Oakes; Middleton, Bean (1950); Manchester, Hubbard 49 (1911).

C. diandra Schrank Bogs; rare. - Danvers, Sears (1878); Beverly, Robinson (1879).

C. prairea Dew. Bogs; rare. - Newburyport, Forbes 827 (1903).

C. stipata Muhl. Low ground; common. - Amesbury, H. 18242 (1958); Merrimac, Bean & H. 8588 (1953); Haverhill, H. 712 (1932); Lawrence, H. 10092 (1955); Andover, Pease 239 (1901); North Andover, H. 9904 (1955); Boxford, M.E. Perley (1882); Groveland, Bean & H. 8473 (1953); West Newbury, Conant (1878); Newbury, White 204 (1913); Georgetown, Horner; Topsfield, H. 17655 (1955); Middleton, H. 18100 (1958); Danvers, H. 7969 (1952); Wenham, Bean & H. 10002 (1955); Rockport, Bishop & Smith 951 (1928); West Manchester, Hubbard (1911); Beverly, Prides Crossing, Hubbard 60 (1911); Salem, Cole (1886); Marblehead, Bean & H. 9746 (1955).

C. disperma Dew. Damp woods; rare. - Amesbury, Huntington (1905); Beverly Farms, Hubbard 78 (1911).

C. trisperma Dew. Mossy woods and bogs; occasional. - Andover, Blake (1883); North Andover, H. 17989 (1958); Georgetown, Horner; Topsfield, Robinson (1877); Manchester, Hubbard 26 (1911); Beverly, Lowell (1849); Lynnfield, Young (1879).

C. trisperma, var. *billingsii* Knight - Andover, H. 17894 (1958); Wenham Swamp, Sears.

C. canescens L., var. *disjuncta* Fern. Swamps and swales; common. - Haverhill, H. 723 (1932); Andover, Lowell (1849); North Andover, H. 9492 (1954); Boxford, H. 17725 (1958); Newbury, Bean & H. 8403 (1953); Ipswich, Middleton, Bean (1950); Wenham, Robinson (1875); West Manchester, Hubbard 14 (1911); Beverly, Robinson (1875).

C. canescens, var. *subloliacea* Laestad. - Andover, H. 17910 (1958).

C. brunnescens (Pers.) Poir., var. *sphaerostachya* (Tuckerm.) KüKenth. Dry woods; rare. - Middleton, H. 18082 (1958); West Manchester, Hubbard 38 (1911), (1913).

C. bromoides Schkuhr. Rich low woods; rare. - South Lawrence, Williams 1 (1900); West Newbury, Conant (1879); Ipswich, Oakes.

C. deweyana Schwein. Rich open woods and banks; rare. - Salem, Great Pastures, Sears (1881).

C. exilis Dew. Bogs and swales; occasional. - Amesbury, Eaton; Andover, Pease 3251 (1902); Boxford, C.E. Faxon (1878); H. 12771 (1957); South Georgetown, Eaton 72 (1896); Ipswich, Oakes; Danvers, Lowell (1848); Hamilton, Robinson (1879); Lynnfield, Young (1879).

C. interior Bailey. Damp Soil; rare. - Methuen, H. 8952 (1953); Boxford, H. 17723 (1958); Essex County, Sears (1878).

C. howei Mackenz. Sphagnous or mossy swamps, thickets or woods; occasional. - Salisbury, Bean & H. 8386 (1953); Andover, Pease 3691 (1904); Ipswich, Oakes; Hamilton-Essex, A.B. Seymour (1908); Manchester, Hubbard 70 (1911).

C. sterilis Willd. Wet soils; rare. - Ipswich, Oakes.

C. atlantica Bailey. Mossy or peaty soils; rare. - Merrimac, Bean & H. 8579 (1953); Methuen, H. 12842 (1957).

C. incomperta Bickn. Swamps; rare. - Merrimac, Bean & H. (1953); Haverhill, H. 724 (1932); Hamilton-Essex, A.B. Seymour (1908).

C. cephalantha (Bailey) Bickn. Acid peaty swales and swamps; frequent. - Andover, H. 17909 (1958); North Andover, H. 17990 (1958); Boxford; M.E. Perley (1882); Groveland, Bean & H. 8090 (1953); Newburyport, Forbes 844+ (1903); Georgetown, Horner (1883); Ipswich, Allen (1912); Middleton, H. 18105 (1958); Danvers, Lowell (1846); Wenham, Robinson (1875); Gloucester, Rand & B.L. Robinson (1896); Rockport, Bishop & Smith 949 (1928); Manchester, Hubbard 44 (1911).

C. angustior Mackenz. Swales and swampy thickets; occasional. - Haverhill, Bradford, Cushman 2955 (1908); Andover, Pease 3087 (1903); Newbury, Bean & H. 8407 (1953); Manchester, Hubbard 45 (1911).

C. seorsa Howe. Wet woods; rare. - South Georgetown, Eaton 47 (1896); Beverly Farms, Hubbard 40 (1911).

C. scoparia Schkuhr. Moist to dry open ground and thickets; common. - Salisbury, D. & E.W. White 7 (1913); Amesbury, Eaton (1897); Merrimac, Bean & H. 8569 (1953); Haverhill, Bradford, Cushman 2944 (1908); Lawrence, H. 10095 (1955); Andover, Pease 3807 (1904); North Andover, H. 9490 (1954); Boxford, H.E. Parkhurst (1882); Groveland, Bean & H. 8490 (1953); Newburyport, Plum Island, H. 13040 (1957); Georgetown, Horner; Rowley, Bean; Ipswich, H. 9573 (1953); Ipswich, Plum Island, H. 12126 (1956); Topsfield, H. 8071 (1952); Wenham, Robinson (1875); Rockport, Cushman (1908); Manchester, Hubbard 65 (1911); Beverly, Lowell (1849); Lynnfield, Young (1879).

C. scoparia, forma *condensa* (Fern.) KüKenth. - Salisbury, White 326 (1913).

C. scoparia, forma *moniliformis* (Tuckerm.) KüKenth. - Andover, Pease 4212 (1904); Rockport, Cushman 3337 (1908); Beverly Farms, Hubbard 77 (1911); Marblehead, Bean & H. 9752 (1955).

C. tribuloides Wahlenb. Swales and low ground; common. - Salisbury, H. 10497 (1955); Merrimac, H. 10853 (1955); Haverhill, H. 719 (1932); Methuen, H. 8933 (1953); Lawrence, H. 12289 (1956); Andover, H. 10062 (1955); North Andover, H. 9913 (1955); West Newbury, Conant (1878); Newburyport, White 190 (1913); Georgetown, Horner; Rowley, H. 12246 (1956); Ipswich, H. 18117 (1958); Essex, H. 10253 (1955); Manchester, H. 10747 (1955); Beverly, H. 10183 (1955); Marblehead, Bean & H. 9748 (1955).

C. projecta Mackenz. Swales; uncommon. - Amesbury, H. 10535 (1955); Haverhill, H. 12635 (1932); Rowley, Plum Island, H. 12257 (1955).

C. cristatella Britt. Swales; rare. - Boxford, M.E. Perley (1882).

C. crawfordii Fern. Damp to dry open ground; frequent. - Lawrence, H. 9860 (1955); Rowley, H. 7919 (1952); Danvers, H. 7973 (1952); Wenham, Bean & H. 10003 (1955); Essex, H. 10256 (1955); Rockport, H. 10310 (1955); West Manchester, Hubbard 61 (1911); Salem, H. 9602 (1955).

C. bebbii Olney. Swales; rare. - Topsfield, H. 17649 (1958); West Manchester, Hubbard 61a (1911).

C. normalis Mackenz. Richwoods and swales; common. - Amesbury, Eaton 57 (1896); Haverhill, H. 709 (1932); Lawrence, H. 10073 (1955); Andover, Pease 3620 (1904); Boxford, Hunnewell 8865 (1923); Groveland, Bean & H. 8549 (1953); West Newbury, Bean & Pease (1940); Ipswich, Oakes; Danvers, Sears (1878); Beverly, Lowell (1847).

C. normalis, forma *perlonga* Fern. - Manchester, Hubbard 47 (1911).

C. tenera Dew. Meadows and moist openings; frequent. - Lawrence, H. 9861 (1955); Andover, Lowell (1849); Boxford, H. 8019 (1952); Plum Island; Rockport, Smith 979 (1928); Manchester, Hubbard 21 (1911); Marblehead, H. 9676 (1955).

C. longii Mackenz. Damp sandy soils; rare. - Merrimac, Bean & H. 8575 (1953); Lynn, Mackintosh (1933).

C. cumulata (Bailey) Mackenz. Acid soils; rare. - West Manchester, Hubbard 195 (1911).

C. brevior (Dew.) Mackenz. Dry open soil; uncommon. - Haverhill, Williams (1896); Boxford, H. 8779 (1953); Rowley, H. 10429 (1955); Topsfield, H. 8853 (1953); Middleton, Bean.

C. bicknellii Britt. Dry ground; rare. - North Andover, H. 8999 (1953); Salem, Morong (1879).

C. silicea Olney. Maritime sands; common. - Newburyport, Plum Island, H. 13058 (1957); Newbury, Plum Island, Pease 2900 (1903); Rowley, Plum Island, H. 12071 (1956); Ipswich, Oakes, East Gloucester, Knowlton (1929); Manchester, Hubbard (1913); Manchester, House Island, Sears 43 (1879); Beverly, Lowell (1849); Salem Willows, Manning (1882); Marblehead, Kennedy (1887).

C. alata T & G. Low ground; rare. - Andover, Robinson (1879); West Newbury, Conant (1879).

C. hormathodes Fern. Brackish to fresh marshes and sands; frequent. - Salisbury, H. 10480 (1955); Newbury, Conant (1878); Ipswich, Allen (1912); Ipswich, Plum Island, H. 12113 (1956); East Gloucester, Knowlton (1929); West Manchester, Hubbard 10 (1911); Salem, Cole (1886).

C. hormathodes, forma *invisa* (W. Boott) Fern. - Salisbury, White 63 (1913); Newbury, Plum Island, White 132 (1913).

C. straminea Willd. Fresh swamps; rare. - Amesbury, Eaton (1903); West Newbury, Conant (1878); Ipswich, Oakes; Lynn, Cushman (1908).

C. leporina L. Swales; uncommon. - Boxford, Pease 13646 (1912); West Newbury, Conant (1879); Wenham, Conant (1885); Essex County, Conant (1881).

C. aenea Fern. Open woods; rare. - Gloucester, Magnolia, Clarke.

C. argyrantha Tuckerm. Dry woods; uncommon. - West Newbury, Conant (1879); Gloucester, Smith 742 (1927); Manchester, Hubbard 51 (1911); West Manchester, Hubbard 34 (1911).

C. leptalea Wahlenb. Wet woods and clearings; uncommon. - Haverhill, H. 12627 (1932); Andover, Pease 1994 (1903); Boxford, H. 17704 (1958); West Newbury, Conant (1878); Georgetown, Horner; Manchester, Hubbard 23 (1911); Beverly Farms, Hubbard (1913); Lynn, Young (1881).

C. caryophyllea Lat. Naturalized from Europe. - Ipswich, Oakes; Salem, Oakes, Russell (1857); Boott (1865), Robinson (1879), Conant (1881), first found by Pickering in 1824.

C. pensylvanica Lam. Open dry soil; common. - Amesbury, H. 9523 (1954); Haverhill, H. 297 (1932); South Lawrence, Williams (1900); Andover, Pease 3033 (1903); North Andover, Morss (1885); Boxford, Cushman 772 (1907); Groveland, Bean & H. 8510 (1953); Newburyport, Conant (1878); Newbury, Plum Island, Pease 13673 (1912); Georgetown, Horner; Rowley, H. 7929 (1952); Rowley, Plum Island, H. 12077 (1956); Topsfield, H. 17642 (1958); Middleton, H. 31915 (1968); Danvers, Bean & H. 9984 (1955); Chebacco, Robinson (1875); Gloucester, Rand & B.L. Robinson (1896); West Manchester, Hubbard 27 (1911); Salem, Robinson (1875).

C. pensylvanica, var. *distans* Peck - Merrimac, Bean & H. 8586 (1953); Manchester, Hubbard 30 (1911); West Manchester, Hubbard 6 (1911).

C. communis Bailey Woods and clearings; occasional. - Salisbury, Bean & H. 8389 (1953); Haverhill, H. 393 (1932); Boxford, R.J. Eaton 5209 (1962); Middleton, H. 31816 (1968); Hamilton, Morong (1875); Manchester, Hubbard 53 (1911).

C. emmonsii Dew. Dry woods; occasional. - Haverhill, H. 342 (1932); Andover, Blake 1714 (1883); Boxford, R.J. Eaton 5215 (1962); Ipswich, Oakes; Hamilton, Rich (1894); Manchester, Hubbard 19 (1911); Peabody, Sears (1881).

C. architecta Mackenz. Dry woods; rare. - Gloucester, H. 12167 (1956); West Manchester, Hubbard (1913).

C. novae-angliae Schwein. Woodlands; rare. - Amesbury, Huntington (1906).

C. umbellata Schkuhr. Dry sandy soil; occasional. - Haverhill, Ward Hill, H. 353 (1932); Andover, Pease 3617 (1904); Boxford, H. 7845 (1952); Hamilton, Rich (1894); Manchester, Hubbard 31 (1911).

C. pedunculata Muhl. Rich woods; rare. - Amesbury, Eaton 83 (1896); Peabody, Sears (1881).

C. aurea Nutt. Meadows; rare. - Haverhill, Bradford, Cushman 2951 (1908); Boxford, Cushman 2951 (1908); Danvers, Robinson (1882); Salem, Robinson (1879).

C. paleacea Wahlenb. Saline or brackish marshes; occasional. - Salisbury, Alcott (1882); Newburyport, Eaton (1895); West Manchester, Hubbard 7 (1911); Beverly Farms, Hubbard (1913); Salem, Kernwood, Russell (1863).

C. salina Wahlenb., var. *kattegatensis* (Fries) Almq. Saline or brackish shores; rare. - Newburyport, Forbes (1903).

C. crinita Lem. Wooded swamps; common. - Salisbury, H. 12210 (1956); Merrimac, H. 10576 (1955); Haverhill, H. 391 (1932); Lawrence, H. 9867 (1955); Andover, Pease 7720 (1905); West Newbury, Conant (1878); Byfield, Sears (1904); Georgetown, M.E. Perley (1882); Topsfield, Porter (1927); Danvers, Sears (1880); Rockport, Swan (1881); Beverly, Lowell (1849); Saugus, H. 10907 (1955).

C. crinita, var. *gynandra* (Schwein.) Schwein. & Torr. - Salisbury, Bean & H. 8385 (1953); Merrimac, Bean & H. 8578 (1953); Groveland, H. 8811 (1953); Ipswich, H. 13905 (1957).

C. aquatilis Wahlenb., var. *altior* (Rydb.) Fern. Fresh marshes; rare. - Georgetown, Horner.

C. nigra (L.) Reichard. Open turf and swales; common. - Andover, Pease 2144 (1903); West Newbury, Conant (1878); Newburyport, White 197 (1913); Newbury, Bean & H. 8397 (1953); Georgetown, Horner; Ipswich, Oakes; Essex Woods, Lowell (1857); Gloucester, Rand & B.L. Robinson (1896); Rockport, Smith 618 (1927); Manchester, Hubbard 11 (1911); Beverly, Lowell (1849); Salem, Boott (1861); Swampscott, Young (1882); Marblehead, Russell (1862).

C. stricta Lam. Swamps and swales, forming "nigger-heads"; common. - Merrimac, Bean & H. 8577 (1953); Haverhill, H. 725 (1932); North Andover, H. 17970 (1958); Boxford, M.E. Perley (1882); Groveland, Bean & H. 8496 (1953); Georgetown, Porter (1939); Topsfield, H. 8069 (1952); Rockport, H. 10274 (1955); West Manchester, Hubbard 17 (1911); Salem, Russell (1852).

C. haydenii Dew. Swales; rare. - Andover, Pease 7721 (1905); Danvers, Lowell (1848).

C. buxbaumii Wahlenb. Wet shores and swamps; occasional. - Haverhill, H. 14194 (1932); Newburyport, Eaton (1897); Ipswich, Oakes; Danvers, Robinson (1879); Lynnfield, E. & C.E. Faxon (1879).

C. scabrata Schwein. Wet woods and meadows; rare. - Andover, Robinson (1879); Newburyport, Schweinfurth (1950's); Gloucester, Clarke (1905); Beverly Farms, Hubbard 79 (1911).

C. lasiocarpa Ehrh., var. americana Fern. Swales and shallow water; occasional. - Andover, Pease 3716 (1904); Boxford, H. 8438 (1953); Newburyport, Deane (1886); Wenham, Oakes; Chebacco, Robinson (1875); Manchester, Hubbard 46 (1911); Beverly, Robinson (1879); Lynnfield, Young (1879).

C. lanuginosa Michz. Rich meadows and swales; frequent. - Haverhill, Bradford, Cushman 2947 (1908); Andover, Lowell (1849); Boxford, H. 8440 (1953); Groveland, Bean & H. 8495 (1953); West Newbury, Conant (1878); Newburyport, Bean & H. 8396 (1953); Newbury, Ogden, Schubert, Smith & Weatherby Pl. Ex. Gray. 826 (1938); Georgetown, Horner; North Salem, Russell (1863).

C. vestita Willd. Dry sandy woods; frequent. - Methuen, H. 18372 (1958); Andover, Blake (1886); Boxford, Cushman 763 (1907); Middleton, Robinson (1877); Danvers, Sears (1880); Beverly, Montserrat, Hubbard 56 (1911); Salem, Robinson (1879).

C. hirta L. Naturalized from Europe. Roadsides and dry fields; occasional. - Haverhill, H. 715 (1932); Danversport, Sears (1883); Manchester, R.J. Eaton (1932).

C. pallescens L., var. *neogea* Fern. Meadows and thickets; common. - Haverhill, 726 (1932); Andover, Blake (1882); Boxford, H. 7865 (1952); West Newbury, Conant (1877); Georgetown, Horner; Rowley, H. 8072 (1952); Wenham, Bean & H. 10028 (1955); West Manchester, Hubbard 1 (1911); Beverly, Robinson (1879); Nahant, S.S. Clark (1916).

C. virescens Muhl. Dry woods; rare. - Boxford, H. 18543 (1958); Essex Woods, Lowell (1848); West Manchester, Hubbard 37 (1911); Essex County, Conant (1878).

C. swanii (Fern.) Mackenz. (Named in honor of Charles Water Swan, 1838-1921.) Woods and thickets; common. - Salisbury, White 88 (1913); Merrimac, Bean & H. 8571 (1953); Haverhill, Bradford, Churchill (1914); Andover, H. 17756 (1958); North Andover, H. 17973 (1958); Boxford, H. 8052 (1952); Ipswich, H. 18175 (1958); Topsfield, H. 8840 (1953); Danvers, Bean & H. 9992 (1955); Wenham, Robinson (1875); Rockport, Smith 4323 (1945); West Manchester, Hubbard 35 (1911); Saugus, Bean & H. 9825 (1955); Lynn, Cushman 2986 (1908).

C. gracillima Schwein. Woodlands and meadows; frequent. - Amesbury, H. 18254 (1958); Haverhill, H. 392 (1932); Boxford, H. 7841 (1952); Groveland, Horner; Ipswich, Oakes; Middleton, Bean (1934); Danvers, Sears; Beverly, Lowell (1849).

C. arctata Boott. Thickets; rare. - Manchester, Hubbard (1913).

C. debilis Michx., var. *rudgei* Bailey. Open Woods and meadows; common. - Salisbury, H. 12221 (1956); Amesbury, Kennedy 52 (1900); Merrimac, Bean & H. 8570 (1953); Haverhill, H. 714 (1932); Andover, Lowell (1849); North Andover, H. 17987 (1958); Boxford, Sears (1878); West Newbury, H. 9537 (1954); Ipswich, H. 18130 (1958); Topsfield, H. 17646 (1958); Danvers, Bean & H. 9993 (1955); Wenham, Robinson (1875); Gloucester, Rand & B.L. Robinson (1896); Rockport, Smith 694 (1927); West Manchester, Hubbard 8 (1911); Lynnfield, Bean (1916); Saugus, Bean & H. 9826 (1955); Lynn, Cushman 3002 (1908).

C. brunnea Thunb. Adventive from Orient. - North Salem, Sears (1905).

C. sprengelii Dew. Alluvial thickets; rare. - Merrimac, Bean & H. 8373 (1953); Haverhill, Bradford, Churchill (1914), Fernald 11322 (1914); West Newbury, H. 31882 (1968).

C. conoidea Schkuhr. Moist grassy places; occasional. - Haverhill, Bradford, Cushman 2949 (1908); Andover, Pease 210 (1901); West Newbury, Conant (1878); Danvers, Sears (1879); West Manchester, Hubbard 15 (1911).

C. livida (Wahlenb.) Willd., var. *grayana* (Dew.) Fern. Meadows and bogs; rare. - Boxford, E. & C.E. Faxon (1878), Robinson (1890); Lynnfield, Forbes.

C. panicea L. Naturalized from Europe. Meadows and grasslands; rare. - Ipswich, Carices Am. Septent. 67; Danvers, Sears (1879).

C. platyphylla Carey. Rich woods; rare. - Hamilton, E. & C.E. Faxon (1888); Salem, Sears (1881).

C. digitalis Willd. Dry hardwoods and glades; occasional. - Amesbury, H. 18273 (1958); Merrimac, Bean & H. 8596 (1953); Andover, H. 17764 (1958); Boxford, R.J. Eaton 5207 (1962); West Newbury, Conant (1879); Georgetown, Horner; Manchester, Hubbard 25 (1911); Beverly, Robinson (1890); Lynn, Cushman 3000 (1908).

C. laxiculmis Schwein. Rich woods and glades; occasional. - Merrimac, H. 18359 (1958); Haverhill, Bradford, Churchill (1914); Andover, Blake (1880); Boxford, H. 8024 (1952); West Newbury, Conant (1879); Danvers, Wright (1852).

C. laxiflora Lam. Rich woods; occasional. - Andover, Pease 2189 (1903); Boxford, Pease & St. John 11938 (1917); Danvers, Sears (1878); Swampscott, Robinson (1877).

C. blanda Dew. Woods and meadows; occasional. - Haverhill, H. 12628 (1932); Boxford, H. 7842 (1952); West Newbury, Pease 28111 (1940); Byfield, H. 7891 (1952); Beverly Farms, Hubbard (1913); Peabody, Forbes (1912).

C. ormostachya Wieg. Sandy or rocky woods; rare. - Manchester, Hubbard (1911).

C. leptonervia Fern. Low woods; rare. - Boxford, Cushman 762 (1907); Beverly Farms, Hubbard 41 (1911).

C. flava L. Meadows and shores; rare. - Haverhill, Bradford, Cushman (1908); Boxford, Williams (1899).

C. flava, var. *fertilis* Peck. Occasional. - Haverhill, Bradford, Cushman 2953a (1908); Andover, Pease 1630 (1902); Groveland, Bean & H. 8478 (1953); West Newbury, Robinson (1879); Georgetown, Horner; Wenham, Lowell (1849); Beverly, Lowell; Lynnfield, Young (1879).

C. viridula Michx. Damp shores and springy places; occasional. - Salisbury, White 258 (1913); Amesbury, Eaton (1896); Topsfield, H. 8854 (1953); Danvers, Lowell (1846); Gloucester, Swan (1881); Rockport, Pigeon Cove, Weir (1875); Beverly, Lowell (1846); Salem, Robinson (1879); Swampscott, Swan (1886); Marblehead, Russell (1866); Nahant, Oakes.

C. typhina Michx. Meadows; rare. - North Andover, Robinson (1879).

C. lacustris Willd. Swamps; occasional. - Methuen, H. 18405 (1958); Boxford, H. 8144 (1952); Newburyport, Eaton (1895); Newbury, Bean & H. 8408 (1953); Danvers, Sears (1879); Hamilton, Conant (1881); Wenham, Conant (1881); Lynnfield, F.C. & L.B. Smith 4254 (1938).

C. comosa Boott. Swamps; common. - Amesbury, Chivers (1902); Haverhill, H. 716 (1932); Methuen, H. 8931 (1953); Lawrence, H. 10088 (1955); Andover, H. 17899 (1958); North Andover, H. 9493 (1954); Boxford, H. 8450 (1953); Groveland, H. 9089 (1953); Rowley, H. 7923 (1952); Ipswich, H. 18121 (1958); Ipswich, Plum Island, St. John & White 534 (1913); Topsfield, Sears (1901); Wenham, Conant (1877); Essex, Bean (1919); Beverly Farms, Hubbard 83 (1911); Lynnfield, Bean (1919); Marblehead, Cole (1928).

C. hystricina Muhl. Swales; occasional. - Methuen, H. 12840 (1957); Groveland, Bean & H. 8547 (1953); Newburyport, Eaton 4 (1896); Beverly, Lowell (1847).

C. lurida Wahlenb. Swamps; very common. - Salisbury, H. 12225 (1956); Amesbury, H. 18233 (1958); Merrimac, Bean & H. 8372 (1953); Haverhill, H. 23 (1931); Lawrence, H. 9862 (1955); Andover, Pease 2645 (1903); North Andover, H. 9908 (1955); Boxford, M.E. Perley (1882); Groveland, Bean & H. 8474 (1953); West Newbury, Robinson (1879); Newburyport, H. 10464 (1955); Georgetown, Horner; Rowley, H. 8118 (1952); Ipswich, H. 14026 (1957); Middleton, H. 18073 (1958); Wenham, H. 13739 (1957); Essex, H. 10255 (1955); Rockport, Cushman 3364 (1908); Manchester, Hubbard 74 (1911); Beverly, Lowell (1848); Lynnfield, Boott (1854).

C. folliculata L. Swampy woods and swales; occasional. - Salisbury, Bean & H. 8388 (1953); Amesbury, H. 18243 (1958); Merrimac, Bean & H. 8581 (1953); Andover, H. 17898 (1958); North Andover, Pease 1524 (1902); Georgetown, Horner; Wenham, Robinson (1875); Rockport, Smith 4325 (1945); Manchester, Hubbard 18 (1911); Beverly, Lowell (1849).

C. intumescens Rudge. Meadows and swales; frequent. - Salisbury, H. 12215 (1956); Amesbury, H. 18249 (1958); Merrimac, H. 9258 (1953); Haverhill, H. 788 (1932); Methuen, H. 18411 (1958); Andover, Pease 3615 (1904); Boxford, H. 8026 (1952); Georgetown, Horner; Ipswich, H. 18144 (1958); Topsfield, H. 8841 (1953); Danvers, Sears (1878); Rockport, Gates & Smith 2393 (1928); Beverly Farms, Hubbard 39 (1911).

C. lupulina Muhl. Swamps; common. - Amesbury, Eaton (1897); Haverhill, H. 717 (1932); Methuen, H. 18386 (1958); Boxford, M.P. Perley (1882); Groveland, H. 9073 (1953); West Newbury, Robinson (1875); Newburyport, White 471 (1913); South Georgetown, Williams (1899); Topsfield, Sears (1901); Middleton, Bean (1940); Danvers, Oakes; Gloucester, Hubbard 103 (1911); Beverly, Lowell (1849); Saugus, Lund (1889).

C. rostrata Stokes, var. *utriculata* (Boott) Bailey. Swamps; occasional. - Haverhill, H. 396 (1932); Newburyport, H. 10374 (1955); Wenham, Lowell (1847); West Gloucester, Hubbard 106 (1911); West Manchester, Hubbard 16 (1911); Beverly, Robinson (1875).

C. bullata Schkuhr. Acid swales and bogs; occasional. - Salisbury, White 265 (1913); Amesbury, Eaton 7 (1896); Haverhill, H. 12634 (1933); Andover, Robinson (1879); Danvers, Sears (1879).

C. vesicaria L. Meadows and swales; occasional. - Amesbury, Eaton (1896); Haverhill, H. 718 (1932); Boxford, H. 8441 (1953); Essex County, Oakes.

C. vesicaria, var. *monile* (Tuckerm.) Fern. - Amesbury, Eaton 10 (1896); Beverly, Sears (1880); Lynnfield, Young (1879).

C. vesicaria, var. *jejuna* Fern. - Andover, Pease 2118 (1903); Boxford, H. 8165 (1952); Ipswich Beach, Hubbard (1913).

ARACEAE (Arum Family)

ARISAEMA Mart. Indian-Turnip.

A. atrorubens (Ait.) Blume, Jack-in-the-Pulpit. Rich woods and thickets; common. - Amesbury, H. 9525 (1954); Merrimac, Bean & H. 8369 (1953); Haverhill, H. 24 (1931); Andover, Pease 2051 (1903); North Andover, Morss (1884); Boxford, Frye (1891); South Groveland, H. 7886 (1952); West Newbury, H. 31883 (1968); Georgetown, Horner (1880); Ipswich, Greenough (1895); Topsfield, V.M. Towne; Middleton, H. 7876 (1952); Gloucester, Cole (1901); Saugus, Bean (1940); Swampscott, Robinson (1872); Marblehead, Cole (1886).

A. atrorubens, forma *viride* (Engler) Fern. - Amesbury, H. 9526 (1954); Boxford, H. 7855 (1952); South Groveland, H. 7885 (1952); West Newbury, H. 7899, (1952); Middleton, H. 7877 (1952); Swampscott, Robinson (1872).

A. stewardsonii Britt. Wet or swampy woods; occasional. - Salisbury, Bean & H. 8390 (1953); Andover, Root (1947); North Andover, H. 17984 (1958); Boxford, H. 7853 (1952); Middleton, H. 8326 (1953); West Manchester, Hubbard (1913).

PELTANDRA Raf. Arrow-Arum.

P. virginica (L.) Schott. Swamps and borders of ponds and streams; common. - Haverhill, Bradford, H. 13160 (1957); Methuen, H. 8921 (1953); Andover, Pease 1996 (1903); North

Andover, H. 13087 (1957); Groveland, H. 12884 (1957); West Newbury, Robinson (1879); Georgetown, H. 12986 (1957); Rowley, H. 13514 (1957); Topsfield, H. 13607 (1957); Middleton, Bean (1950); Essex H. 13346 (1957); Peabody, H. 12758 (1957); Saugus, Ripley (1915); Lynn, H. 13176 (1957).

P. virginica, forma *latifolia* (Raf.) Blake - Boxford, H. 12770 (1957); Essex, Fernald, Hunnewell & Long 9101 (1913).

P. virginica, forma *rotundata* Blake - Boxford, H. 8451 (1953).

P. virginica, forma *hastifolia* Blake - North Andover, H. 9497 (1954); Georgetown, Horner (1880); Hamilton, H. 9415 (1953); Chebacco, Robinson (1875).

P. virginica, forma *brachyota* Blake - Boxford, H. 12776 (1957); Topsfield, H. 13625 (1957); Lynnfield, Young (1879).

P. virginica, forma *heterophylla* (Raf.) Blake - Peabody, H. 12757 (1957).

CALLA L. Water-Arum.

C. palustris L. Wild Calla. Bogs and pond margins; local. - Haverhill, H. 510 (1932); Andover, Sutton 69 (1874); Boxford, Horner (1873); Wenham, Sears (1884); Beverly, Robinson (1872).

SYMPLOCARPUS Salisb. Skunk-Cabbage.

S. foetidus (L.) Nutt. Meadows and swampy woods; common. - Merrimac, Bean & H. 8370 (1953); Haverhill, Bradford, H. 290 (1932); Lawrence, H. 10079 (1955); Andover, Pease 1846 (1903); North Andover, Morss (1885); Boxford, Frye (1880); Groveland, Bean & H. 8558 (1953); West Newbury, H. 31884 (1968); Ipswich, Oakes; Topsfield, H.R. Towne; Danvers, Sears (1885); Wenham, Robinson (1875); Gloucester, H. 10157 (1955); Salem, Cole (1902); Swampscott, Robinson (1875); Marblehead, Cole (1886).

ACORUS L. Sweetflag.

A. calamus L. Wet places; frequent. - Haverhill, H. 469 (1932); North Andover, Morss (1884); West Newbury, N.E.B.C. (1951); Newburyport, Forbes 883 (1903); Rowley, H. 13483 (1957); Topsfield, Robinson (1875); Wenham, Bean & H. 10000 (1955); Rockport, Smith 625 (1927); Beverly, Lowell (1847); Salem, Cole (1899).

LEMNACEAE (Duckweed Family)

SPIRODELA Schleid.

S. polyrhiza (L.) Schleid. Water-flaxseed. Pools and pond- and stream- margins; frequent. - Amesbury, H. 18191 (1958); Haverhill, Bradford, H. 13126 (1957); Methuen, H. 12829 (1957); Andover, Pease 273 (1901); North Andover, H. 12915 (1957); Boxford, H. 9632 (1954); West Newbury, Bean & H. 7727 (1951); Rowley, H. 13481 (1957); Topsfield, H. 8892 (1953); Swampscott, Robinson (1875).

LEMNA L. Duckweed

L. valdiviana Phil. - Byfield, Little Crane Pond, Milliken (1915).

L. minor L. Quiet waters; common. - Haverhill, Bradford, H. 13125 (1957); Methuen, Bean & H. 7755 (1951); Andover, Pease 30430 (1943); North Andover, H. 13059 (1957); Boxford, Duncan & H. 7661 (1951); Groveland, H. 8806 (1953); West Newbury, H. 876 (1933); Georgetown, Bean & H. 7729 (1951); Ipswich, H. 18129 (1958); Ipswich, Plum Island, H. 12382 (1956); Topsfield, H. 25578 (1963); Danvers, Duncan & H. 9016 (1953); Wenham, H. 13902 (1957); Gloucester, Robinson & Sears (1880); Peabody, H. 9292 (1953); Lynnfield, H. 31787 (1967); Lynn, H. 13167 (1957); Marblehead, Sears (1882).

WOLFFIA Horkel Water-meal.

W. columbiana Karst. Quiet waters; rare but apparently becoming more frequent. - Haverhill, Bradford, H. 13127 (1957); Methuen, Bean & H. 7754 (1951); Boxford, recently made farm pond, H. 14074 (1957); Topsfield, H. 25579 (1963); Danvers, Mill Pond, Duncan & H. 9015 (1953).

XYRIDACEAE (Yellow-eyed Grass Family)

XYRIS L. Yellow-eyed Grass.

X. torta Sm. Damp to dryish sandy peaty depressions; occasional. - Newbury, Plum Island, Pease 25762 (1936); Ipswich, Prof. T.E.N. Eaton (1880); Furber (1911); Essex County, Lowell.

X. caroliniana Walt. Wet peaty or sandy soil; occasional. - Merrimac, H. 9214 (1953); Andover, Bean (1945); Boxford, Robinson (1876); Georgetown, Horner; Middleton, Lowell (1845); Chebacco, Robinson (1876); Essex, Lowell (1846); Salem, Sears (1905); Lynnfield, (1839).

X. congdoni Small. Peaty shores; local. - Middleton, Oakes; Wenham, Purdie (1894); Essex, Fernald, Hunnewell & Long 9136 (1913); Peabody, Sears (1905).

ERIOCAULACEAE (Pipewort Family)

ERIOCAULON L. Pipewort.

E. septangulare With. White-buttons. Shores of ponds and streams; common. - Salisbury, White 276 (1913); Amesbury, H. 13019 (1957); Merrimac, H. 13728 (1957); Haverhill, H. 511 (1932); Andover, Blake (1883); North Andover, Root (1943); Boxford, M.P. Perley (1883); Groveland, H. 9091 (1953); West Newbury, Robinson (1879); Georgetown, Horner; Ipswich, H. 13236 (1957); Topsfield, A.F. Towne; Middleton, Lowell (1845); Wenham, H. 13113 (1957); Hamilton, H. 12819 (1957); Essex, Lowell (1846); Gloucester, H. 13820 (1957); Lynnfield, H. 14135 (1957); Lynn, Sears (1885).

E. parkeri Robins. Tidal mud and estuaries; rare. - Newbury, Glück & Fassett 2321 (1924).

COMMELINACEAE (Spiderwort Family)

COMMELINA L. Dayflower.

C. communis L. Naturalized from Asia. Dooryards, roadsides and waste places; frequent. - Haverhill, H. 11207 (1955); Methuen, H. 11755 (1955); Lawrence, H. 11777 (1955); North Andover, Root (1945); Newburyport, H. 20735 (1959); Newbury, H. 20710 (1959); Gloucester, Magnolia, H. 12373 (1956); Beverly, H. 11684 (1955); Salem, H. 11516 (1955); Peabody, H. 11519 (1955); Lynnfield, H. 14112 (1957); Saugus, H. 11851 (1955); Lynn, H. 11829 (1955); Marblehead, H. 11461 (1955).

TRADESCANTIA L. Spiderwort.

T. virginiana L. Escape from gardens, native farther South. Roadsides and waste places; occasional. - Methuen, H. 12271 (1956); West Boxford, Root & Zink (1945); Salem, Robinson (1879).

PONTEDERIACEAE (Pickerelweed Family)

PONTEDERIA L. Pickerelweed.

P. cordata L. Shores and shallow water; common. - Amesbury, H. 13012 (1957); Merrimac, H. 13713 (1957); Methuen, H. 12849 (1957); Andover, H. 12794 (1957); North Andover, H. 12939 (1957); Boxford, M.E. Perley (1882); Groveland, H. 12878 (1957); Georgetown, H. 13979 (1957); Rowley, H. 13525 (1957); Ipswich, (1912); flowers white; Topsfield, H. 13626 (1957); Middleton, H. 12900 (1957); Wenham, 13118 (1957); Hamilton, H. 12816 (1957); Essex, H. 13358 (1957); Gloucester, H. 13810 (1957); Manchester, H. 10770 (1955); Salem, Forbes 915 (1903); Peabody, Mackintosh (1933); Lynnfield, Young (1879); Lynn, H. 13178 (1957).

P. cordata, forma *latifolia* (Raf.) House - Haverhill, H. 470 (1932); North Andover, Morss (1885); Byfield, White 393 (1913); Ipswich, Topsfield, G.F. Dow; Middleton, Bean (1950); Gloucester, Magnolia, Rogers (1898); Lynnfield, Young (1879).

P. cordata, forma *angustifolia* (Pursh) Solms - Amesbury, H. 13013 (1957); Haverhill, Bradford, H. 13144 (1957); North Andover, Root (1945); Boxford, H. 8975 (1953).

P. cordata, forma *taenia* Fassett - Amesbury, H. 13007 (1957); Merrimac, H. 13730 (1957); Boxford, H. 13662 (1957); Topsfield, H. 13610 (1957); Middleton, H. 12903 (1957); Gloucester, H. 13812 (1957).

HETERANTHERA R. & P. Mud-Plantain.

H. dubia (Jacq.) MacM. Water-Stargrass. Streams and quiet waters; rare. - Wenham, Robinson (1875).

JUNCACEAE (Juncus Family)

JUNCUS L. Bog-rush.

J. bufonius L. Open ground and waste places; common. - Haverhill, H. 700 (1932); Lawrence, Jenks, Nickerson & Swan (1884); Groveland, H. 78380 (1865); Newbury, Plum Island, White 143

(1913); Georgetown, Horner; Ipswich, Hubbard (1913); Ipswich, Plum Island, H., McGregor & Schweinfurth 12706 (1957); Danvers, Lowell (1846); Gloucester, Prince (1896); Rockport, Fernald, Hunnewell & Long 9141 (1913).

J. bufonius, var. *halophilus* Buchenau & Fern. Saline soils; occasional. - Salisbury, White 4 (1913); Newbury, Plum Island, Pease 13670 (1912).

J. gerardi Loisel. Black Grass. Salt marshes; common. - Newbury, White 226 (1913); Newbury, Plum Island, White 115 (1913); Rowley, E.C. Morse (1885); Ipswich, Oakes "5 or 6 years before 1833"; Essex, H. 9390 (1953); Beverly, Lowell (1846); Salem, Robinson (1877); Peabody, Mackintosh (1933); Marblehead Neck, Kennedy (1887); Nahant, Deane (1885).

J. secundus Beauv. Dry open sterile soil; uncommon. - Boxford, H. 8785 (1953); Topsfield, H. 8850 (1953); Essex County, Sears (1878).

J. tenuis Willd. Open soils and waste places; common. - Salisbury, White 356 (1913); Haverhill, H. 25 (1931); Andover, Pease 2665 (1903); North Andover, H. 18044 (1958); Boxford, M.P. Perley (1882); West Newbury, Bean & Pease (1940); Newburyport, Plum Island, H. 13039 (1957); Georgetown, Horner (1885); Ipswich; Topsfield, R.J. Eaton (1931); Middleton, Bean & H. 7782 (1951); Wenham, Robinson (1875); Gloucester, Magnolia Fitch (1883); Rockport, Robinson (1875); Beverly, Lowell (1846); Marblehead, R.R., Thoreau.

J. tenuis, var. *anthelatus* Wieg. Occasional. - Andover, Pease 4897 (1904); Wenham, Weatherby (1915).

J. tenuis, var. *williamsii* Fern. Occasional. - Andover, Pease 3808 (1904); Boxford, Hopkins & Pease 23168 (1932); Rowley, H. 12247 (1956).

J. platyphyllus (Wieg.) Fern. Acid soil; uncommon. - West Newbury, Pease 28067 (1940); Danvers, Mackintosh (1924).

J. greenei Oakes & Tuckerm. Sandy areas; common. - Salisbury, D. & E.W. White 22 (1913); Haverhill, H. 373 (1932); Newburyport, Plum Island, H. 13050 (1957); Newbury, Plum Island, White 106 (1913); Rowley, Plum Island, H. 12568 (1956); Ipswich, Oakes; Ipswich, Plum Island, H. 12438 (1956); Topsfield, H. 8076 (1952); Gloucester, Conant & Sears (1877); Rockport, Cushman 3370 (1908); Marblehead, Bean & H. 9744 (1955); Nahant.

J. filiformis L. Wet places; rare. - Lawrence, Pease 2131 (1903).

J. effusus L., var. *compactus* Leg. & Court. Swales; occasional. - Andover, Pease 2390 (1903); Ipswich, Oakes; Wenham, Weatherby (1916).

J. effusus var. *decipiens* Bucheneau Swales; uncommon. - Haverhill, H. 26 (1931); Topsfield, G.F. Dow.

J. effusus, var. *dylaei* (LaHarpe), Fern. & Wieg. Wet places; frequent. - Haverhill, Bradford, Cushman 2960 (1908); North Andover, H. 12928 (1957); Boxford, M.E. Perley (1882); Ipswich, Oakes; Rockport, Cushman 3370 (1908); Beverly, Lowell (1846).

J. effusus, var. *solutus* Fern. & Wieg. Damp sandy soil; frequent. - Amesbury, H. 18245 (1958); Haverhill, H. 27 (1931); Boxford, H. 8782 (1953); Newbury, H. 8240 (1952); Georgetown, Horner; Ipswich, Oakes; Middleton, H. 9423 (1953); Salem, Cole (1886); Saugus, H. 10874 (1955).

J. balticus Willd., var. *littoralis* Engelm. Sandy brackish to fresh shores; frequent, - Newbury, Kents Island, Sears (1904); Newbury, Plum Island, White 111 (1913); Georgetown, Horner; Rowley, H. 8281 (1952); Ipswich, Oakes; Rockport, Thoreau.

J. marginatus Rostk. Moist soil; common. - Merrimac, H. 9208b (1953); Lawrence, Robinson (1877); Andover, Blake (1882); Boxford, Robinson (1890); Byfield, Sears (1904); Rowley, H. 8126 (1952); Ipswich, Oakes; Topsfield, R.J. Eaton (1931); Danvers, Lowell (1849); Essex, H. 13333 (1957); Gloucester, Swan (1881); Beverly, Robinson (1879); Salem, Pickering.

J. canadensis J.Gay. Marshy places; common. - Salisbury, White 260 (1913); Merrimac, H. 9224 (1953); Haverhill, H. 29 (1931); Methuen, H. 8930 (1953); Andover, Churchill (1903); Boxford, Williams (1899); Groveland, H. 9066 (1953); West Newbury, N.E.B.C. (1951); Georgetown, Horner: Rowely, H. 13510 (1957); Rowley, Plum Island, H. 12559 (1956); Ipswich, Prince; Ipswich, Plum Island, H. 12389 (1956); Middleton, H. 9422 (1953); Wenham, Robinson (1877); Chebacco, Robinson (1877); Essex, H. 13366 (1957); Gloucester, H. 13831 (1957); Rockport, Smith 912 (1927); Manchester, H. 10723 (1955); Beverly, Lowell (1847); Salem, Cole (1886); Saugus, Bean (1936).

J. canadensis, forma *conglobatus* Fern. Brackish to saline marshes; occasional. - Newbury, Kents Island, Sears (1904); Newbury, Plum Island, Eaton (1896); Ipswich, Little Neck, Forbes (1913); Ipswich, Plum Island, White 550 (1913).

J. brevicaudatus (Engelm.) Fern. Wet places; frequent. - Merrimac, H. 9219 (1953); Haverhill, H. 28 (1931); Methuen, H. 8938 (1953); Andover, Pease 2137 (1903); Boxford, H. 9289 (1953); Ipswich, Forbes (1913); Gloucester, Fernald, Hunnewell & Long 9199 (1913); Beverly, Lowell (1849).

J. acuminatus Michx. Damp soil; frequent. - Salisbury, White 263 (1913); Haverhill, H. 471 (1932); North Andover, H. 14054 (1957); Groveland, H. 9071 (1953); West Newbury, Bean (1913); Newbury, White 216 (1913); Ipswich, H. 13442 (1957); Danvers, Lowell (1846); Wenham Swamp, Oakes; Gloucester, Prince; Beverly, Lowell (1846); Peabody, Mackintosh (1933).

J. militaris Bigel. Margins of ponds; frequent. - Andover, Pease 283 (1900); North Andover, H. 13065 (1957); Boxford, H. 8974 (1953); Ipswich, H. 13237 (1957); Topsfield, H. 12321 (1957); Wenham, Lowell (1847); Hamilton, H. 9413 (1953); Essex, Fernald, Hunnewell & Long 9219 (1913); Gloucester, H. 13837 (1957).

J. militaris, forma *subnudus* Fern. - Andover, Young (1882).

J. articulatus L. Wet Ground; occasional. - West Newbury, Conant (1879); Plum Island, St. John 686 (1913); Rowley, H. 8127 (1952); Ipswich, H. 13253 (1957); Wenham, Conant (1877); Essex, H. 13332 (1957); Gloucester, Robinson (1877).

J. articulatus, var. *obtusatus* Engelm. Scarce. - Newbury, Plum Island, Forbes (1930); Rockport, Fernald, Hunnewell & Long 9227 (1913).

J. pelocarpus Mey. Damp shores; common. - Amesbury, H. 9152 (1953); Merrimac, H. 9228 (1953); Haverhill, H. 30 (1931); Andover, Blake (1872); North Andover, H. 14064 (1957); Boxford, H. 8190 (1952); Ipswich, Wenham, Robinson (1879); Essex, Lowell (1848); Beverly, Lowell (1849); Saugus, H. 10866 (1955).

J. pelocarpus, forma *submersus* Fassett - Gloucester, H. 13814 (1957).

LUZULA DC. Woodrush.

L. acuminata Raf. Woods and clearings; rare. - Essex County, Oakes.

L. multiflora (Retz.) Lejeune. Fields and open woods; very common. - Amesbury, H. 10522 (1955); Merrimac, Bean & H. 8587 (1953); Haverhill, H. 31 (1931); Andover, Pease 291 (1901); North Andover, Robinson (1884); Boxford, M.E. Perley (1882); Groveland, Bean & H. 8483 (1953); West Newbury, H. 9536 (1954); Georgetown, Horner; Ipswich, Prince (1896); Topsfield, Mrs. L.T. Brown; Middleton, Porter (1927); Danvers, Bean & H. 9990 (1955); Wehnham, Bean & H. 10026 (1955); Hamilton, H. 10802 (1955); Gloucester, Magnolia, Sears (1884); Rockport, Smith 2636 (1932); Manchester, Hubbard (1913); Beverly, H. 10213 (1955); Lynnfield, Bean (1919); Saugus, Bean & H. 9809 (1955); Marblehead, H. 9683 (1955).

LILIACEAE (Lily Family)

VERATRUM L. False Hellebore.

V. viride Ait. Indian Poke. Swamps and along brooks; frequent; - Merrimac, Bean & H. 8366 (1953); Haverhill, H. 8345 (1953); Andover, Pease 1997 (1903); North Andover, H. 9909 (1955); Boxford, Frye (1880); Groveland, Bean & H. 8559 (1953); Georgetown, M.E. Perley (1884); Ipswich, Oakes; Topsfield, O. Floyd; Danvers, Sears (1885); Wenham, Sears (1899); Swampscott, Cole (1895).

UVULARIA L. Bellwort.

U. perfoliata L. Under hardwood trees; occasional. - North Andover, Carleton (1885); Boxford, M.E. Perley (1882); Georgetown, Horner; Topsfield, Oakes; Middleton, Bean (1934); Danvers (1847); Wenham, Sears (1899); Salem, Phippen.

U. sessilifolia L. Wild-oats. Open woods; common. - Salisbury, Eaton (1882); Haverhill, Ward Hill, Cushman (1908); Andover, Sutton 38 (1874); North Andover, Morss (1885); Boxford, Frye (1881); Newbury, E.N. Cole (1861); Ipswich, Oakes; Topsfield, B.B. Towne; Danvers, Sears (1885); Rockport, Anderson, Shafer & Smith (1931); Beverly Farms, Hubbard (1913); Salem, Cole (1886); Lynn, Phippen; Swampscott, Cole (1899).

ALLIUM L.

A. canadense L. Wild Garlic. Low woods and meadows; occasional. - Amesbury, H. 10536 (1955); Merrimac, H. 10573 (1955); Haverhill, Bradford, Collins (1910); Lawrence, H. 10086 (1955); Andover, Robinson (1879); North Andover, Morss (1900); West Boxford, Root & Zink (1945); Groveland, Bean & H. 8553 (1953); Newburyport, Cole (1898); Georgetown, H. 7881 (1952); Ipswich; Topsfield, M.Clarke; Danvers, Hathorne, St. John 600 (1912).

A. vineale L. Field-garlic. Naturalized from Europe; not common. - Ipswich, H. 13871 (1957); Topsfield, M.L. Pierce.

A. schoenoprasum L. Chives. Introduced from Eurasia. Escape from cultivation; occasional. - Boxford, Horner; Georgetown, Horner; Topsfield, G.T. Dow.

A. tricoccum Ait. Wild Leek. Rich woods; rare. - Andover, Pease 1888 (1903); Boxford, H. 12642 (1957).

A. cepa L. Onion. Introduced from Old World; occasional on dumps. - Boxford, H. 12198 (1956); Ipswich, H. 13870 (1957); Salem, Hall (1925).

A. cepa, var. *viviparum* Metz - Boxford, H. 8628 (1953); Newburyport, H. 10359 (1955).

HEMEROCALLIS L. Day-lily.

H. fulva L. Common Orange Day-lily. Naturalized from Eurasia; rapidly spreading by vegetative means only and frequent by roadsides and borders of fields. - Merrimac, H. 10572 (1955); Haverhill, H. 473 (1932); Metheun, H. 9613 (1954); North Andover, Root (1945); Boxford, M.E. Perley (1882); Rowley, H. 12239 (1956); Ipswich, H. 18160 (1958); Topsfield, Mrs. Peabody; Wenham, Robinson (1875); [Rockport ArW, (1969).] Beverly, Montserrat-Prides, Hubbard (1913).

H. flava L. Yellow Day-lily. Introduced from Asia. Escape from cultivation; occasional. - Georgetown, H. 17681 (1958).

LILIUM L. Lily.

L. philadelphicum L. Wood-lily. Frequent in open, rather dry places. - Amesbury, H. 18233 (1958); Andover, Pease 303 (1899); North Andover, Morss (1883); Boxford, H.E. Parkhurst (1882); Georgetown, Williams (1906); Ipswich, Oakes; Topsfield, H.R. Towne; Middleton, Williams (1906); Danvers, Frost; [Gloucester, Rockport, A & W (1968)] Gloucester, Prince (1883); Rockport, Gates & Smith 998 (1928); Beverly, Montserrat, Hubbard (1913); Salem, Mackintosh (1933).

L. canadense L. Canada Lily. Meadows and low thickets; frequent. - Haverhill, H. 474 (1932); Methuen, H. 18373 (1958); Andover, Pease 304 (1901); North Andover, Morss (1885); Boxford, M.E. Perley (1885); West Newbury, H. 11108 (1955); Georgetown, Williams (1906); Ipswich, Oakes; Topsfield, P.E. Kneeland; Middleton, Cole (1919); Danvers, Frost (1875); Wenham, Cole (1911); Salem, Phippen (1839); West Peabody, Mackintosh (1933); Marblehead, Kennedy (1887).

L. tigrinum L. Tiger-lily. Introduced and naturalized from eastern Asia; occasional escape by roadsides. - Merrimac, H. 10590 (1955); Boxford, H. 13271 (1957); Topsfield, H.R. Towne; Wenham, Robinson (1879); Marblehead, Cole (1930).

ERYTHRONIUM L. Dog's-tooth-violet.

E. americanum Ker. Yellow Adder's-tongue. Moist rich woods; frequent. - Haverhill, H. 300 (1932); Andover, Sutton 32 (1874); North Andover, Morss (1885); West Newbury, Horner; Ipswich, Oakes; Topsfield, L. Dow; Danvers, Robinson (1872); Gloucester, Magnolia, Newhall (1899); Rockport, Smith 937 (1928); Salem, Phippen (1838).

E. americanum, forma *castaneum* Smith. Known only from type collection. - Rockport, Smith 938 (1928).

SCILLA L. Squill.

S. sibirica Andr. Introduced and naturalized from Eurasia. - Topsfield, H. 24217 (1962).

ORNITHOGALUM L. Star-of-Bethlehem.

O. umbellatum L. Introduced and naturalized from Europe; occasional along roadsides and in thickets. - Andover, Sutton 117 (1874); Byfield, Horner; Topsfield, Mrs. W. Towne; Danvers, Russell (1852); Salem, Cole (1924); Marblehead, Cole (1930).

MUSCARIA Mill. Grape-Hyacinth.

M. botryoides (L.) Mill. Introduced and naturalized from Europe; a rare escape. - North Andover, Sargent (1885); Danvers, Robinson (1879).

YUCCA L. Spanish-bayonet.

Y. smalliana Fern. Introduced from the South; waif on dump. - Marblehead, H. 20960 (1959).

ASPARAGUS L. Asparagus.

A. officinalis L. Garden Asparagus. Introduced from Europe. In fields and along roadsides where seed has been spread by birds; common. - Amesbury, H. 10527 (1955); Haverhill, H. 33 (1931); Lawrence, H. 12296 (1956); Andover, Morss (1900); North Andover, Cole (1909); Boxford, H. 11974 (1955); Newbury; Rowley, H. 11926 (1955); Wenham, Robinson (1875); Gloucester, J.M. Hunnewell (1901); Beverly Farms, Hubbard (1913); Salem, Cole (1887).

CLINTONIA Raf.

C. borealis (Ait.) Raf. In old deep woods; occasional. - Haverhill, H. 397 (1932); Andover, Morss (1884); Boxford, Fernald (1909); Newburyport, Moulton (1887); Georgetown, Horner (1881); Topsfield, Arimoto; Danvers, Sears (1885); Wenham, Cole (1908); Essex, Robinson (1872); Gloucester, Magnolia, Sears (1884); Rockport, Smith 4326 (1945); Manchester, Hubbard (1913); Beverly, Cole (1886).

SMILACINA Desf. False Solomon's-seal.

S. racemosa (L.) Desf. False Spikenard. Woods and thickets; occasional. - Haverhill, H. 398 (1932); Boxford, M.E. Perley (1882); Manchester, Hubbard (1913); Lynnfield, Cole (1891).

S. racemosa, var. *cylindrata* Fern. Common. - Lawrence, Cole (1928); North Andover, Robinson (1872); Boxford, H.E. Parkhurst (1882); Georgetown, M.E. Perley (1882); Rowley, H. 7928 (1952); Topsfield, L.T. Brown (1912); Danvers, Sears (1885); Hamilton, H. 10777 (1955); Chebacco, Cole (1887); Manchester, Hubbard (1913); Beverly, Lowell (1846); Saugus, Bean & H. 9831 (1955); Lynn, Sears (1885).

S. stellata (L.) Desf. Sandy areas and open meadows; common near the coast, occasional elsewhere. - Haverhill, Ward Hill, Bean (1916); Lawrence, Morss (1900); Andover, Swan (1880); North Andover, Pease 2661 (1903); Newbury, Knowlton (1913); Newbury, Plum Island, H. 12006 (1956); Georgetown, Horner (1874); Rowley, H. 11928 (1955); Ipswich, Plum Island, H. 12018 (1956); Danvers, Frost (1874); Gloucester, Salt Island, Morss (1927); Beverly, Lowell (1846); Salem, Cole (1898); Saugus, Chute; Lynn, Marblehead, Cole (1900).

S. stellata var. *crassa* Vict. Ipswich, Deane (1911) Essex, Knowlton (1909).

S. trifolia (L.) Desf. Bogs and mossy woods; rare. - Andover, Sutton 68 (1874); Pease 1583 (1901); Gloucester, Magnolia, J.W. Robbins, Robinson (1877).

MAIANTHEMUM Weber.

M. canadense Desf. Wild Lily-of-the-valley. Woods and clearings; very common. - Salisbury, Bean & H. 8383 (1953); Amesbury, H. 9519 (1954); Haverhill, Ward Hill, Cushman (1908); North Andover, Root (1943); Boxford, M.E. Perley (1881); Newbury, E.N. Cole (1861); Georgetown, Horner; Rowley, H. 7924 (1952); Rowley, Plum Island, H., McGregor & Schweinfurth 12716

(1957); Ipswich, Oakes; Topsfield, L.T. Brown (1912); Middleton, H. 8329 (1953); Danvers, Sears (1885); Gloucester, Annisquam, Ripley 15913 (1915); Rockport, Anderson, Shafer & Smith (1931); Manchester, Hubbard (1913); Beverly, Robinson (1872); Saugus, Bean & H. 9820 (1955); Marblehead, Cole (1886).

STREPTOPUS Michx. Twisted-stalk.

S. roseus Michx., var. *perspectus* Fassett. Moist woods; rare. - Newburyport, Moulton (1887); Rockport, Babson (1866).

POLYGONATUM Mill. Solomon's-seal.

P. pubescens (Willd.) Pursh. Woods; frequent. - Haverhill, H. 400 (1932); Andover, Sutton 63 (1874); Boxford, Frye (1881); Georgetown, Horner; Plum Island, Williams (1899); Topsfield, Mrs. S. Conley; Middleton, Manning (1884); Danvers, Robinson (1874); Manchester, Hubbard (1913); Beverly Farms, Sears (1885); Salem, Cole (1887); Peabody, Sears (1885).

P. latifolium (Jacq.). Desf. Introduced from Europe; rare escape from gardens. - Topsfield; H.R. Towne.

CONVALLERIA L. Lily-of-the-valley.

C. majalis L. Introduced from Europe; spreading from cultivation. - Danvers, Frost; Salem, Robinson (1879).

MEDEOLA L. Indian Cucumber.

M. virginiana L. Rich woods; frequent. - Salisbury, Bean & H. 8384 (1953); Haverhill, Ward Hill, Cushman (1908); Methuen, H. 18388 (1958); Boxford, Frye (1881); Newburyport, E.N. Cole (1861); Topsfield, G.T. Dow; Danvers, Frost; Chebacco Woods, Robinson (1890); Rockport, Smith 730 (1927); Manchester, Hubbard (1913); Beverly, Lowell (1846); Lynn, Sears (1885).

TRILLIUM L. Wakerobin.

T. cernuum L. Nodding Trillium. Damp woods; frequent. - Haverhill, H. 354 (1932); Andover, Pease 2038 (1903); Boxford, M.E. Perley (1883); Groveland, Paine (1884); Newburyport, E.N. Cole (1861); Georgetown, Horner; Ipswich, H. 18167 (1958); Topsfield, H.S. Porter; Middleton, Pease 27766 (1940); Danvers, Sears (1885); Peabody, Cole (1900).

T. erectum L. Purple Trillium. Rich woods; occasional. - Andover, Sutton 27 (1874); Groveland, Paine (1884); Topsfield, B.B. Towne; Danvers, (1859); Swampscott, Andrews (1879).

T. erectum, forma *albiflorum* R. Hoffm. - Groveland, Paine (1884).

T. undulatum Willd. Painted Trillium. Acid woods; rare. - Newburyport, Moulton (1887); Manchester, Babson (1866).

SMILAX L. Greenbrier.

S. herbacea L. Carrion-flower. Low woods and meadows; common. - Haverhill, Bradford, Fernald 11331 (1914); Methuen, H. 9622 (1954); Lawrence, H. 10074 (1955); North Andover, Root (1945); Boxford, M.E. Perley (1885); Groveland, H. 8798 (1953); Topsfield, G.F. Dow;

Danvers, Lowell (1846); Wenham, Robinson (1875); Hamilton, H. 10780 (1955); Rockport, Forbes 1050 (1902); Beverly, Lowell (1849); Salem, Phippen (1839); Peabody, Mackintosh (1925); Lynn, Cushman (1907); Marblehead, Cole (1943).

S. rotundifolia L. Common Greenbrier. Thickets and woods; too common. - Amesbury, H. 9435 (1954); Haverhill, H. 34 (1931); Andover, Pease 315 (1902); Boxford, M.E. Perley (1881); Newburyport, Deane (1886); Georgetown, Horner; Rowley, Plum Island, H., McGregor & Schweinfurth 12731 (1957); Ipswich, Townsend (1922); Topsfield, H.R. Towne; Middleton, Hopkins & Pease 24055 (1934); Danvers, Sears (1886); Wenham, Robinson (1875); Hamilton, Hubbard (1913); Essex, Hubbard (1913); Gloucester, Magnolia, Forbes 1054 (1902); Rockport, Smith 956 (1928); Beverly, Cole (1902); Salem, Cole (1908); Lynnfield, Forbes (1912); Lynn, Robinson (1890).

AMARYLLIDACEAE (Amaryllis Family)

NARCISSUS L. Narcissus.

N. pseudo-narcissus L. Daffodil. Introduced from Europe; rarely naturalized in open places. - Topsfield, Mrs. Still.

HYPOXIS L. Stargrass.

H. hirsuta (L.) Coville. Open woods and meadows; frequent. - Haverhill, Ward Hill, Cushman (1908); North Andover, H. 9488 (1954); Boxford, Frye (1881); Groveland, Bean & H. 8491 (1953); Newburyport, H. 11286 (1955); Ipswich, Oakes; Topsfield, H.R. Dow; Middleton, H. 9485 (1954); Danvers, Frost; Beverly, Robinson (1879); Salem, Cole (1886); Saugus, Bean & H. 9787 (1955); Marblehead, Cole (1889).

IRIDACEAE (Iris Family)

SISYRINCHIUM L. Blue-eyed Grass.

S. montanum Greene, var. *crebrum* Fern. Damp open soil; common. - Haverhill, Ward Hill, H. 374 (1932); Lawrence, H. 9859 (1955); Andover, Pease 3676 (1904); North Andover, Root (1943); Boxford, A.W. & C.A. Cheever (1913); Newburyport, E.N. Cole (1861); Georgetown, Horner; Ipswich, Allen (1912); Topsfield, Bean & H. 9937 (1955); Middleton, M.E. Perley (1884); Danvers, Sears (1884); Essex, Hubbard (1913); West Manchester, Hubbard (1913); Beverly, Robinson (1879); Salem, Sears (1899); Peabody, Mackintosh (1933); Marblehead, Batchelder (1911).

S. arenicola Bickn. Dry sandy soil; rare. - Lynnfield, Bean (1916).

S. angustifolium Mill. Meadows and low woods; frequent. - Salisbury, Fernald (1913); North Andover, Bean (1947); Boxford, M.E. Perley (1884); Groveland, Bean & H. 8494 (1953); Newbury, Bean & H. 8411 (1953); Rowley, H. 7941 (1952); Topsfield, B.B. Towne; Middleton, M.E. Perley (1884); Danvers, Sears (1885); Hamilton, Hubbard (1913); Marblehead, Bean & H. 9723 (1955).

S. atlanticum Bickn. Damp to dry meadows; frequent. - Merrimac, Bean & H. 8599 (1955); Andover, Pease 2152 (1903); North Andover, H. 9919 (1955); Boxford, H. 8005 (1952); Rowley, Bean (1940); Topsfield, Sears (1899); Hamilton, Hubbard (1913); Rockport, Smith 622 (1927); Manchester, Williams (1896); Saugus, Ripley 15525 (1915); Swampscott, Swan (1886); Marblehead, Bean & H. 9736 (1955).

IRIS L. Iris.

I. prismatica Pursh. Slender Blue Flag. Brackish to fresh marshes or meadows; frequent. - Salisbury, Cole (1901); Haverhill, H. 1294 (1934); Andover, Robinson (1879); North Andover, Morss (1885); Georgetown, Williams (1906); Rowley, Bean (1940); Ipswich, Oakes; Ipswich, Plum Island, H., McGregor & Schweinfurth 12701 (1957); Topsfield, P. Kneeland; Middleton, Frost; Rockport, Forbes 1074 (1903).

I. versicolor L. Blue Flag. Meadows and marshes; common. - Haverhill, H. 402 (1932); Methuen, H. 18401 (1958); Lawrence, H. 10087 (1955); Andover, Pease 329 (1902); North Andover, Root (1943); Boxford, Frye (1880); Newbury, H. 8241 (1952); Byfield, Robinson (1875); Rowley, Plum Island, H., McGregor & Schweinfurth 12712 (1957); Ipswich, H. 13933 (1957); Ipswich, Plum Island, H. 12119 (1956); Topsfield, C. Sweeney; Danvers, Frost (1873); Wenham, Robinson (1884); Rockport, Smith 749 (1927); West Manchester, Hubbard (1913); Salem, Cole (1902).

I. pseudacorus L. Introduced from Europe. Escape from cultivation, meadows and streams; occasional. - Boxford, H. 17633 (1958); Topsfield, H. 17637 (1958).

I. germanica L. Introduced from Europe. Escape from cultivation; occasional on dumps. - Haverhill, Bradford, H. 18816 (1958); Georgetown,H. 17680 (1958); Saugus, H. 18641 (1958).

I. sibirica L. Introduced from Europe. Escape from cultivation; occasional. - Boxford, H. 19466 (1959).

ORCHIDACEAE (Orchis Family)

CYPRIPEDIUM L. Lady's-slipper.

C. calceolus L., var. *parviflorum* (Salisb.) Fern. Small Yellow Lady's-slipper. Mossy swamps and woods; formerly rare in County, no recent collections. - West Boxford, Wilmarth; Danvers, Blind Hole Swamp, Sears (1875).

C. calceolus, var. *pubescens* (Willd.) Correll. Large Yellow Lady's-slipper. Andover, Gutterson (1901); Boutwell & Root (1951).

C. reginae Walt. Showy Lady's-slipper. Swamps and woodland glades; formerly rare in County, no recent collections. - Salisbury, True (1897); Danvers, Sears (1877); Wenham, Leach's Swamp, Sears (1885).

C. acaule Ait. Common Lady's-slipper. Coniferous woods; frequent. - Haverhill, H. (1932); Boxford, Frye (1881); Newburyport, E.N. Cole (1861); Rowley, Horner; Ipswich, Oakes; Topsfield, B.B. Towne; Hamilton, Cole (1887); Gloucester, Magnolia, Sears (1896); Manchester, Hubbard (1913); Beverly, Robinson (1872); Lynnfield, Lund.

C. acaule, forma *albiflorum* Rand & Redfield - Boxford, H. 7950 (1952); Georgetown, Horner (1875); Manchester, Hart (1882).

HABENARIA Willd. Rein-Orchis.

H. clavellata (Michx.) Spreng. Green Woodland Orchis. Moist mossy woods; rare. - Essex, Sears (1877); Gloucester, Magnolia, Kidder (1884); Salem, Robinson (1877); Lynnfield, Chute; Essex County, Russell.

H. clavellata, var. *ophioglossoides* Fern. - Middleton, H. 18084 (1958).

H. flava (L.) R.Br., var. *herbiola* (R.Br.) Ames & Correll. Pale Green Orchis. Swampy woods and swales; occasional. - Andover, Pease 2205 (1903); Boxford, M.E. Perley (1881); Newbury, Bean & H. 8401 (1953); Georgetown, Horner; Ipswich, Oakes; Topsfield, B.B. Towne; Wenham, Sears (1887).

H. hookeri Torr. Hooker's Orchis. Rare, no recent collections. - Topsfield, G.T. Dow. "Exhibited from Georgetown at Hort. Soc. Boston 1875 by Mrs. Horner!" Note without specimen at Peabody Museum.

H. macrophylla Goldie. Round-leaved Orchis. Mixed woods; rare. - Salisbury, Eaton (1896); Andover, Mrs. G.W. Chandler; Boxford, Mrs. J.F. Boynton; Georgetown, Horner (1882); Gloucester, Mrs. Frover (1875).

H. blephariglottis (Willd.) Hook. White Fringed Orchis. Bogs; occasional. - Andover, Pease 360 (1901); Gloucester, Jenks (1882); Manchester, Sears (1882); Essex County, Oakes.

H. lacera (Michx.) Lodd. Ragged Orchis. Meadows; frequent. - Merrimac, H. 18329 (1958); Haverhill, H. 512 (1932); Andover, Pease 4257 (1904); North Andover, Morss (1885); Boxford, M.E. Perley (1883); West Newbury, H. 20773 (1959); Newbury, Williams (1898); Georgetown, Horner (1895); Rowley, H. 8119 (1952); Middleton, Robinson (1874); Wenham, Sears (1887); Gloucester, Magnolia, J.M. Hunnewell (1898); Rockport, Gates & Smith 1000 (1928); Beverly, Robinson (1874); Marblehead, Cole (1886).

H. psycodes (L.) Spreng. Small Purple Fringed Orchis. Meadows; occasional. - Haverhill, H. 36 (1931); Boxford, Williams (1899); Newbury, Cushing; Georgetown, Horner (1880); Ipswich; Topsfield, A. Jacobs; Danvers (1845); Peabody, Robinson (1874).

H. fimbriata (Ait.) R.Br. Large Purple Fringed Orchis. Not as frequent as the last. - Andover, Pease 366 (1902); Georgetown, Horner (1880); Ipswich, Oakes; Topsfield, A. Jacobs; Middleton, H. 18070 (1958); Wenham, Pickering; Manchester, Hubbard (1913); [Gloucester, Rockport, A & W (1968)].

POGONIA Juss. Pogonia.

P. ophioglossoides (L.) Ker. Meadows and shores of ponds; occasional. - Haverhill, Palfrey (1890); Andover, Pease 334 (1899); Boxford, M.E. Perley (1882); Georgetown, Horner; Ipswich, Oakes; Danvers, Frost (1873); Hamilton, Cole (1910); Gloucester, Prince (1883); Rockport, Forbes 1114 (1904); Manchester, Oakes; Beverly, Lowell (1846); Marblehead, Cole (1886).

ISOTRIA Raf. Whorled Pogonia.

I. verticillata (Willd.) Raf. Acid woodlands; rare, no recent collections. - Lynnfield, Hitchings (1890); Lynn, Fuller (1896); Essex County (1901).

CALOPOGON R.Br. Swamp-pink.

C. pulchellus (Salisb.) R.Br. Meadows and bogs; occasional. - Haverhill, Palfrey (1890); Andover, Robinson (1879); North Andover, Morss (1885); Boxford, M.E. Perley (1882); Georgetown, Horner; Topsfield, Mrs. W. Herrick; Danvers, Frost (1873); Gloucester, Prince; Rockport, Smith 738 (1927); Manchester, Oakes; Beverly, Robinson (1874); Lynnfield, Bean (1919); Marblehead, Cole (1886).

ARETHUSA L. Arethusa.

A. bulbosa L. Meadows and bogs; occasional. - Andover, Pease 333 (1899); West Boxford, Zink (1943); Newburyport, E.N. Cole (1861); Georgetown, Horner (1881); Ipswich; Topsfield, Thayer; Danvers, Frost (1881); Wenham, Cole (1899); Hamilton, Robinson (1879); Gloucester, Rand & B.L. Robinson (1896); Beverly, Cooke & Robinson (1872); Lynnfield, Bean (1919).

EPIPACTIS Sw. Helleborine.

E. helleborine (L.) Introduced and naturalized from Europe. Woods and roadsides, formerly rare, now more common. Boxford, H. 25142 (1962); [Ipswich, R. Goodale (1968)] ; Middleton, H. Lewis (1968); [A & W (1969)] Rockport.

SPIRANTHES Richard Ladies'-tresses.

S. lacera Raf. Northern Slender Ladies'-tresses. Dry to moist meadows; occasional. - Haverhill, H. 1843 (1935); North Andover, Morss (1885); Georgetown, Horner; Gloucester, Annisquam, Ripley 15889 (1915); Lynn, Fuller (1897).

S. gracilis (Bigel.) Beak. Southern Slender Ladies'-tresses. Rather dry pastures; occasional. - Boxford, M.E. Perley (1883); Georgetown, Horner; Ipswich, Oakes; Topsfield, G.T. Dow; Danvers, Sears (1887); Gloucester, Prince (1883); Beverly, Robinson (1875); Salem, Forbes 1127 (1903); Peabody, Mackintosh (1933).

S. cernua (L.) Richard. Common Ladies'-tresses. Low ground; frequent in late summer. - Merrimac, H. 9222 (1953); Andover, Bean (1945); Boxford, H. 18606 (1958); Georgetown, Robinson (1875); Ipswich, H. & Wade 18792 (1958); Topsfield, A.M. Dow; Danvers, Phippen; Gloucester, Magnolia, J.M. Hunnewell (1898); Rockport, Smith 1082 (1928); Salem, Cole (1928); Peabody, Sears (1884).

S. cernua, var. *ochroleuca* (Rydb.) Ames - Boxford, M.E. Perley (1884).

S. lucida (H.H. Eat.) Ames. Damp meadows; rare. - Newbury, Bean & H. 8405 (1953).

GOODYERA R.Br. Rattlesnake-plantain.

G. repens (L.) R.Br., var. *ophioides* Fern. Dwarf Rattlesnake-plantain. Damp woods; rare. - Manchester, Oakes.

G. tesselata Lodd. Dry to moist woodlands; frequent. - Andover, Pease 1769 (1903); Boxford, M.E. Perley (1882); Georgetown, Horner (1881); Manchester, Williams (1892); Beverly, Lowell (1846).

G. pubescens (Willd.) R.Br. Downy Rattlesnake-plantain. Dry or moist woodlands; frequent. - Andover, Pease 1768 (1903); North Andover, Cole (1898); Boxford, M.E. Perley (1885); Georgetown, Horner (1881); Ipswich, H. 18143 (1958); Topsfield, G.F. Dow; Middleton, H. 18068 (1958); Essex, Robinson (1873); Gloucester, Magnolia, J.M. Hunnewell (1899); Manchester, Williams (1895); Beverly, Cole (1898).

LISTERA R.Br. Twayblade.

L. cordata (L.) R.Br. Heartleaf Twayblade. Mossy woods; rare. - Gloucester, Magnolia, Sprague (1880), Goose Cove, Lufkin (1891), Mount Ann Park, Cole (1905).

CORALLORHIZA Chatelain Coral-root.

C. trifida Chatelain, var. *verna* (Nutt.) Fern. Early Coral-root. Damp woods; rare. - [Boxford, Miliotis (1968)] Gloucester, Magnolia, Sears (1884); Manchester, Oakes; Saugus, Bean (1937); Lynn, Fuller (1897).

C. maculata Raf. Spotted Coral-root. Dry woods; occasional. - Haverhill, H. 28390 (1965); North Andover, Cole (1898); Boxford, Alcott (1884); Georgetown, Horner (1873); Topsfield, G.F. Dow; Middleton, Oakes; Danvers, Cole (1906); Chebacco Woods, Robinson (1874); Gloucester, Magnolia, J.M. Hunnewell (1898); Rockport, Anderson & Smith 1152 (1930); Manchester, Cole (1910); Peabody, Mackintosh (1933).

MALAXIS Sw. Adder's-mouth.

M. unifolia Michx. Green Adder's-mouth. Dry or moist woods; rare. - [Rockport, A & W (1968), spec. PMSH] Salem, Oakes; Lynn, Fuller (1896) "found by Ella".

LIPARIS Richard Twayblade.

L. lilifolia (L.) Richard. Loamy or sandy woods; rare. - Amesbury, Eaton (1896); Nahant.

L. loeselii (L.) Richard. Bog-twayblade. Bogs and peaty meadows; occasional. - Amesbury, Eaton (1896); Lawrence, W. Bailey; Andover, Pease 377 (1902); Boxford, H. 17714 (1958); East Gloucester, Sears (1888); Rockport, Smith (1918); Beverly, Robinson (1874); West Peabody, H. 11068 (1955).

SALICACEAE (Willow Family).

SALIX L. Willow.

S. nigra Marsh. Black Willow. Banks of streams, shores and low rich woods; frequent. - Merrimac, H. 13720 (1957); Methuen H. 18390 (1958); North Andover, Root (1951); Boxford, H. 8736 (1953); Ipswich, H. 13899 (1957); Topsfield, Sears (1877); Essex, H. 13360 (1957); Manchester, Hunnewell (1913); Beverly, Lowell (1848); Lynnfield, H. 8342 (1953); Marblehead Neck, Kennedy (1887).

S. pentandra L. Bay-leaved willow. Introduced from Europe. Spread from cultivation; occasional. - North Andover, Root (1949); South Groveland, H. 13959 (1957); Ipswich, St. John 101 (1911); H. 14001 (1957); Middleton, H. 12895 (1957).

S. lucida Muhl. Shining willow. Low grounds; frequent. - Salisbury, H. 10496 (1955); Merrimac, H. 9206 (1953); North Andover, Robinson (1884); Boxford, Robinson & Sears (1877); Newbury, Plum Island, White 120 (1913); Ipswich, Wiegand 2088 (1911); Topsfield, Sears (1878); Salem, Sears (1880).

S. lucida, var. *angustifolia* Anderss. - Andover, Lowell Junction, H. 17922 (1958).

S. fragilis L. Crack-Willow. Introduced from Europe. Spread from cultivation; occasional. - West Newbury, H. & Wade 18803 (1958); Ipswich, Oakes; Wenham, H. 13104 (1957); Salem, Robinson (1891).

S. babylonica L. Weeping Willow. Introduced from Eurasia. Spread from cultivation; occasional. - North Andover, Root (1947); Wenham, Robinson (1875); Salem, Cole (1933).

S. alba L., var. *vitellina* (L.) Stokes. White Willow. Introduced from Europe. Naturalized through much of county; frequent. - Haverhill, H. 14155 (1957); Methuen, H. 18375 (1958); Boxford, M.E. Perley (1881); Rowley, H. 10452 (1955); Ipswich, Oakes; Topsfield, H.R. Towne; Middleton, H. 8319 (1953); Danvers, Robinson (1875); Essex, Hubbard (1913); Salem, Russell (1859).

S. rigida Muhl. Banks of streams, pond shores and low thickets; frequent. - Amesbury, H. 10557 (1955); Lawrence, Morss (1900); Andover, Pease 3602 (1904); North Andover, Robinson (1875); Rowley, H. 10434 (1955); Ipswich Beach, Hubbard (1913); Topsfield, Robinson (1875); Middleton, Bean (1940); Beverly, H. 10201 (1955); Salem, Cole (1929).

S. bebbiana Sarg. Long-beaked Willow. Moist to dry thickets; common. - Haverhill 1267 (1931); Methuen, H. 8908 (1953); Andover, Pease 3714 (1904); North Andover, Morss (1900); Boxford, H. 7822 (1952); Georgetown, H. 13592 (1957); Rowley, H. 10438 (1955); Ipswich, Oakes; Ipswich, Plum Island, H., McGregor & Schweinfurth 12673 (1957); Topsfield, Bean & H. 9932 (1955); Danvers, Robinson (1875); Wenham, Robinson (1875); Hamilton, Hubbard (1913); Chebacco, Robinson (1875); Rockport, H. 10306 (1955); Beverly, Robinson (1877); [MNS, Snyder (1971)].

S. pedicellaris Pursh, var. *hypoglauca* Fern. Acid bogs and sphagnous shores; occasional. - Methuen, H. 18380 (1958); Andover, Pease 426 (1902); Boxford, C.E. Faxon (1877); Groveland, Paine (1884); Topsfield, Emerson (1838); Danvers, Sears (1877); Lynnfield, Young (1879).

S. discolor Muhl. Large Pussy-willow. Damp thickets and shores; common. - Merrimac, H. 9207 (1955); Haverhill, H. 344 (1932); Methuen, H. 18376 (1958); Andover, Pease 3698 (1904); North Andover, Watson (1875); Boxford, M.E. Perley (1886); Rowley, Plum Island, H. 12563 (1956); Topsfield, Mrs. Benson; Middleton, H. 8305 (1953); Danvers, Sears (1875); Wenham, Robinson (1875); Gloucester, Lanesville, Robinson (1877); Beverly, H. 10176 (1955); Salem, Cole (1899).

S. humilis Marsh. Small Pussy-willow. Dry thickets and openings; occasional. - Andover, Lowell Junction, H. 17930 (1958); North Andover, Root (1951); Boxford, H. 7809 (1952); Wenham, Robinson (1875); Gloucester, Lanesville, Robinson (1875); Beverly, Lowell.

S. humilis, var. *microphylla* (Anderss.) Fern. Dwarf Gray Willow. Dry barrens and slopes; occasional. - Andover, Pease 2006 (1903); North Andover, H. 11045 (1955); Boxford, H. 7807 (1952); Ipswich, Oakes; Middleton, Robinson (1877); Danvers, Frost (1881); Peabody, Mackintosh (1925); Lynnfield, Robinson (1875).

S. humilis X S. *gracilis,* var. *textoris* - Boxford, Forbes (1909).

S. gracilis Anderss., var. *textoris* Fern. Meadows and swales; frequent. - Andover, Pease 24089 (1943); North Andover, Robinson (1875); Georgetown, Horner (1875); Topsfield; Middleton, H. 8306 (1953); Danvers, Sears (1890); Wenham, Robinson (1875); Chebacco, Robinson (1875); Beverly, H. 10164 (1955); Peabody, H. 8323 (1953); Lynnfield, H. 9361 (1953); Marlbehead, Cole (1930).

X *S. subsericea* (Anderss.) Schneid. - Newbury, Plum Island, H., McGregor & Schweinfurth 12741 (1957); Wenham, H. 13741 (1957).

S. sericea Marsh. Silky Willow. Low thickets and banks of streams; occasional. - Andover, Lowell Junction, H. 17928 (1958); Newbury, Bean & H. 8400 (1953); Georgetown, Kennedy (1906);

Ipswich, Oakes; Topsfield, Robinson (1877); Middleton, H. 8320 (1953); Danvers, Robinson (1879); Rockport, Smith 930 (1928); Peabody, H. 8322 (1953).

S. candida Flügge. Hoary Willow. Calcareous bogs and thickets; rare. - Boxford, Robinson (1875); C.E. Faxon, Robinson & Sears (1877), C.E. & E. Faxon (1882, 1889). The only station in Massachusetts east of Berkshire County.

S. purpurea L. Basket-willow. Introduced from Europe for basket making. Low grounds; occasional. - Boxford, Alcott (1882); Newburyport, Robinson (1875); Rowley, H. & Wade 18900 (1958); Salem, Robinson (1875); Nahant, Knowlton (1916).

POPULUS L. Poplar. Aspen.

P. tremuloides Michx. Quaking Aspen. Dry open woods; common. - Salisbury, White 300 (1913); Haverhill, H. 291 (1932); Andover, Pease 3426 (1904); North Andover, Morss (1885); West Boxford, Horner; Newbury, H. 10387 (1955); Rowley, H. 10453 (1955); Rowley, Plum Island, H. 12548 (1956); Ipswich, St. John 1104 (1914); Topsfield, Robinson (1875); Wenham, Robinson (1875); Hamilton, H. 10819 (1955); Rockport, H. 10267 (1955); West Manchester, Hubbard (1913); Beverly, H. 10210 (1955); Lynnfield, H. 8340 (1953); Marblehead, Cole (1900).

P. tremuloides, forma *reniformis* Tidestr. - Swampscott, Cole (1924).

P. grandidentata Michx. Large-toothed Aspen. Dry woods and openings; common. - Haverhill, H. 404 (1932); Metheun, H. 9604 (1954); Lawrence, Morss (1900); Andover, Ballardvale, Pease 1834 (1903); North Andover, Root (1957); Boxford, M.E. Perley (1884); West Newbury, H. 10966 (1955); Rowley, H. 10450 (1955); Ipswich, H. 13991 (1957); Ipswich, Plum Island, St. John & White 540 (1913); Topsfield, Bean & H. 9949 (1955); Middleton, H. 8302 (1953); Wenham, Robinson (1875); Hamilton, Hubbard (1913); West Manchester, Hubbard (1913); Peabody, Sears (1887); Lynnfield, H. 8339 (1953); Saugus, Bean & H. 9781 (1955); Swampscott, Cole (1924).

P. alba L. White Poplar. Introduced from Europe. Roadsides, occasional. - Andover, Ballardvale, Pease 2165 (1903); Boxford, H. 11159 (1955); Wenham, Robinson (1875); West Manchester, Hubbard (1913).

P. canescens (Ait.) Sm. Gray Poplar. Introduced from Europe. Spreading from cultivation; occasional. - North Andover, Root (1951).

P. deltoides Marsh. Cottonwood. River-banks and low woods; rare. - Middleton, H. 8304 (1953).

P. nigra L. Introduced from Europe. Spread from cultivation; occasional. - Newbury, H. 18452 (1958).

P. nigra, var. *italica* Muenchh. Lombardy Poplar. Introduced from Europe. Spread from cultivation; occasional. - Groveland, M.E. Perley (1885); Danvers, Sears (1875).

P. balsamifera L. Balsam-Poplar. River-banks or -gravels; rare. - Andover and North Andover, Root (1950), tree on boundary and in both towns.

P. balsamifera, var. *subcordata* Hylander - Haverhill, Batchelder (1885).

P. balsamifera X P. *nigra* ? - Merrimac, Bean & H. 7733 (1951).

X *P. gileadensis* Rouleau. Balm-of-Gilead. Origin unknown. Spread from cultivation; occasional. - West Andover, Root (1947); North Andover, Morss (1885); Boxford, M.E. Perley (1884); Rowley, H. 8249 (1952); Ipswich, St. John 465 (1912); Gloucester, Magnolia, J.M. Hunnewell (1898); Beverly, Robinson (1875); Salem, Cole (1912).

MYRICACEAE (Wax-Myrtle Family)

MYRICA L.

M. gale L. Sweet Gale. Shallow water and swamps; common. - Salisbury - Amesbury, White 256 (1913); Amesbury, H. 9529 (1954); Merrimac, H. 13677 (1957); Haverhill, H. 476 (1932); Methuen, H. 8939 (1953); South Lawrence, Morss (1885); Boxford, H. 13642 (1957); Groveland, H. 9063 (1953); Rowley, H. 11946 (1955); Ipswich, Oakes; Topsfield, M. Clarke; Middleton, H. 18104 (1958); Wenham, Lowell (1849); Hamilton, H. 9406 (1953); Essex Woods. Lowell (1847); Gloucester, H. 13835 (1957); Beverly, Robinson (1879); Peabody, Mackintosh (1925); Lynnfield, H. 14111 (1957); Lynn, Young (1880); Marblehead, Cole (1889).

M. gale, var. *subglabra* (Chev.) Fern. - Andover, Pease 420 (1901); Georgetown, Horner.

M. pensylvanica Loisel. Bayberry. Dry or wet sterile soil; common. - Salisbury, H. 10492 (1955); Lawrence, H. 9854 (1955); North Andover, Root (1943); Boxford, M.E. Perley (1882); West Newbury, H. 9112 (1953); Newbury, H. 10401 (1955); Newbury, Plum Island, White 126 (1913); Rowley, H. 8264 (1952); Rowley, Plum Island, H. 12049 (1956); Ipswich, Deane & Fernald (1908); Ipswich, Plum Island, H. 12118 (1956); Topsfield, H.R. Towne; Danvers, Frost (1881); Essex, Hubbard (1913); Gloucester, Hoar (1858); Rockport, H. 10279 (1955); Beverly, H. 10204 (1955); Salem, Cole (1889); Salem, Little Misery Island, H. 13293 (1957); Swampscott, H. 10642 (1955); Marblehead, Bean & H. 9730 (1955).

COMPTONIA L'Her. Sweet-Fern.

C. peregrina (L.) Coult. Open sterile woodlands, clearings and pastures; common. - Salisbury, H. 10474 (1955); Haverhill, H. 37 (1931); Lawrence, Pease 2931 (1903); Andover, H. 10039 (1955); North Andover, Morss (1885); Boxford, M.E. Perley (1881); Newbury, H. 10394 (1955); Rowley, H. 10454 (1955); Topsfield, G.T. Conley; Middleton, Bean & H. 7769 (1951); Danvers, Bean & H. 9986 (1955); Wenham, Bean & H. 10030 (1955); Hamilton, H. 10824 (1955); Gloucester, Smith 746 (1927); Beverly, Robinson (1879); Salem, Cole (1889); Lynnfield, Young (1878); Saugus, Bean & H. 9776 (1955); Lynn, H. 10921 (1955).

JUGLANDACEAE (Walnut Family)

JUGLANS L. Walnut.

J. cinerea L. Butternut. Rich woods; frequent. - Haverhill, H. (1932); Andover, Pease 416 (1901); North Andover, Morss (1900); Boxford, M.E. Perley (1884); Groveland, Bean & H. 8523 (1953); Topsfield, Robinson & Sears (1875); Danvers, Sears (1875); Hamilton, H. 10774 (1955); Salem, Cole (1900).

J. nigra L. Black Walnut. Introduced from farther West. Occasionally spreading from cultivation. - Andover, Root (1949); Gloucester, Annisquam, Ripley 15887 (1915).

CARYA Nutt. Hickory.

C. cordiformis (Wang.) K. Koch. Pignut. Wet to dry woods; occasional. - Andover, Kennedy (1908); North Andover, Robinson (1890); West Boxford, Root & Zink (1945); Ipswich, Kennedy 1603 (1914); Topsfield, Robinson (1875); Danvers, Sears (1888); Wenham, Robinson (1875); Salem, Robinson (1891); Swampscott, Buck (1897).

C. ovata (Mill.) K. Koch. Shagbark-Hickory. Rich woods; common. - Haverhill, H. 642 (1932); Lawrence, H. 10084 (1955); Andover, Pease 415 (1902); North Andover, Robinson (1890); Boxford, M.P. Perley (1882); Ipswich, Oakes; Topsfield, Robinson (1875); Danvers, Frost (1881); Wenham, Robinson (1875); Essex, Hubbard (1913); North Beverly, Robinson (1890); Salem, Cole (1909); Saugus, Bean & H. 9801 (1955).

C. tomentosa Nutt. Mockernut. Dry to moist woodlands; rare. - Ipswich, Oakes; Danvers, Sears (1877).

C. glabra (Mill.) Sweet. Pignut. Dry woods; frequent. - Andover, Pease 2005 (1903); North Andover, Root (1957); Boxford, M.P. Perley (1882); Ipswich, Morong (1875); Topsfield, Robinson (1875); Danvers, Sears (1885); Chebacco, Robinson (1875); Manchester, Deane (1887); Beverly, Robinson (1890); Salem, Cole (1889); Swamscott, Buck 70 (1896).

C. ovalis (Wang.) Sarg., var. *odorata* (Marsh.) Sarg. Rich woodlands; rare. - North Andover, Root (1951); West Boxford, Root (1951).

CORYLACEAE (Hazel Family)

CORYLUS L. Hazelnut

C. americana Walt. American Hazelnut. Thickets; common. - Haverhill, H. 789 (1931); Andover, Pease 3428 (1904); North Andover, Morss (1885); Boxford, M.E. Perley (1881); Groveland, Bean & H. 8508 (1953); Newbury, Oak Island, White 236 (1913); Ipswich, H. 18417 (1957); Topsfield, G.T. Dow; Middleton, Bean (1950); Wenham, Robinson (1875); Chebacco, Robinson (1875); Peabody, Robinson (1876).

C. americana, forma *missouriensis* (A. DC.) Fern. - Lynnfield, Rich (1892).

C. cornuta Marsh. Beaked Hazelnut. Rich thickets and borders or woods; frequent. - Haverhill, H. 547 (1932); Andover, Pease 450 (1902); Boxford, M.E. Perley (1881); Topsfield, Gloucester, Magnolia, Forbes 1210 (1902); Beverly, Prides Crossing, Sears (1887); Salem, Cole (1889).

OSTRYA Scop. Hop-Hornbeam.

O. virginiana (Mill.) K. Koch. Leverwood. Rich woods; frequent. - Merrimac, H. 18337 (1958); North Andover, Root (1957); Boxford, M.E. Perley (1883); Groveland, Bean & H. 8519 (1953); Topsfield, M. Spofford; Danvers, Sears (1887); Salem, Conant & Robinson (1877); Peabody, Mackintosh (1933); Swampscott, Buck (1895); Marblehead, Cole (1944).

CARPINUS L. Hornbeam.

C. caroliniana Walt., var. *virginiana* (Marsh.) Fern. Rich woods and swamps; occasional. - Andover, Pease 447 (1902); North Andover, Root (1960); Boxford, M.E. Perley (1885); West

Newbury, Robinson (1879); Georgetown, Robinson (1875); Topsfield, H. 10113 (1955); Middleton, H. 12638 (1957); Danvers, Sears (1886); Wenham, Robinson (1876); Peabody, H. 9669 (1955).

BETULA L. Birch.

B. lenta L. Black Birch. Rich woods; common. - Salisbury, White 271 (1913); Haverhill, H. 1295 (1934); Andover, Pease 441 (1901); North Andover, Root & Zink (1945); Boxford, M.E. Perly (1885); West Newbury, H. 10338 (1955); Newburyport, White 462 (1913); Newbury, H. 8237 (1952); Georgetown, H. 10703 (1955); Topsfield, Mrs. S. Conley; Hamilton, Morong (1874); Essex, H. 11984 (1955); Gloucester, H. 12153 (1956); Rockport, H. 19030 (1958); Manchester, Oakes; Salem, Cole (1924); Saugus, Bean & H. 9774 (1955).

B. lutea Michx. f. Yellow Birch. Rich woods; frequent. - Salisbury, H. 12220 (1956); North Andover, Root (1957); Boxford, Zink (1945); Ipswich, Oakes; Wenham, Bean & H. 10019 (1955); Chebacco, Robinson (1874); Essex, Robinson (1876); Gloucester, Annisquam, Ripley (1914); West Manchester, Hubbard (1913); Beverly, Robinson (1876); Swampscott, Buck (1895).

B. nigra L. River-Birch, Red Birch. Bottonlands and borders of streams; common. - Amesbury, Rich (1906); Merrimac, H. 13705 (1957); Haverhill, Batchelder (1885); Methuen, Churchill (1909); South Lawrence, Williams (1900); North Andover, Robinson (1875); Boxford, H. 7887 (1952); South Groveland, H. 7882 (1952); West Newbury, Bean & Pease (1940); Georgetown, Horner (1875); Ipswich, St. John 11879 (1915) [and Hollow in dunes, Robert L. Goodale, (1969)]; Topsfield, H. 8316 (1953); Middleton, H. 8319 (1953); Danvers, H. 8325 (1953); Wenham, Cole (1919); Beverly, Fernald & Svenson 887 (1928); Peabody, H. 8324 (1953); Lynnfield, H. 9355 (1953). Formerly thought to be restricted to the Merrimack Valley but now much more widespread.

B. populifolia Marsh. Gray Birch. Sterile dry to wet soil; common. - Haverhill, H. 790 (1931); Lawrence, H. 9863 (1955); Andover, Sutton 57 (1874); North Andover, Morss (1885); Boxford, M.E. Perley (1885); Newbury, Plum Island, H., McGregor & Schweinfurth 12742 (1957); Rowley, H. 10447 (1955); Rowley, Plum Island, H. 12566 (1956); Ipswich, H. 13941 (1957); Topsfield, H.R. Towne; Danvers, Frost (1881); Wenham, Robinson (1874); Chebacco, Robinson (1874); Gloucester, H. 10143 (1955); Rockport, H. 10278 (1955); Manchester, Hubbard (1913); Beverly, H. 10175 (1955); Salem, Cole (1899); Lynnfield, H. 9357 (1953); Saugus, Bean & H. 9794 (1955).

B. papyrifera Marsh. Paper-Birch. Wooded slopes; frequent. - Haverhill, Bradford, Fernald 11334 (1914); North Andover, Root (1949); Boxford, H. 10609 (1955); Groveland, M.E. Perley (1883); West Newbury, H. 11148 (1955); Newbury, Kents Island, Sears (1904); Topsfield, Oakes; Wenham, Robinson (1875); Hamilton, Hubbard (1913); Gloucester, H. 10138 (1955); Saugus, Bean & H. 9815 (1955).

B. papyrifera, var. *commutata* (Regel) Fern. - Gloucester, Johnston (1944).

B. papyrifera, var. *cordifolia* (Regel) Fern. - Boxford, M.E. Perley (1885).

ALNUS B. Ehrh. Alder.

A. rugosa (Du Roi) Spreng. Speckled Alder. Low Ground; common. - Haverhill, H. 292 (1932); Boxford, H.M.E. Perley (1881); Rowley, Plum Island, H. 12561 (1956); Ipswich, Lowell; Ipswich, Plum Island, St. John & White 543 (1913); Topsfield, Pease 7479 (1905); West Danvers, Manning (1886); Rockport, Gates & Smith 1009 (1928); West Manchester, Hubbard (1913); Marblehead, Phippen (1857).

A. rugosa, forma *emersoniana* Fern. - Lawrence, H. 9887 (1955); North Andover, Morss (1885); Groveland, Bean & H. 8560 (1953); West Newbury, H. 10948 (1955); Middleton, H. 8681 (1953); Danvers, Sears (1884); Hamilton, H. 10843 (1955); Beverly, Robinson (1875); Salem, Cole (1929).

A. serrulata (Ait.) Willd. Common Alder. Swamps and stream margins; common. - Merrimac, H. 13723 (1957); Methuen, H. 11005 (1955); Andover, Pease 3432 (1904); Boxford, M.E. Perley (1881); Groveland, E. 10988 (1955); West Newbury, NEBC. (1951); Ipswich, H. 13923 (1957); Topsfield, Pease 7480 (1905); Middleton, Bean (1950); Danvers, Sears (1885); Wenham, Robinson (1875); Rockport, H. 10292 (1955); Salem, Cole (1929); Saugus, Bean & H. 9778 (1955); Swampscott, H. 10687 (1955).

A. serrulata, forma *noveboracensis* (Britt.) Fern. - Topsfield, H. 10135 (1955); Beverly, H. 10169 (1955).

A. serrulata, var. *subelliptica* Fern. - North Andover, H. 8663 (1953); Topsfield, H.R. Towne; Lynnfield, H. 9358 (1953).

A. serrulata, var. *subelliptica,* forma *mollescens* Fern. - Rowley, H. 12245 (1956); Plum Island, St. John 837 (1913).

FAGACEAE (Beech Family)

FAGUS L. Beech.

F. grandifolia Ehrh. Rich woods; frequent. - Salisbury, H. 18974 (1958); Merrimac, Bean & H. 7702 (1951); Lawrence, Morss (1901); North Andover, Root (1957); Boxford, Pease & St. John 11945 (1917); Groveland, Horner; West Newbury, H. 10853 (1958); Ipswich, H. 18437 (1958); Topsfield, H. 10128 (1955); Danvers, Frost (1881); Chebacco, Robinson (1875); Hamilton, Cole (1900); Essex, H. 11975 (1955); Gloucester, Magnolia, Robinson (1877); Rockport, Sears (1885); West Manchester, Hubbard (1913); Salem, Cole (1909); Saugus, Bean & H. 9804 (1955); Swampscott, Buck (1896).

CASTANEA Mill. Chestnut.

C. dentata (Marsh.) Borkh. Dry rocky or gravelly, acid soil; formerly a frequent tree but wiped out by the Chestnut Blight in the past half century. Sucker growth still appears from the old roots and sometimes attains sufficient size to produce fruit but eventually succumbs to the Blight. - Amesbury, H. 18203 (1958); Merrimac, Bean & H. 7705 (1951); Haverhill, H. 405 (1932); Andover, Young (1882); North Andover, Morss (1885); Boxford, H. 8631 (1953); Groveland, H. 10987 (1955); Newburyport, H. 10340 (1955); Topsfield, H.R. Towne; Danvers, Sears (1877); Wenham, Cole (1928); Hamilton, Hubbard (1913); Beverly, Prides Crossing, Kidder (1885); West Peabody, Sears (1885); Lynnfield, Robinson (1874).

QUERCUS L. Oak.

Q. alba L. White Oak. Dry woods; common. - Salisbury, Bean & H. 8277 (1953); Haverhill, H. 684 (1932); Lawrence, H. 9875 (1955); Andover, Pease 2608 (1903); North Andover, Robinson (1875); Boxford, M.E. Perley (1885); Groveland, Bean & H. 8506 (1953); West Newbury, Horner; Rowley, H. 8267 (1952); Ipswich, H. 18429 (1958); Topsfield, Robinson (1875); Middleton, Sears (1885); Wenham, Bean & H. 10023 (1955); Hamilton, H. 10845 (1955); Chebacco, Robinson (1874); Salem, Cole (1928); Peabody, Sears (1885); Lynnfield, Robinson (1877); Saugus, Bean & H. 9805 (1955).

Q. robur L. English Oak. Introduced from Europe. Spread from cultivation; occasional. - Salem, Vanderbilt; Salem, Little Misery Island, H. 13289 (1957); Marblehead, H. 12601 (1956); Nahant, Bean & H. 9838 (1955); [MNS, Snyder 1950's, common].

Q. robur, var. *pedunculata* DC. - Andover, Pease 2809 (1908); Topsfield, H. 8890 (1953).

Q. macrocarpa Michx. Mossy-cup Oak. Probably introduced from the West but the tress are located in a spot where they are apparently wild. - Newbury, Downfall Street, H. 8226 (1952), Bean & H. 8398 (1953).

Q. bicolor Willd. Swamp-White Oak. Bottomlands and swamps; common. - Amesbury, H. 10546 (1955); Methuen, H. 9626 (1954); Andover, Pease 1942 (1903); North Andover, Bean (1935); Boxford, M.E. Perley (1885); Groveland, H. 10983 (1955); West Newbury, Robinson (1879); Newbury, H. 8224 (1952); Ipswich, H. 18153 (1958); Topsfield, Robinson (1891); Danvers, Robinson (1877); Wenham, Oakes; Hamilton, St. John 11888 (1916).

Q. prinoides Willd. Chinquapin-Oak. Dry thickets and borders of woods; occasional, more frequent in northern and western portion of County. - Andover, Pease 2609 (1903); North Andover, Morss (1885); Boxford, Robinson & Sears (1877); Groveland, Bean & H. 8503 (1953); Georgetown, Robinson (1875); Topsfield, Robinson (1875); Middleton, Robinson (1887); South Danvers, Russell (1857); Peabody, Sears (1886).

Q. prinus L. Chestnut-Oak. Dry or rocky woods; rare. - Saugus, Bean (1908).

Q. rubra L. Red Oak. Dry woods; common. - Salisbury, Bean & H. 8378 (1953); Lawrence, Morss (1900); Andover, Pease 4550 (1902); Boxford, M.E. Perley (1881); West Newbury, Robinson (1877); Newbury, Plum Island, H., McGregor & Schweinfurth 12740 (1957); Rowley, H. 8268 (1952); Rowley, Plum Island, H., McGregor & Schweinfurth 12733 (1957); Middleton, Bean & H. 7768 (1951); Wenham, Bean & H. 10029 (1955); Gloucester, H. 12154 (1956); Rockport, H. 10287 (1955); West Manchester, Hubbard (1913); Beverly, Robinson (1872); Saugus, Bean & H. 9830 (1955); Swampscott, H. 10648 (1955).

Q. palustris Muenchh. Pin-Oak. Introduced from farther South. Spread from cultivation. - Newbury, H. 18445 (1958); Ipswich, Robinson (1890).

Q. coccinea Meunchh. Scarlet Oak. Dry light soil; frequent. - Lawrence, H. 9874 (1955); Andover, Sutton 83 (1874); North Andover, Root (1949); Boxford, Robinson & Sears (1877); Groveland, Bean & H. 8504 (1953); Georgetown, Robinson (1876); Rowley, H. 8265 (1952); Middleton, Robinson (1884); Danvers, Sears (1885); Gloucester, H. 12156 (1956); Lynnfield, H. 9348 (1953); Saugus, Bean & H. 9831 (1955); Swampscott, Cole (1924).

Q. velutina Lam. Black Oak. Dry woods; common. - Salisbury, Bean & H. 8376 (1953); Methuen, H. 9612 (1954); Lawrence, H. 9873 (1955); Andover, Pease 455B (1902); North Andover, Robinson (1875); Boxford, H. 8135 (1952); Newbury, Bean & H. 8418 (1953); Rowley Plum Island, H., McGregor & Schweinfurth 12722 (1957); Ipswich, H. 18428 (1958); Topsfield, Robinson (1875); Middleton, H. 8680 (1953); Wenham, Robinson (1875); Hamilton, H. 10805 (1955); Gloucester, Robinson & Sears (1877); Rockport, Robinson (1875); West Manchester, Hubbard (1913); Peabody, Mackintosh (1933); Lynnfield, Robinson (1874); Swampscott, Robinson (1875).

Q. ilicifolia Wang. Scrub-Oak. Dry rocky, gravelly or sandy barrens; common. - Methuen, H. 12256 (1956); Lawrence, H. 9872 (1955); Andover, Pease 458 (1902); North Andover, H. 8661

(1953); Boxford, Robinson (1875); Groveland, Bean & H. 8502 (1953); Georgetown, Robinson (1876); Ipswich, H. 9575 (1954); Topsfield, Bean & H. 9936 (1955); Middleton, Robinson (1887); Salem, Cole (1926); Peabody, H. 9295 (1953); Lynnfield, Robinson (1874); Saugus, H. 7793 (1951).

ULMACEAE (Elm Family)

Ulmus L. Elm.

U. rubra Muhl. Slippery Elm. Rich soil; rare. - Boxford, Robinson & Sears (1877); Georgetown, Horner (1875); Topsfield, B.B. Towne; Wenham, Robinson (1875); Salem, Robinson (1891); Marblehead, Cole (1939).

U. procera Salisb. English Elm. Introduced from Europe. Probably present only as a cultivated tree. - Danvers, Sears (1886); [MNS, Snyder (1963)] ; Salem, Robinson (1879).

U. americana L. American Elm. Rich soil; common but rapidly being killed by the Dutch Elm Disease. - Andover, Pease 4373 (1904); North Andover, H.E. Parkhurst (1883); Boxford, M.E. Perley (1886).

U. americana, forma *pendula* (Ait.) Fern. - Boxford, M.E. Perley; Groveland, Bean & H. 8531 (1953); Newbury, H. 10407 (1955); Ipswich, H. 14006 (1957); Topsfield, B.B. Towne.

U. americana, forma *laevior* Fern. - Methuen, H. 18385 (1958); Georgetown, H. 10701 (1955); Ipswich, H. 14003 (1957); Danvers, Sears (1885); Swampscott, H. 10681 (1955); Essex County, Oakes.

U. americana, forma *alba* (Ait.) Fern. - North Andover, Root (1949); Boxford, H. 8612 (1953); Groveland, H. 10985 (1955); West Newbury, H. 10327 (1955); Newbury, H. 10400 (1955); Georgetown, H. 10700 (1955); Rowley, H. 10436 (1955); Ipswich, H. 14004 (1957); Topsfield, H. 10132 (1955); Middleton, H. 8694 (1953); Wenham, Robinson (1875); Hamitlon, H. 10813 (1955); Saugus, Bean & H. 9790 (1955).

U. americana, forma *intercedens* Fern. - Merrimac, Bean & H. 8590 (1953); Boxford, Robinson (1877); Groveland, H. 10984 (1955); West Newbury, H. 11101 (1955); Georgetown, H. 10702 (1955); Ipswich, H. 13906 (1957).

CELTIS L. Hackberry.

C. occidentalis L., var. *pumila* (Pursh) Gray. Damp to dry woodlands and shores; rare as a wild tree but sometimes cultivated. - Salisbury, Bean & Stockman (1953); Newburyport, Plum Island, Robinson & Sears (1876); Ipswich, Oakes; Salem Pastures, Robinson (1879).

MORACEAE (Mulberry Family)

MORUS L. Mulberry.

M. rubra L. Red Mulberry. Rich woods; probably present only as cultivated trees. - Ipswich, Robinson (1876).

M. alba L. White Mulberry. Introduced and naturalized from Asia; most specimens are probably from cultivated trees. - Andover, Pease 3802 (1904); Boxford, H. 9551 (1954), cultivated; Groveland, H. 12583 (1956); Topsfield, M.E. Perley (1882); Danvers, Sears (1882); Salem, Cole (1901).

BROUSSONETIA L'Her. Paper-Mulberry.

B. papyrifera (L.) Vent. Introduced from Asia. - Gloucester, Dogtown Common, H. 12174 (1956). A single tree growing by cellar-hole #18; first noticed by John Kieran.

MACLURA Nutt. Osage Orange.

M. pomifera (Raf.) Schneid. Introduced from the South. - Danvers, Kennedy (1907), escaped, 25 feet high.

CANNABINACEAE (Hemp Family)

CANNABIS L. Hemp. Marijuana.

C. sativa L. Adventive from Asia. Waste ground; occasional but locally abundant. - Haverhill, H. 407 (1932); Lawrence, Sears (1885); Boxford, M.E. Perley (1885); Newburyport, White 95 (1913); Beverly, Robinson (1875); North Salem, Hall (1926); Peabody, Osborn (1930).

HUMULUS L. Hop.

H. lupulus L. Common Hop. Introduced from Europe. Waste places and fencerows; common. - Haverhill, H. 643 (1932); Methuen, H. 11718 (1955); North Andover, Root (1945); Boxford, M.E. Perley (1885); Georgetown, Horner; Ipswich, H. 13867 (1957); Topsfield, Mrs. Conley; Wenham, Robinson (1875); Hamilton, H. 10784 (1955); Salem, H. 11493 (1955); Peabody, H. 11531 (1955); Marblehead, H. 11426 (1955).

H. japonicus Sieb. & Zucc. Japanese Hop. Introduced from Asia. - Beverly, H. 18902 (1958); Salem, Sears (1904); Saugus, H. 18710 (1958).

URTICACEAE (Nettle Family)

URTICA L. Nettle.

U. procera Muhl. Thickets, waste places and roadsides; common but seldom collected. - North Andover, Root (1945); Boxford, M.E. Perley (1884); Groveland, Seymour 4861 (1936); Newburyport, White 421 (1913); Topsfield, H.R. Towne; Danvers, Sears (1884); Rockport, L.B. & R.C. Smith 4337 (1945); Beverly, Prince.

U. dioica L. Naturalized from Eurasia. Waste places; occasional. - Topsfield, H.R. Towne; Wenham, Robinson (1875); Salem, Cole (1899); Nahant, Bean & H. 9840 (1955).

U. urens L. Burning Nettle. Naturalized from Europe. Waste places; occasional. - Andover, Blake (1881); Danvers, Oakes, Sears (1885); West Gloucester, Robinson & Sears (1877).

LAPORTEA Gaud. Wood-nettle.

L. canadensis (L.) Wedd. Low woods and banks of streams; occasional. - Amesbury, Bean (1950); Groveland, Sears (1883); West Newbury, H. 11125 (1955); Byfield, Horner; Salem, Robinson (1877).

PILEA Lindl. Richweed.

P. pumila (L.) Gray. Cool, moist shaded places; frequent. - Merrimac, H. 9279 (1953); Haverhill, H. 28394 (1965); Andover, Pease 2611 (1903); North Andover, Root (1945); Boxford, M.P. Perley (1883); West Newbury, H. 11130 (1955); Georgetown, Robinson (1875); Topsfield, H.R. Towne; Danvers, Oakes; Chebacco, Robinson (1876); Salem, Cole (1899).

BOEHMERIA Jacq. False Nettle.

B. cylindrica (L.) Sw. Bog-hemp. Moist or shady ground; common. - Haverhill, Ward Hill, Bean (1916); Methuen, H. 18410 (1958); West Andover, Morss (1885); Boxford, M.E. Perley (1885); Groveland, H. 8819 (1953); Rowley, H. 13495 (1957); Ipswich, H. 13478 (1957); Topsfield, Mackintosh (1924); Middleton, H. 18091 (1958); Danvers, Sears (1885); Wenham, Lowell (1847); Beverly, Robinson (1874); Salem, Cole (1899).

B. cylindrica, var. *drummondiana* Wedd. - Beverly, Knowlton (1907).

PARIETARIA L. Pellitory.

P. pensylvanica Muhl. Rocky shaded ground; rare. - Swampscott, Sears (1885).

SANTALACEAE (Sandalwood Family)

COMANDRA Nutt. Bastard-Toadflax.

C. umbellata (L.) Nutt. Dry sterile ground; common. - Haverhill, H. 408 (1932); Andover, Pease 464 (1902); Boxford, H.E. Parkhurst (1882); Georgetown, Horner; Topsfield, L.M. Dow; Middleton, Pease 27768 (1940); Danvers, Robinson (1877); Wenham, Robinson (1879); Gloucester, Lanesville, Collins (1918) [John Kieran (1971)]; Peabody, Sears (1896); Lynnfield, H. 8338 (1953); Lynn, Cushman (1908); Marblehead, Cole (1888); [MNS, Snyder 1950 on].

LORANTHACEAE (Mistletoe Family)

ARCEUTHOBIUM Bieb.

A. pusillum Peck. Dwarf Mistletoe. Parasite on *Picea mariana;* rare. - Andover, Pease 407 (1902), 1845 (1903), 2092 (1904).

ARISTOLOCHIACEAE (Birthwort Family)

ASARUM L. Wild Ginger.

A. canadense L. Rich woods; rare. - Andover, Robinson (1879) "field near house where it has become thoroughly naturalized", Pease 2022 (1903) "copse, Woburn Road", Root (1947).

POLYGONACEAE (Buckwheat Family)

RUMEX L. Dock. Sorrel.

R. verticillatus L. Water-Dock. Swamps and margins of ponds; rare. - West Newbury, H. 9137 (1953).

R. mexicanus Meisn. Adventive. Rich soil; occasional. - North Andover, H. 9900 (1955); Newburyport, Bean (1940); Topsfield, H. 9973 (1955); Gloucester, Bay View, Fernald, Long & Hunnewell 9384 (1913); Beverly, H. 11696 (1955).

R. mexicanus, var. *fueginus* (Phil.) Dusen. Ipswich, Plum Island, R.J. Eaton (1965).

R. pallidus Bigel. Seabeach-Dock. Saline or brackish marshes; occasional. - Newburyport, Bean (1945); Plum Island, St. John 1106 (1913); Danvers, Russell (1850); Gloucester, Lanesville, Sears (1884); Rockport, Sears (1886).

R. patientia L. Patience-Dock. Naturalized from Eurasia. Waste places; occasional. - Andover, Pease 3161 (1903); Gloucester, Lanesville, Collins (1915); West Manchester, Hubbard (1913).

R. orbiculatus Gray. Water-Dock. Wet meadows, swamps and shores; frequent. - Methuen, Boott (1868), H. 12848 (1957); Andover, Pease 494 (1901); North Andover, H. 13062 (1957); West Newbury, Robinson (1879); Wenham, Lowell (1847); Beverly, Lowell (1849); Swampscott, Robinson (1877); Marblehead, Cole (1930).

R. crispus L. Yellow-Dock. Naturalized from Europe. Waste land and old fields; common. - Salisbury, H. 10502 (1955); Merrimac, H. 10589 (1955); Haverhill, H. 38 (1931); Andover, Sutton (1874); North Andover, Root (1945); Groveland, Bean & H. 8470 (1953); Newburyport, H. 10385 (1955); Newbury, H. 10389 (1955); Newbury, Plum Island, H., McGregor & Schweinfurth 12744 (1957); Georgetown, M.E. Perley (1884); Rowley, H. & Wetherall 9561 (1954); Ipswich, Plum Island, H., McGregor & Schweinfurth 12679 (1957); Topsfield, H.R. Towne; Wenham, Robinson (1875); Essex, H. 10246 (1955); Rockport, Cushman 3343 (1908); West Manchester, Hubbard (1913); Beverly, Lowell (1847); Salem, Cole (1886); Swampscott, H. 10662 (1955); Nahant, Bean & H., 9834 (1955).

R. obtusifolius L. Bitter-Dock. Naturalized from Europe. Waste places; frequent. - Merrimac, H. 10568 (1955); Haverhill, H. 39 (1931); North Andover, H. 9912 (1955); Boxford, M.E. Perley (1885); Groveland, Cole (1930); West Newbury, H. 11107 (1955); Topsfield, H.R. Towne; Danvers, H. 7977 (1952); Wenham, Robinson (1875); Hamilton, H. 10837 (1955); Gloucester, Williams (1897); West Manchester, Hubbard (1913); Lynnfield, Young (1879); Saugus, H. 18635 (1958).

R. acetosella L. Sheep-Sorrel. Naturalized from Europe. Sour soils; very common. - Salisbury, H. 10495 (1955); Amesbury, H. 10556 (1955); Haverhill, H. 409 (1932); Lawrence, H. 9884 (1955); Andover, Pease 3525 (1904); North Andover, Root (1943); Boxford, M.E. Perley (1882); Groveland, H. 10601 (1955); Newbury, H. 10402 (1955); Newbury, Plum Island, H. 12016 (1956); Georgetown, H. 10706 (1955); Rowley, H. 10428 (1955); Rowley, Plum Island, H. 12603 (1956); Ipswich, Plum Island, H. 12127 (1956); Topsfield, H.R. Towne; Wenham, Robinson (1875); Hamilton, H. 10812 (1955); Gloucester, H. 10159 (1955); Rockport, Smith 2637 (1943); Manchester, Hubbard (1913); Beverly, Lowell (1847); Salem, H. 9708 (1955); Salem, Little Misery Island, H. 13302 (1957); Saugus, Bean & H. 9784 (1955); Marblehead, H. 9696 (1955).

R. hastatulus Baldw. Sandy soil; rare. - Salisbury, Eaton (1902).

R. acetosa L. Garden-Sorrel. Introduced and naturalized from Europe. Meadows and roadsides; occasional. - Boxford, H. 9925 (1955), cultivated ?; West Newbury, A. W. & C. A. Cheever & Sanford (1933); Newbury, H. 10386 (1955); Georgetown, Sears (1905).

POLYGONUM L. Knotweed. Smartweed.

P. exsertum Small. Saltmarsh; rare. - Newbury, H. 8219 (1952).

P. prolificum (Small) Robins. Saltmarsh; occasional. - Newbury, Plum Island, Pease 3206 (1903); Rowley, Plum Island, H. 12558 (1956); Ipswich, Horner; Essex, H. 9381 (1953); Beverly, Robinson (1875); Salem, Cole (1928); Lynn, Robinson (1875).

P. ramosissimum Michx. Roadsides; rare. - Haverhill, H. 702 (1932); Marblehead Neck, H. 7829 (1933).

P. ramosissimum, forma *atlanticum* Robins. - Salisbury, Bean (1945).

P. erectum L. Waste places; rare. - Ipswich, Robinson (1876); Danvers, Sears (1902); Essex County, Russell.

P. aviculare L. Knotweed. Naturalized from Europe. Waste places; common. - Amesbury, H. 9153 (1953); Merrimac, H. 9238 (1953); Haverhill, H. 701 (1932); Lawrence, H. 11798 (1955); Andover, Pease 481 (1900); North Andover, Morss (1885); Boxford, M.E. Perley (1885); West Newbury, H. 11110 (1955); Georgetown, H. 13576 (1957); Topsfield, H. 11366 (1955); Danvers, H. 11619 (1955); Gloucester, Williams (1897); Beverly, H. 11669 (1955); Salem, Cole (1886); Salem, Little Misery Island, H. 13309 (1957); Peabody, Mackintosh (1890); Lynnfield, H. 14131 (1957); Marblehead, Robinson (1876).

P. aviculare, var. *vegetum* Ledeb. - Boxford, H. 8189 (1952).

P. aviculare, var. *littorale* (Link) W.D.J. Koch. Native; seashore; occasional. - Ipswich, H. 13878 (1957); Ipswich, Plum Island, H. 12399 (1956); Salem, North Gooseberry Island, H. 13282 (1957).

P. tenue Michx. Dry open soil; occasional. - Danvers, Lowell 70 (1847); Russell (1859); Lynnfield, Chute, H. 14149 (1957).

P. bistorta L. Adventive from Europe. - Boxford, Alcott (1890), H. 7905 (1952).

P. amphibium L., var. *stipulaceum* (Coleman) Fern. Water-smartweed. Shores and ditches; occasional. - Haverhill, H. 28393 (1965); Methuen, H. 18371 (1958); Groveland, H. 5699 (1933); Ipswich 126y; Topsfield, H.R. Towne; Peabody, Robinson (1874).

P. amphibium, var. *stipulaceum,* forma *simile* Fern. - Peabody, Mackintosh (1932); Lynnfield, Kingman (1898).

P. amphibium, var. *stipulaceum,* forma *fluitans* (Eat.) Fern. Aquatic; occasional. - Haverhill, Lunt (1912); North Andover, H. 12936 (1957); Boxford, H. 13198 (1957); Danvers, Oakes; Peabody, Robinson (1874).

P. amphibium, var. *stipulaceum* (Farw.) Fern. Boxford H. 30776 (1966).

P. coccineum Muhl. Shores and margins of ponds; frequent. - Methuen, Boott (1865); Andover, Pease 471 (1902); North Andover, Morss (1885); Boxford, Brown (1885); West Newbury, H. 9136 (1953); Danvers, H. 9045 (1953); Wenham Pond, Oakes; Ipswich River, Robinson (1876); Marblehead, Cole (1930).

P. pensylvanicum L., var. *laevigatum* Fern. Pinkweed. Damp shores, clearings and disturbed soil; common. - Amesbury, H. 9162b (1953); Haverhill, H. 40 (1931); Metheun, H. 11754 (1955); Lawrence, H.11247 (1955); North Andover, Morss (1885); Boxford, H. 11562 (1955); Groveland, H. 11183 (1955); Newburyport, H. 20747 (1959); Newbury, H. 20716 (1959); Rowley, H. 18488 (1958); Rowley, Plum Island, H. 12451 (1956); Ipswich, H. 11603 (1955); Ipswich, Plum Island, H. 12392 (1956); Topsfield, H. 11405 (1955); Danvers, H. 11611 (1955); Wenham, H. 9420 (1953); Gloucester, Magnolia, H. 12338 (1956); Manchester, H. 11352 (1955); Salem, H. 11489 (1955); Peabody, Mackintosh (1933); Lynnfield, Robinson (1874); Saugus, H. 10891 (1955); Lynn, Young (1879).

P. lapathifolium L. Shores, damp clearings and disturbed soil; common. - Amesbury, H. 9162a (1953); Haverhill, H. 11208 (1955); Lawrence, H. 11240 (1955); North Andover, H. 14063 (1957); Groveland, H. 11180 (1955); Georgetown, H. 13580 (1957); Rowley, H. 18511 (1958); Topsfield, H. 11404 (1955); Danvers, H. 18570 (1958); Essex, H. 18854 (1958); Gloucester, Magnolia, H. 12339 (1956); Manchester, H. 11350 (1955); Salem, Sears (1883); Peabody, Mackintosh (1933); Saugus, H. 11863 (1955); Swampscott, H. 10694 (1955); Marblehead, H. 11483 (1955).

P. lapathifolium, var. *salicifolium* Sibth. - Andover, H. 14091 (1957); West Newbury, H. 9132 (1953); Lynnfield, H. 14142 (1957).

P. scabrum Moench. Naturalized from Europe. Waste places; occasional. - Haverhill, H. 78385 (1965); Boxford, H. 8648 (1953); Newburyport, H. 10365 (1955); Rowley, H. 11967 (1955); Salem, H. 10627 (1955).

P. orientale L. Kiss-me-over-the-garden-gate. Introduced from Europe. Garden escape, waste places; frequent. - Haverhill, H. 645 (1932); Methuen, H. 11763 (1955); Boxford, H. 13632 (1957); Newburyport, M.E. Perley (1884); Georgetown, Horner (1882); Rowley, H. 18501 (1958); Ipswich, H. 13860 (1957); Topsfield, H.R. Towne; Wenham, Robinson (1877); Peabody, H. 18591 (1958); Saugus, H. 18663 (1958); Lynn, Young (1879).

P. careyi Olney. Low thickets and swamps; occasional. - North Andover, H. 13061 (1957); Groveland, H. 11075 (1955); Danvers, Sears (1880); Manchester, Robinson & Sears (1876).

P. hydropiper L. Water-Pepper. Damp soils; common. - Amesbury, H. 9156 (1953); North Andover, Root (1945); Boxford, H.E. Parkhurst (1882); Groveland, H. 11185 (1955); West Newbury, H. 9133 (1953); Ipswich, H. 14035 (1957); Topsfield, H.R. Towne; Danvers, H. 11631 (1955); Chebacco, Robinson (1874); Rockport, Smith 2720 (1933); Manchester, H. 11327 (1955); Salem, Cole (1888); Peabody, Mackintosh (1933); Saugus, Robinson (1874); Marblehead, H. 12597 (1956).

P. dubium Slein. Naturalized from Europe. Waste places; rare. - Boxford, H. 8651 (1953).

P. cespitosum Blume, var. *longisetum* (DeBruyn) Stewart. Naturalized from Europe. Waste places; rare. - Topsfield, H. 11372 (1955).

P. persicaria L. Lady's Thumb. Naturalized from Europe. Roadsides and waste places; common. - Haverhill, H. 646 (1932); Methuen, H. 11759 (1955); Lawrence, H. 11783 (1955); Andover, H. 12297 (1956); North Andover, Morss (1885); Boxford, M.E. Perley (1885); West Newbury, H. 11103 (1955); Newburyport, H. 20742 (1959); Newbury, H. 18460 (1958); Rowley, H. 18498 (1958); Topsfield, H.R. Towne; Danvers, H. 11632 (1955); Wenham, Robinson (1874); Hamilton, H. 9403 (1953); Essex, H. 10247 (1955); Gloucester, H. 13846 (1957); Rockport, H. 10271

(1955); Beverly, Lowell 134 (1846); Salem, H. 10630 (1955); Peabody, H. 8752 (1953); Lynnfield, H. 14141 (1957); Saugus, H. 10890 (1955); Lynn, H. 11819 (1955); Marblehead, H. 11456 (1955).

P. persicaria, var. *ruderale* (Salisb.) Neisn. - Newbury, H. 12354 (1956); Essex, H. 11301 (1955); Gloucester, Magnolia, H. 12342 (1956).

P. punctatum Ell. Water-Smartweed. Wet places; occasional. - Salisbury, White 247 (1913); Merrimac, H. 9268 (1953); Haverhill, H. 41 (1931); Andover, Pease 2644 (1903); North Andover, Root (1945); Groveland, H. 9070 (1953); West Newbury, H. 9127 (1953); Chebacco, Robinson (1874); Beverly, H. 11694 (1955).

P. punctatum, var. *leptostachyum* (Meisn.) Small - Haverhill, H. 28391 (1965); Lawrence, H. 11252 (1955); Boxford, M.E. Perley (1886); Salem, Lowell (1847).

P. punctatum, var. *parvum* Vict. & Rousseau - Amesbury, H. 9174 (1953); West Newbury, H. 11124 (1955); Manchester, H. 10741 (1955).

P. robustius (Small) Fern. Margins of ponds; occasional. - North Andover, Morss (1885); Boxford, H. 8209 (1952); West Newbury, H. 9128 (1953); Rowley, H. 13494 (1957); Topsfield, H. 8837 (1953).

P. hydropiperoides Michx. Shallow water or wet shores; common. - Andover, Pease 486 (1901); North Andover, H. 13075 (1957); Boxford, H. 8726 (1953); Groveland, H. 10981 (1955); Ipswich, H. 13461 (1957); Topsfield, H. 13612 (1957); Hamilton, H. 10815 (1955); Essex, Robinson (1876); Saugus, H. 10889 (1955); Swampscott, H. 10693 (1955).

P. hydropiperoides, forma *strigosum* (Small) Stanford - Lynn. H. 13173 (1957).

P. sagittatum L. Arrow-leaved Tearthumb. Low ground; common. - Amesbury, H. 9160 (1953); Merrimac, H. 18432 (1958); Haverhill, H. 549 (1932); Methuen, H. 8927 (1953); North Andover, Morss (1885); Boxford, M.E. Perley (1883); Groveland, H. 8797 (1953); West Newbury, H. 11102 (1955); Rowley, H. 12255 (1956); Ipswich, H. 14039 (1957); Topsfield, H.R. Towne; Middleton, H. 18097 (1958); Danvers, H. 11610 (1955); Hamilton, H. 10783 (1955); Manchester, Hubbard (1913); Peabody, Mackintosh (1933); Saugus, Robinson (1874).

P. sagittatum, forma *chloranthum* Fern. - Amesbury, Fassett 2333 (1924).

P. arifolium L., var. *pubescens* (Keller) Fern. - Halberd-leaved Tearthumb. Wet places; frequent. - Amesbury, Bean (1949); Merrimac, H. 18343 (1958); Haverhill, H. 548 (1932); Andover, Morss (1885); West Boxford, Root (1945); Ipswich, R.J. Eaton (1931); Topsfield, Mrs. Conley; Hamilton, Robinson (1875); Beverly, Lowell (1846); Peabody, Mackintosh (1933).

P. cilinode Michx. Dry thickets; occasional. - Danvers, Oakes; Gloucester, Sears (1890); Rockport, Robinson (1874); Manchester, Oakes; Beverly, Robinson (1877); Peabody, Fuller (1897).

P. gonvolvulus L. Naturalized from Europe. Thickets and clearings; common. - Haverhill, H. 647 (1932); Methuen, H. 11741 (1955); Andover, Pease 485 (1901); North Andover, Root (1945); Boxford, M.E. Perley (1885); Newburyport, H. 10357 (1956); Topsfield, H.R. Towne; Wenham, Robinson (1875); West Gloucester, Robinson (1877); North Beverly, Robinson (1875); Salem, H. 11492 (1955); Salem, North Gooseberry Island, H. 13280 (1957); Saugus, H. 10868 (1955); Lynn, H. 11831 (1955).

P. cristatum Engelm. & Gray. Thickets and clearings; occasional. - Haverhill, H. 649 (1932); North Andover, Morss (1885); Rowley, W.S. Horner; Peabody, Mackintosh (1933).

P. scandens L. Climbing False Buckwheat. Damp thickets; common. - Salisbury, Morss (1925); Haverhill, H. 648 (1932); Lawrence, E. 11794 (1955); North Andover, Bean (1916); Groveland, H. 11076 (1955); West Newbury, H. 11146 (1955); Newbury, Plum Island, H. 124 (1930); Ipswich, Plum Island, H. 12441 (1956); West Gloucester, Robinson (1877); Beverly, Lowell 198 (1847); Peabody, H. 11552 (1955); Lynnfield, Rich (1886); Marblehead, H. 11427 (1955).

P. auberti (L.) Henry. Saugus, T.W. Wells (1967).

P. cuspidatum Sieb. & Zucc. Japanese Knotweed. Introduced from east Asia. Waste places; common. - Salisbury, H. 10477 (1955); Haverhill, H. (1932); Methuen, H. 11720 (1955); Lawrence, H. 11790 (1955); North Andover, Root (1951); Boxford, H. 8315 (1953); Newburyport, H. 10350 (1955); Topsfield, H. 11403 (1955); Danvers; Kennedy (1907); Salem, Cole (1903); Lynn, H. 11815 (1955); Swampscott, H. 10685 (1955); Marblehead, H. 9692 (1955).

P. sachalinense F. Schmidt. Giant Knotweed. Introduced from east Asia. Roadsides and waste places; occasional. - Haverhill, Bradford, H. 7831 (1952); Boxford, M.E. Perley (1915); Wenham, H. 7832 (1952); Manchester, W.N. Bullard (1908); Marblehead, Bean (1955).

FAGOPYRUM Mill. Buckwheat.

F. sagittatum Gilib. Introduced from Asia. Waste places and old fields; occasional. - Merrimac, H. 9236 (1953); North Andover, Root (1945); Boxford, H. 8775 (1953); Newburyport, Cole (1899); Newbury, Plum Island, H. 12525 (1956); Topsfield, Sears (1875); Middleton Pond, Lowell (1845); Beverly, Lowell (1846); Marblehead, H. 11463 (1955).

POLYGONELLA Michx. Jointweed.

P. articulata (L.) Meisn. Dry sand; common. - Haverhill, Bradford, Fernald 11341 (1914); Andover, Pease (1906); North Andover, Carpenter (1952); Groveland, Paine (1885); Newburyport, B.L. Robinson (1903); Newburyport, Plum Island, Cole (1901); Newbury, Plum Island, H. 125 (1930); Georgetown, Horner (1881); Rowley, Plum Island, H. 12477 (1956); Ipswich, Oakes; Ipswich, Plum Island, H. 12435 (1956); Topsfield, Mrs. Conley; Essex, Bean (1938); Gloucester, Ripley (1914); Peabody, Mackintosh (1933).

CHENOPODIACEAE (Goosefoot Family)

KOCHIA Roth

K. scoparia (L.) Roth. Summer-Cypress. Introduced from Eurasia. Escaped from cultivation; rare. - Lawrence, H. 18755 (1958); Beverly, F.H. Peabody (1902); Marblehead, Deane (1911).

BASSIA All.

B. hirsuta (L.) Aschers. Naturalized from Eurasia. Saline soil; occasional. - Ipswich, H. 11592 (1955); Ipswich, Plum Island, H. 12385 (1956); Essex, H. 9383 (1953); Saugus, H. 11855 (1955).

CHENOPODIUM L. Goosefoot. Pigweed.

C. murale L. Naturalized from Europe. Waste places; rare. - Newburyport, Knowlton (1912).

C. urbicum L. Adventive from Europe. Waste places; occasional. - Andover, Pease 3102 (1903); Beverly, Lowell (1847); Lynn, Young (1879).

C. hybridum L., var. *gigantospermum* (Aellen) Rouleau. Maple-leaved Goosefoot. Waste places; frequent. - Lawrence, H. 18772 (1958); Andover, Pease 501 (1901); North Andover, Root (1945); Boxford, H. 30744 (1966); Newburyport, Knowlton (1912); Georgetown, Horner; Rowley, H. 18509 (1958); Ipswich, H. 13865 (1957); Topsfield, Robinson (1877); Lynn, Robinson (1875).

C. leptophyllum Nutt. Sandy coast, adventive elsewhere; occasional. - Haverhill, H. 7286 (1932); Rowley, Plum Island, H. 12478 (1956); Plum Island, Williams (1899); Peabody, H. 9298 (1953).

C. boscianum Moq. Roadsides and waste places; rare. - Groveland, F.C. Seymour 4862 (1936).

C. lanceolatum Muhl. Naturalized from Europe. Waste or cultivated ground: rare. - Boxford, M.E. Perley (1883).

C. album L. Pigweed. Lamb's-quarters. Naturalized from Europe. Weed of cultivated and waste ground; common. - Salisbury, White 285 (1913); Haverhill, H. 7286 (1932); Methuen, H. 11722 (1955); Lawrence, H. 11771 (1955); Boxford, H. 8901 (1953); Groveland, H. 11192 (1955); Newbury, Plum Island, H. 12528 (1956); Rowley, Plum Island, H. 12479 (1956); Ipswich, H. 11597 (1955); Topsfield, H.R. Towne; Danvers, H. 11628 (1955); Wenham, Robinson (1875); Hamilton, H. 10785 (1955); Essex, H. 11309 (1955); Gloucester, Kennedy (1907); Manchester, H. 11335 (1955); Beverly, H. 11693 (1955); Salem, Pickering; Salem, Little Misery Island, H. 8755 (1953); Saugus, H. 10898 (1955); Lynn, Robinson (1875); Marblehead, H. 11433 (1955).

C. paganum. Reichenb. Boxford H. 30746 (1966).

C. botrys L. Jerusalem-oak. Introduced from Eurasia. Waste places; occasional. - Georgetown, H.E. Perley (1886), cultivated; Topsfield, H.R. Towne; Salem, Robinson (1875); Mackintosh (1924); Marblehead, H. 18930 (1958).

C. ambrosioides L. Mexican-tea. Naturalized from tropical America. Waste places and cultivated ground; common. - Haverhill, H. 727 (1932); Methuen, H. 11751 (1955); Lawrence, H. 11793 (1955); Groveland, H. 11194 (1955); Newburyport, H. 20740 (1959); Topsfield, H. 11209 (1955); Gloucester, Magnolia, H. 11995 (1955); Beverly, H. 11686 (1955); Salem, Sears (1889); Peabody, H. 11526 (1955); Saugus, H. 18667 (1958); Lynn, Robinson (1877); Marblehead, H. 11428 (1955).

C. ambrosioides, var. *anthelminticum* (L.) Gray. Wormseed. Waste places; occasional. - Haverhill, Bradford, H. 11206 (1955); Andover, Pease 2853 (1903); Rowley, H. 11941 (1955); Gloucester, Magnolia, 12328 (1956); Saugus, H. 11862 (1955); Lynn, H. 11823 (1955).

C. pumilio R.Br. (*C. carinatum* R.Br.) Lawrence, H. 18743 (1958) fide H.A. Wahl.

C. glaucum L. Oak-leaved Goosefoot. Naturalized from Old World. Waste places; rare. - Salem, Robinson (1879).

C. rubrum L. Coast-Blite. Salt-marshes; rare. - Newburyport, Bean (1940); Salem, Robinson (1877); Essex County, Richardson.

C. bonus-henricus L. Good King Henry. Naturalized from Europe. - Salem, Little Misery Island, H. 13305 (1957).

BETA L. Beet.

B. vulgaris L. Introduced from Europe. Escape from cultivation. Peabody, H. 27072 (1964).

B. vulgaris var. *cicula* L. Swiss Chard. Introduced from Europe. Escape from cultivation. - Saugus, H. 18608 (1958).

ATRIPLEX L. Orach.

A. patula L. Saline soils; occasional. - Newburyport, Bean (1941); Essex, H. 9369 (1953); Gloucester, Fernald, Hunnewell & Long 9436 (1913); Salem, Cole (1886); Nahant, H. 12588 (1956).

A. patula, var. *hastata* (L.) Gray. Common. - Amesbury, H. 18959 (1958); Newburyport, White 407 (1913); Newbury, Bean & Stockman (1954); Newbury, Plum Island, H. 12527 (1956); Rowley, H. 8251 (1952); Ipswich, H. 11576 (1955); Ipswich, Plum Island, H. 12403 (1956); Essex, H. 11312 (1955); Beverly, Robinson (1875); Salem, Russell; Peabody, H. 11534 (1955); Saugus, H. 11868 (1955); Marblehead Neck, Merrill (1849); Nahant, Sanford (1944).

A. patula, var. *littoralis* (L.) Gray - Newbury, Plum Island, Pease 3075 (1903); Rockport, Sears (1886).

A. glabriuscula Edmondston. Seacoast; occasional. - Newbury, H. 8218 (1952); Rowley, H. 8269 (1952); Ipswich Beach, Sears (1884); Essex, H. 9387 (1953); Marblehead, H. 11913 (1955); Nahant, Williams (1894).

A. arenaria Nutt. Borders of salt-marshes; occasional. - Newbury, Plum Island, H. 12522 (1956); Lynn, Tracy (1859).

AXYRIS L.

A. amaranthoides L. Russian Pigweed. Adventive from Siberia and Russia. - Haverhill, H. 729 (1932).

SALICORNIA L. Glasswort. Samphire.

S. bigelovii Torr. Dwarf Saltwort. Salt-marshes; frequent. - Rowley, H. 8259 (1952); Ipswich, Plum Island, Furber (1911); Essex, Bean (1938); Saugus, H. 11856 (1955); Lynn, Robinson (1875).

S. europea L. Samphire. Salt-marshes; common. - Newbury, Bean & Stockman (1954); Newbury, Plum Island, Pease 3097 (1903); Rowley, H. 8258 (1952); Ipswich, H. 11587 (1955); Ipswich, Plum Island, Furber (1911); Essex, Bean (1938); Gloucester, Cushman 3490 (1908); Salem, Cole (1908); Peabody, Mackintosh (1933); Saugus, H. 11857 (1955); Lynn, Robinson (1875); Marblehead, H. 11912 (1955); Nahant, Beal (1887).

S. virginica L. Perennial Saltwort. Borders of salt-marshes; occasional. - Ipswich, H. 13890 (1957); Gloucester, Annisquam, Sears (1892); Lynn, Robinson (1875); Essex County, Oakes.

SUEDA Forsk. Sea-Blite.

S. maritima (L.) Dumort. Salt-marshes; frequent. - Newbury, White 424 (1913); Ipswich, Williams (1899); Ipswich, Plum Island, St. John & White 586 (1913); Essex, Knowlton (1908); Gloucester, Prince (1883); Nahant, Rich (1899).

S. richii Fern. Salt-marshes; frequent. - Newbury, H. 11288 (1955); Newbury, Plum Island, H. 12524 (1955); Ipswich, Steyermark 8000 (1931); Ipswich, Plum Island, H. 12384 (1956); Gloucester, Bass Rocks, Cushman 3485 (1908); Marblehead, H. 11906 (1905).

S. americana (Pers.) Fern. Salt-marshes; rare. - Essex, H. 9386 (1953).???

S. linearis (Ell.) Moq. Salt-marshes; common. - Newbury, Plum Island, Pease 2645 (1903); Rowley, H. 8257 (1952); Ipswich, H. 11590 (1955); Essex, H. 9384 (1953); West Gloucester, Swan (1881); Beverly, Lowell (1847); Salem, Russell (1859); Saugus, H. 11864 (1955); Marblehead, H. 11907 (1955); Nahant, Kennedy (1894).

SALSOLA L. Saltwort.

S. kali L. Common-Saltwort. Sandy-seashores; frequent. - Newbury, Plum Island, Pease 2640 (1903); Ipswich, H. 13888 (1957); Beverly, Lowell (1846); Nahant, 1838, Merrill (1850).

S. kali, var. *caroliniana* (Walt.) Nutt. Gloucester, Tiffney 28396 (1965); Manchester, Prince (1895); Lynn, Robinson (1875); Nahant, Williams (1894).

S. kali, var. *tenuifolia* Tausch. Naturalized from Europe. Roadsides and waste places; occasional. - Lawrence, H. 18475 (1958); Gloucester, Salem, Cole (1908); Saugus, H. 18686 (1958); Lynn, H. 18839 (1958); Marblehead, H. 18934 (1958).

AMARANTHACEAE (Amaranth Family)

AMARANTHUS L. Amaranth.

A. palmeri S. Wats. Adventive from southwest. Waste places and wool-waste; rare. - South Lawrence, Pease 510 (1902); Lawrence, H. 11785 (1955); Salem, Hall (1926).

A. cruentus L. Purple Amaranth. Adventive from Asia. Waste places; occasional. - Topsfield, H.R. Towne; Peabody, H. 11533 (1955); Saugus, H. 18677 (1958); Lynn, Young (1880).

A. hybridus L. Pigweed. Cultivated fields and waste places; frequent. - Haverhill, Bradford, H. 730 (1932); Lawrence, H. 11802 (1955); Groveland, H. 11187 (1955); Newburyport, Bean (1940); Topsfield, M.E. Perley (1885); Danvers, Sears (1904); North Salem, Hall (1926); Saugus, H. 18682 (1958).

A. retroflexus L. Green Amaranth. Waste or cultivated ground; common. - Haverhill, Bradford, H. 731 (1932); Methuen, H. 11740 (1955); Andover, Pease 509 (1900); North Andover, Bean (1947); Boxford, Robinson (1877); Groveland, H. 11195 (1955); Newburyport, Bean (1940); Newbury, H. 20706 (1959); Rowley, H. 18487 (1958); Topsfield, H.R. Towne; Danvers, H. 11630 (1955); Essex, H. 18861 (1958); Gloucester, Magnolia, J.M. Hunnewell (1899); Salem, H. 11505 (1955); Salem, North Gooseberry Island, H. 13277 (1957); Salem, Little Misery Island, 13308 (1957); Peabody, H. 11529 (1955); Lynn, H. 11832 (1955).

A. powellii S. Wats. Adventive from southeast. Waste places; rare. - Boxford, Hopkins & Pease 23167 (1932).

A. albus L. Tumbleweed. Disturbed or waste ground; common. - Boxford, M.E. Perley (1885); Newbury, Plum Island, H. 12526 (1956); Rowley, H. 18490 (1958); Ipswich, Oakes; Topsfield, H.R. Towne; Danvers, Kennedy (1905); Peabody, H. 11541 (1955); Saugus, H. 18664 (1958); Lynn, H. 11805 (1955).

A. graecizans L. Disturbed or waste ground; occasional. - Andover, Pease 505 (1902); Boxford, H. 30748 (1966); Rowley, H. 11940 (1955); Ipswich, H. 13873 (1957).

A. spinosus L. Thorny Amaranth. Adventive from tropics. Waste places; rare. - Lynn, Young (1880).

ACNIDA L. Water-Hemp.

A. cannabina L. Salt-marshes; frequent; - Salisbury, H. 10507 (1955); Amesbury, H. 18965 (1958); Newburyport, Moulton (1886); Newbury, Sears (1890); Rowley, Sears (1887).

NYCTAGINACEAE (Four-o'clock Family).

MIRABILIS L. Four-o'clock.

M. nyctaginea (Michx.) MacM. Adventive from West. Waste places; occasional. - Lawrence, Lewis 19051 (1958); Newburyport, H. 20745 (1959); Saugus, H. 18709 (1958).

M. hirsuta (Pursh) MacM. Adventive from West. Waste places; rare. - Andover, Ballardvale, E.W. Thompson (1924).

PHYTOLACCACEAE (Pokeweed Family).

PHYTOLACCA L. Pokeweed.

P. americana L. Recent clearings, waste ground and roadsides; common. - Amesbury, H. 12993 (1957); Merrimac, Bean & H. 7701 (1951); Haverhill, H. 513 (1932); Methuen, H. 11728 (1955); Andover, Pease 512 (1900); North Andover, H. 17959 (1958); Boxford, M.E. Perley (1881); Groveland, H. 10982 (1955); Newburyport, H. & McGregor 12310 (1956); Newbury, H. 18458 (1958); Georgetown, Horner (1881); Rowley, H. 18483 (1958); Rowley, Plum Island, H. 12456 (1956); Topsfield, M. Clark; Hamilton, H. 10826 (1955); Gloucester, Magnolia H. 18867 (1958); Rockport, H. 19017 (1958); Manchester, Hubbard (1913); Beverly, H. 11677 (1955); Salem, Sears (1884); Peabody, Mackintosh (1933); Lynnfield, Kennedy (1862); Saugus, H. 18668 (1958).

AIZOACEAE (Carpetweed Family).

MOLLUGO L. Indian Chickweed.

M. verticillata L. Carpetweed. Immigrant from Tropical America. Open ground; common. - Salisbury, White 297 (1913); Haverhill, H. 703 (1932); Andover, Pease 513 (1902); Boxford, Paine (1885); Newburyport, Plum Island, H. 13048 (1957); Newbury, H. 10404 (1955); Newbury, Plum Island, H. 12536 (1956); Georgetown, Horner; Rowley, H. 8125 (1952); Ipswich, Oakes; Ipswich, Plum Island, H. 12439 (1956); Topsfield, H.R. Towne; Gloucester, Annisquam, Ripley 15897

(1915); Manchester, H. 10712 (1955); Beverly, Lowell (1846); Salem, Robinson (1874); Peabody, Mackintosh (1933); Saugus, H. 10892 (1955); Swampscott, H. 10652 (1955).

TETRAGONIA L.

T. tetragonioides (Pall.) Ktze. New Zealand Spinach. Introduced from east Asia. Garden escape; rare. - Beverly, H. 11682 (1955).

PORTULACACEAE (Purslane Family).

PORTULACA L. Purslane.

P. oleracea L. "Pusley". Naturalized from Europe. Cultivated and waste ground; much more common than collections indicate. - Haverhill, Ward Hill, H. 377 (1932); Methuen, H. 11737 (1955); Lawrence, H. 11801 (1955); North Andover, H. 13072 (1957); Boxford, M.E. Perley (1886); Newbury, H. 20711 (1959); Rowley, H. 18492 (1958); Topsfield, M. Todd: Gloucester, Magnolia, H. 12377 (1956); Salem, Robinson (1879); Salem, North Gooseberry Island, H. 13279 (1957); Salem, Little Misery Island, 13291 (1957); Peabody, H. 11560 (1955); Saugus, H. 10893 (1955); Lynn, H. 11812 (1955).

P. grandiflora Hook. Portulaca. Introduced from South America. Garden escape; occasional. - Methuen, H. 11739 (1955); Lawrence, H. 18752 (1958); Saugus, H. 18610 (1958).

MONTIA L.

M. sibirica (L.) Howell. Introduced from Far West. Rare waif. - Manchester, Farlow (1916). See *Rhodora* 18:240 (1916).

CARYOPHYLLACEAE (Pink Family).

SCLERANTHUS L. Knawel.

S. annuus L. Naturalized from Europe. Waste places; common. - Haverhill, H. 191 (1931); Methuen, H. 9618 (1954); Andover, Blake (1879); Boxford, H. 7911 (1952); Groveland, Paine (1885); Rowley, H. 11963 (1955); Ipswich, H. 10244 (1955); Topsfield, H. 8059 (1952); Danvers, H. 7979 (1952); Beverly, Lowell (1847); Salem, Forbes 1524 (1903); Salem, Bakers Island, Lowell (1848); Lynnfield, H. 14119 (1957); Marblehead, Hall (1925); Nahant, Robinson (1875).

PARONYCHIA Mill. Whitlow-wort.

P. argyrocoma (Michx.) Nutt., var. *albimontana* Fern. Silverling. Bare granitic slopes; rare. - Newburyport, Carrs Island, Osgood, Castelhans (1884), McGregor (1944), Bean (1945). A remarkable station far removed from the mountains of New Hampshire and Maine where it is more common. Discovered by Dr. Osgood in 1884 and rediscovered by Frank J. McGregor in 1944.

P. canadensis (L.) Wood. Forked Chickweed. Thin dry woods and sandy openings; scarce. - Methuen, Boott (1845); Andover, Blake (1882); Boxford, Sears (1875); Danvers, Lowell (1848); Lynn, Rodman (1886); Essex County, Oakes.

P. fastigiata (Raf.) Fern. Forked Chickweed. Dry woods and sandy openings; rare. - Newbury, Palmer 46293 (1942); Danversport, Sears (1885).

SPERGULARIA J. & C. Presl Sand-Spurrey.

S. rubra (L.) J. & C. Presl. Dry sandy or sterile soils; common. - Haverhill, Ward Hill, H. 376 (1932); Andover, Pease 536 (1903); North Andover, Morss (1885); Boxford, M.P. Perley (1882); Ipswich, H. 14016 (1957); Ipswich, Plum Island, H. 12111 (1956); Topsfield, B.B. Towne; Danvers, H. 7962 (1952); Gloucester, H. 10145 (1955); Rockport, Rich (1897); Manchester, Hubbard (1913); Beverly, Montserrat, Hubbard (1913); Salem, Robinson (1879); Marblehead, Cole (1927).

S. marina (L.) Griseb. Saline or brackish soils; common. - Newbury, Plum Island, H. 884 (1933); Rowley, H. 8260 (1952); Ipswich, Morss (1921); Ipswich, Plum Island, H., McGregor & Schweinfurth 12705 (1957); Danvers, Robinson (1875); Gloucester, Swan (1881); Rockport, Robinson (1874); Beverly, Robinson (1875); Marblehead Neck, H. 885 (1933); Nahant, Blankinship (1895).

S. marina, var. *leiosperma* (Kindb.) Gürke - Salisbury, White 11 (1913); Newburyport, Bean (1940); Newbury, Williams (1898); Newbury, Plum Island, Pease 3076 (1902); Rowley, H. 11925 (1955); Ipswich Beach, Hubbard (1913); Ipswich, Plum Island, H. 12400 (1956); Essex, Bean (1938); Beverly, Lowell (1847); Salem, Cole (1910); Salem, North Gooseberry Island, H. 13281 (1957); Salem, Little Misery Island, H. 13310 (1957); Marblehead, Kennedy (1887); Nahant, Blankinship (1895).

S. canadensis (Pers.) Don. Saline and brackish soils; apparently rare. - Cape Ann, Brinton (1884); Marblehead, Kennedy (1887).

SPERGULA L. Spurrey.

S. arvensis L. Naturalized from Europe. Weed of cultivated ground; common. - Amesbury, H. 10517 (1955); Andover, Pease 4239 (1904); North Andover, Morss (1885); Boxford, M.E. Perley (1883); Ipswich, Oakes; Topsfield, H.R. Towne; Rockport, Forbes 1513 (1903); Manchester, H. 10768 (1955); Beverly, Lowell (1846); Swampscott, Sears (1885).

SAGINA L. Pearlwort.

S. procumbens L. Birdseye. Damp open soil and shores; common. - Haverhill, Ward Hill, H. 671 (1932); Andover, Pease 1993 (1903); Boxford, M.E. Perley (1884); Ipswich, Plum Island, H., McGregor & Schweinfurth 12668 (1957); Gloucester, Eastern Point, Rand & B.L. Robinson (1896); Rockport, Robinson (1874); Salem, Sears (1899); Salem, Bakers Island, Robinson (1890); Salem, Little Misery Island, H. 13311 (1957); Saugus, Manning (1881); Swampscott, Deane (1913); Nahant, Farlow (1878).

S. nodosa (L.) Fenzl, var. *pubescens* Mert. & Koch. Rocky soil near coast; rare. - Rockport, Pigeon Cove, Robinson (1875); Sears (1885), Williams (1897), Forbes (1903).

ARENARIA L. Sandwort.

A. lateriflora L. Grove-Sandwort. Shores, meadows and borders of woods; common. - Salisbury, White 35 (1913); Haverhill, H. 410 (1932); Lawrence, H. 9855 (1955); West Andover, Robinson (1872); North Andover, Morss (1884); Boxford, Frye (1881); Groveland, H. 10598 (1955); Newburyport, Forbes 1510 (1903); Newbury, Schweinfurth (1950); Newbury, Plum Island, White

137 (1913); Georgetown, M.E. Perley (1883); Rowley, Plum Island, H., McGregor & Schweinfurth 12726 (1957); Ipswich, Plum Island, Mackintosh (1925); Topsfield, Paine (1884); Middleton, H. 8328 (1953); Wenham, Bean & H. 10020 (1955); Gloucester, Eastern Point, Rand & B.L. Robinson (1896); Rockport, Bishop & Smith 950 (1928); Beverly, Lowell (1846); Salem, Cole (1889); Salem, Bakers Island, Robinson (1890); West Peabody, E.L. Mackintosh (1924); Marblehead, Cole (1886).

A. serpyllifolia L. Thyme-leaved Sandwort. Naturalized from Europe. Sterile fields and roadsides; occasional. - Newburyport, H. & Wetherall 9566 (1954); Danvers, Lowell (1848); Salem, Robinson (1877); Marblehead, Sears (1885).

A. peploides L., var. *robusta* Fern. Seabeach-Sandwort. Maritime sands; probably more abundant than collections indicate. - Newbury, Plum Island, Pease 3099 (1903); Ipswich, Oakes; Ipswich, Plum Island, Mackintosh (1915); Gloucester, Hewitt (1901); Rockport, Lufkin (1878).

STELLARIA L. Chickweed.

S. media (L.) Cyrillo. Common Chickweed. Naturalized from Europe. Everywhere; blossoms throughout the year. - Haverhill, H. 301 (1932); North Andover, Morss (1885); Boxford, M.P. Perley (1882); West Newbury, White 508 (1913); Newbury, H. 10416 (1955); Goergetown, Horner; Ipswich, Oakes; Topsfield, H.R. Towne; Gloucester, Magnolia, H. 11991 (1955); Rockport, Forbes 1465 (1902); Salem, Robinson (1875); Peabody, Mackintosh (1913); Marblehead, Bean & H. 9735 (1955).

S. graminea L. Common Stitchwort. Naturalized from Europe. Grasslands; common. - East Salisbury, Moulton (1888); Amesbury, H. 10529 (1955); Haverhill, H. 672 (1932); Andover, Pease 538 (1901); North Andover, Root (1943); Groveland, H. 10607 (1955); Newbury, Schweinfurth (1950); Ipswich, Oakes; Topsfield, Mary Williams; Middleton, H. 17692 (1958); Wenham, Cole (1910); Essex, Hubbard (1913); Gloucester, Lanesville, Collins (1915); Rockport, Pigeon Cove, Lufkin (1886); Beverly Farms, Hubbard (1913); Salem, Mackintosh (1915); Peabody, Mackintosh (1933); Lynnfield, Bean (1909); Saugus, Bean & H. 9817 (1955).

S. longifolia Muhl. Damp thickets and meadows; frequent. - Methuen, H. 7987 (1952); Andover, Jenks (1880); Newburyport, Moulton (1886); Ipswich, Sears (1888); Ipswich, Plum Island, H. 12082 (1956); Danversport, Sears (1904); East Gloucester, Jenks (1881); Salem, Robinson (1879).

S. alsine Grimm. Springheads and wet places; occasional. - Haverhill, H. 132 (1932); Andover, Blake (1883), Pease 2086 (1903); Newburyport, Moulton (1886).

S. calycantha (Ledeb.) Bong., var. *isophylla* Fern. Wet places; occasional. - Amesbury, H. 18209 (1958); Andover, H. 17762 (1958); Wenham, Leach's Swamp, Sears (1883).

CERASTIUM L. Mouse-ear Chickweed.

C. vulgatum L. Common Mouse-ear. Naturalized from Eurasia, Fields and cultivated ground; common. - Merrimac, Bean & H. 8574 (1953); Haverhill, Ward Hill, H. 355 (1932); Andover, Pease 520 (1901); North Andover, Root (1945); Boxford, M.P. Perley (1882); Newburyport, White 463 (1913); Ipswich, Plum Island, H. 12079 (1956); Topsfield, H.R. Towne; Essex, H. 10236 (1955); Chebacco, Robinson (1875); Rockport, Forbes 1486 (1902); West Manchester, Hubbard (1913); Beverly, H. 10195 (1955); Salem, Robinson (1875); Lynnfield, Kennedy (1911); Marblehead, Hall (1927).

C. arvense L. Field-Chickweed. Gravelly soils and grasslands; local. - Salisbury, White 75 (1913); Essex, Buxton (1925); Lynn, Forbes (1912); Swampscott, M.K.R.; Nahant, Oakes; Robinson (1877); Morss (1919).

HOLOSTEUM L. Jagged Chickweed.

H. umbellatum L. Naturalized from Europe. Cultivated ground; rare. - Haverhill, Ward Hill, H. 356 (1932).

AGROSTEMMA L. Corn-Cockle.

A. githago L. Purple Cockle. Naturalized from Europe. Fields and roadsides; occasional. - Boxford, Frye (1881); Georgetown, Horner; Ipswich, Topsfield, Mrs. S. Conley; Salem, Robinson (1878); Peabody, Mackintosh (1926); Lynn, Sprague; Marblehead, Cole (1900).

LYCHNIS L. Campion.

L. coronaria (L.) Desr. Mullein-Pink. Introduced from southern Europe. Garden escape; occasional. - Methuen, H. 12257 (1956); Boxford, H. 12890 (1957); Georgetown, H. 17673 (1958); Topsfield, H. 18307 (1958); Danvers, Sears (1885); Essex County, Robinson (1879).

L. dioica L. Red Campion. Naturalized from Eurasia. Waste places; frequent. - Methuen, H. 7990 (1952); North Andover, H. 18018 (1958); Groveland, Bean & H. 9556 (1954); West Newbury, H. 11122 (1955); Newburyport, White 152 (1913); Newbury, H. 20723 (1959); Georgetown, H. 13590 (1957); Rowley, H. & Wetherall 9556 (1954); Danvers, H. 11660 (1955); Wenham, H. 9417 (1953); Gloucester, Magnolia, H. 12350 (1956); Rockport, L.B. & R.C. Smith (1934); Swampscott, Fernald, Hunnewell & Long 9493 (1913); Marblehead, H. 11414 (1955).

L. alba Mill. White Campion. Naturalized from Eurasia. Roadsides and waste places; common. - Haverhill, H. 573 (1932); Methuen, H. 7991 (1952); North Andover, H. 18043 (1958); West Newbury, H. 11151 (1955); Newburyport, Bean (1940); Topsfield, M.L. Pierce; Danvers, H. 7961 (1952); Wenham, Bean & H. 10015 (1955); Gloucester, Magnolia, H. 12351 (1956); Rockport, Smith 4328 (1945); Manchester, H. 11331 (1955); Salem, Sears (1883); Peabody, Mackintosh (1933); Lynnfield, H. 14105 (1957); Saugus, H. 10884 (1955); Lynn, H. 11807 (1955); Swampscott, Kidder (1886); Marblehead, Lindsay (1884).

L. chalcedonica L. Scarlet Lychnis. Introduced from Asia. Garden escape; occasional. - Methuen, H. 12267 (1956); Boxford, Kennedy (1906), H. 12195 (1956).

L. flos-cuculi L. Naturalized from Europe. Fields and meadows; occasional except for the Newbury area where it is common. - North Andover, McCoy & Schiff (1951); Boxford, H. 9484 (1954); Groveland, H. 10595 (1955); Newburyport, Cole (1901); Newbury, Pease 28690 (1941); Topsfield, E. Morissey; Danvers, Cole (1954); Gloucester, Cole (1935).

L. viscaria L. German Catchfly. Introduced from Eurasia. Garden escape; rare. - Topsfield, H. 9513 (1954).

SILENE L. Catchfly.

S. cucubalus Wibel. Bladder-Campion. Naturalized from Eurasia. Roadsides and borders of fields; common. - Salisbury, White 378 (1913); Methuen, H. 7985 (1952); North Andover, Morss (1885); Boxford, H. 9671 (1955); Rowley, H. 7939 (1955); Topsfield, H.R. Towne; Wenham, H. 13742

(1957); Essex, H. 10251 (1955); Rockport, A & W (1969); Salem, M.E. Perley (1885); Saugus, Ripley 15527 (1915); Lynn, Oakes; Swampscott, H. 10665 (1955); Marblehead, Chute.

S. cserei Baumg. Naturalized from Europe. Waste places; rare. - Saugus, H. 11852 (1955).

S. antirrhina L. Sleepy Catchfly. Open woods and fields; common. - Salisbury, Eaton; Haverhill, H. 8460 (1953); Methuen, H. 12263 (1956); North Andover, Root (1945); Groveland, Bean & H. 8528 (1953); Newbury, Plum Island, White 97 (1913); Rowley, Plum Island, H. 12076 (1956); Gloucester, Magnolia, Robinson (1877); Salem, Phippen (1875); Salem, Jeggle Island, Cole (1888); Lynnfield, Bean (1934); Marblehead, Bean & H. 9759 (1955); Nahant, Murdock 1666 (1904).

S. antirrhina, forma *deaneana* Fern. - Andover, Pease 1592 (1901); Boxford, H. 10216 (1955).

S. armeria L. Sweet-William-Catchfly. Introduced from Eurasia. Garden escape; frequent. - Lawrence, Pease 527 (1902); Andover, Pease 3100 (1903); Boxford, H. 10614 (1955); Newburyport, Bean (1945); Georgetown, H. 12799 (1957); Topsfield, H.R. Towne; Danvers, Cole (1954); Gloucester, Magnolia, H. 12356 (1956); Salem, Robinson (1896); Saugus, H. 18636 (1958); Swampscott, Robinson (1877); Marblehead, Bean & H. 9737 (1955).

S. dichotoma Ehrh. Forking Catchfly. Naturalized from Europe. Waste places; rare. - Peabody, H. 8749 (1953).

S. noctiflora L. Night-flowering Catchfly. Naturalized from Europe. Garden escape; occasional. - Georgetown, Horner (1881); Ipswich, Oakes; Topsfield, H.R. Towne; Salem, Robinson (1875); Lynnfield, Bean (1932).

SAPONARIA L.

S. officinalis L. Bouncing-Bet. Naturalized from Europe. Roadsides; common. - Haverhill, Ward Hill, H. 550 (1932); Methuen, H. 12262 (1956); Lawrence, H. 11792 (1955); North Andover, Root (1945); Boxford, M.E. Perley (1885); Groveland, M.E. Perley (1885); West Newbury, H. 11126 (1955); Newburyport, White 159 (1913); Newburyport, Plum Island, H. 13047 (1957); Rowley, H. 18521 (1958); Ipswich, Oakes; Topsfield, H. Goodale; Danvers, Cushman 4843 (1908); Essex, H. 11319 (1955); Gloucester, Annisquam, Williams (1897); Rockport, Robinson (1875); Salem, Sears (1884); Peabody, H. 11556 (1955); Saugus, H. 18655 (1958); Lynn, H. 11844 (1955); Marblehead, H. 11441 (1955).

S. vaccaria L. Adventive from Europe. Waste places; rare. - Topsfield, A.W. Conant.

GYPSOPHILA L.

G. elegans Bieb. Introduced from Europe. Garden escape; rare. - Salem, Hall (1927).

DIANTHUS L. Pink.

D. armeria L. Deptford Pink. Naturalized from Europe. Dry fields and roadsides; common. - Methuen, H. 9630 (1954); West Newbury, H. 10330 (1955); Newburyport, H. 10343 (1955); Byfield, M.E. Perley (1881); Ipswich, H. & Wade 18790 (1958); Topsfield, H. 8056 (1952); Middleton, Bean (1950); Danvers, Oakes; Essex, H. 11320 (1955); Salem, Cole (1886); Peabody, H. 8750 (1953); Lynnfield, Bean (1934); Lynn, Lowell (1846); Swampscott, H. 10643 (1955); Marblehead, Clifton, Morss (1924); Nahant, Preble (1917).

D. barbatus L. Sweet William. Introduced from Europe. Garden escape; rare. - Danvers, Sears (1878); Beverly Farms, Hubbard (1913).

D. deltoides L. Maiden-Pink. Naturalized from Europe. Roadsides and dry fields; occasional. - Haverhill, H. 678 (1932); North Andover, Root (1945); South Groveland, H. 8608 (1953); West Newbury, Conant (1879); Gloucester, H. 12152 (1956); Marblehead, Ives (1927).

D. plumarius L. Introduced from Europe. Garden escape; rare. - Topsfield, Bean & H. 9959 (1955).

CERATOPHYLLACEAE (Hornwort Family)

CERATOPHYLLUM L. Hornwort.

C. demersum L. Quiet waters; common. - Andover, Pease 2612 (1903); Boxford, H. 8990 (1953); Groveland, H. 12879 (1957); West Newbury, H. 9145 (1953); Newburyport, Sears (1886); Ipswich, H. 13255 (1957); Wenham, Robinson (1875); Chebacco, Robinson (1874); Peabody, H. 12765 (1957).

C. echinatum Gray. Quiet waters; occasional. - North Andover, H. 9500 (1954); Hamitlon, H. 12805 (1957).

NYMPHAEACEAE (Water-Lily Family).

NUPHAR Sm. Cow-Lily. Spatter-Dock.

N. variegatum Engelm. Bullhead-Lily. Slow streams and ponds; common. - Amesbury, H. 13015 (1957); Merrimac, H. 13691 (1957); Haverhill, H. 477 (1932); Methuen, H. 12855 (1957); Andover, Sutton 74 (1874); North Andover, Morss (1885); Boxford, M.E. Perley (1882); Groveland, H. 12877 (1957); West Newbury, H. 9117 (1953); Georgetown, H. 12984 (1957); Ipswich, H. 13458 (1957); Topsfield, B.B. Towne; Middleton, H. 12905 (1957); Wenham, Lowell (1849); Hamilton, H. 12804 (1957); Essex, H. 13375 (1957); North Beverly, Deane (1886); Peabody, Sears (1885); Saugus, H. 13029 (1957); Lynn, H. 13183 (1957).

NYMPHAEA L. Water-Lily.

N. odorata Ait. Slow streams and ponds; common. - Amesbury, H. 13016 (1957); Haverhill, H. 478 (1932); Methuen, H. 12856 (1957); Andover, H. 12790 (1957); North Andover, Root (1951); Boxford, H.E. Parkhurst (1883); Groveland, H. 12883 (1957); Georgetown, H. 12984 (1957); Rowley, H. 13517 (1957); Ipswich, Oakes; Topsfield, B.B. Towne; Middleton, H. 12901 (1957); Danvers, H. 9054 (1953); Wenham, H. 13103 (1957); Hamilton, H. 12823 (1957); Essex, Watters (1883); Gloucester, H. 13811 (1957); Manchester, Phippen (1857); Salem-Marblehead, small boy (1891); Peabody, H. 12760 (1957).

N. odorata, forma *rubra* Guillon - Danvers, H. 9055 (1953); Wenham, H. 13094 (1957).

N. odorata, var. *gigantea* Tricker - Ipswich, H. 13463 (1957); Danvers, H. 9056 (1953); Lynn, H. 13184 (1957).

NELUMBO Adans. Sacred Bean.

N. lutea (Willd.) Pers. Introduced from the South. - West Peabody, Devil's Dishfull, Larrime (1913), Mackintosh (1933); H. 9337 (1953). No evidence of the plant 6 August 1957, H., [Wenham, Pleasant Pond, Hellquist (1971)].

BRASENIA Schreb. Water-shield.

B. schreberi Gmel. Purple Wen-dock. Ponds and slow streams; common. - Haverhill, Bradford, H. 13137 (1957); Methuen, H. 12850 (1957); Andover, Pease 2606 (1903); North Andover, H. 9502 (1954); West Boxford, Spofford, Groveland, H. 12889 (1957); Georgetown, H. 12978 (1957); Rowley, H. 13520 (1957); Ipswich, H. 8623 (1953); Topsfield, A.F. Towne; Middleton, H. 12907 (1957); Wenham, Oakes; Hamilton, H. 9395 (1953); Chebacco, Robinson (1879); Essex, H. 13343 (1957).

CABOMBA Aubl.

C. caroliniana Gray. Fanwort. Naturalized from the South. - Andover, Foster's Pond, H. 12797 (1957). Introduced at some time after 1903 and now abundant.

RANUNCULACEAE (Crowfoot Family).

RANUNCULUS L. Crowfoot. Buttercup.

R. trichophyllus Chaix., var. *calvescens* W.B. Drew. White Water-Crowfoot. Fresh to brackish waters; rare. - Danvers, Sears (1879); Wenham Pond, Lowell (1847), Robinson (1875); Saugus, Collins 730 (1909).

R. cymbalaria Pursh. Seaside Crowfoot. Saline shores; occasional. - Salisbury, Alcott; Salisbury-Amesbury, White 252 (1913); Newburyport, Forbes 1631 (1903); Plum Island, Mackintosh (1924); Ipswich Beach, Sears (1897); Ipswich, Plum Island, Sears (1886); Danvers, Sears (1885); Salem, Cole (1899); Lynn, Lowell (1847); Marblehead Neck, Sears (1883); Nahant, Merrill (1850).

R. ficaria L. Lesser Celandine. Introduced from Europe. Garden escape; rare. - Topsfield, M. Clark; Danvers, Morss (1924).

R. flabellaris Raf. Yellow Water-Crowfoot. Quiet waters; common. - Amesbury, H. 13004 (1957); Andover, Pease 1986 (1903); North Andover, Fernald (1909); Boxford, Robinson (1877); Georgetown, Horner; Rowley, H. 13501 (1957); Ipswich, H. 13460 (1957); Topsfield, Sears (1885); Middleton, Robinson (1885); Danvers, St. John 515 (1912); Marblehead, Sears (1897).

R. flabellaris, forma *riparius* Fern. Damp shores; frequent. - Merrimac, H. 13715 (1957); Boxford, Alcott (1882); Groveland, H. 9083 (1953); West Newbury, H. 887 (1933); Middleton, Bean & H. 7776 (1951); Marblehead, Sears (1885).

R. ambigens S. Wats. Water-Plantain-Spearwort. Ditches and muddy swamps; rare. - Newburyport, Sears (1886); Newbury, Sears (1896); Byfield, Sears (1902).

R. reptans L., var. *ovalis* (Bigel.) T. & G. Creeping Spearwort. Damp shores; common. - Merrimac, H. 13734 (1957); Haverhill, Bradford, H. 13129 (1957); North Andover, Sears (1872);

Boxford, Williams (1899); Groveland, Horner; Topsfield, Oakes; Wenham, (1891); Hamilton, Sears (1896); Essex, Fernald, Hunnewell & Long 9512 (1913); Lynn, Young (1880).

R. scleratus L. Cursed Crowfoot. Sloughs; rare. - Salem, Robinson (1879), Sears (1897); Nahant, Deane (1885).

R. allegheniensis Britt. Rich woods; rare. - Danvers, Sears (1897); North Beverly, Sears (1897).

R. abortivus L. Kidneyleaf-Buttercup. Low woods and clearings; common. - Amesbury, H. 9520 (1954); Merrimac, Bean & H. 8364 (1953); Haverhill, Palfrey (1893); Andover, Pease 584 (1902); North Andover, Morss (1884); Boxford, M.E. Perley (1882); Groveland, H. 7901 (1952); Rowley, H. 11952 (1955); Ipswich, H. 12008 (1956); Topsfield, Sears (1902); Danvers, Sears (1897); Wenham, Sears (1899); Beverly, Manning (1882); Salem, Sears (1883); Swampscott, Robinson (1879); Marblehead, Kingman (1899).

R. abortivus, var. *acrolasius* Fern. - West Boxford, Root & Zink (1945); West Newbury, H. 31886 (1968); Danvers, Sears (1896); Salem, Sears (1897); Swampscott, Cole (1887).

R. abortivus, var. *eucyclus* Fern. - Andover, Pease 3638 (1904); Boxford, R.J. Eaton 5233 (1962); Topsfield, Sears (1899).

R. micranthus Nutt. Rich woods; rare. - Peabody, rocky hill near Spring Pond, Sears (1888, 1896), Forbes (1912).

R. recurvatus Poir. Damp rich woods; occasional. - West Haverhill, Pease 3664 (1904); Andover, Pease 2077 (1903); North Andover, Morss (1900); Boxford, H. 8355 (1953); Ipswich, Oakes; Topsfield, M. Clark; Danvers, (1879).

R. pensylvanicus L. f. Bristly Crowfoot. Alluvium; rare. - Lawrence, above the dam, north side, Robinson (1879).

R. fascicularis Muhl. Early Crowfoot. Exposed hills and ledges; rare. - Danvers, Oakes; Swampscott, Fenollosa & Cole (1889).

R. repens L. Creeping Buttercup. Naturalized from Europe. Wet open ground; frequent. - Amesbury, H. 18211 (1958); Merrimac, Bean & H. 8363 (1953); Lawrence, H. 10093 (1955); Andover, H. 17759 (1958); North Andover, H. 9911 (1955); Boxford, M.E. Perley (1886); Groveland, Bean & H. 8479 (1953); Rowley, H. 7918 (1952); Topsfield, H.R. Towne; Danversport, Sears (1887); Manchester, Hubbard (1913); Beverly, Lowell (1847); Saugus, Bean & H. 9821 (1955); Marblehead, Cole (1888).

R. repens, var. *villosus* Lamotte - Andover, Root & Zink (1951); Boxford, Alcott (1909); Groveland, H. 10605 (1955); Ipswich, H. 18146 (1958); Wenham, Bean & H. 10001 (1955); Lynnfield, Rodman (1887).

R. repens, var. *glabratus* DC. - Danversport, Sears (1887); Wenham, Sears (1899); Salem Neck, Sears (1898).

R. repens, var. *erectus* DC. - Danvers, Robinson (1879); Beverly Farms, Hubbard (1913).

R. acris L. Tall Buttercup. Naturalized from Europe. Fields and clearings; common. - Amesbury, H. 18215 (1958); Haverhill, Ward Hill, H. 379 (1932); Andover, Pease 1989 (1903); North

Andover, Root (1943); Boxford, Frye (1880); West Newbury, H. 10333 (1955); Newburyport, H. 10356 (1955); Ipswich, Sears (1890); Ipswich, Plum Island, H., McGregor & Schweinfurth 12688 (1957); Topsfield, H.R. Towne; Wenham, Sears (1905); Rockport, Smith 460 (1923); West Manchester, Hubbard (1913); Beverly, Robinson (1879); Salem, Sears (1899); Saugus, Bean & H. 9824 (1955); Swampscott, Deane (1913); Nahant, S.S. Clark 28 (1916).

R. bulbosus L. Bulbous Buttercup. Naturalized from Europe. Fields and roadsides; common. - Haverhill, H. 358 (1932); Andover, Pease 3612 (1904); North Andover, Morss (1885); Boxford, Frye (1880); West Newbury, H. 31898 (1968); Newburyport, H. 10353 (1955); Ipswich, Oakes; Topsfield, H.R. Towne; Danvers, Sears (1872); Wenham, Robinson (1879); Gloucester, H. 12175 (1956); Rockport, H. 10301 (1955); Manchester, Hubbard (1913); Beverly, Robinson (1879); Salem, Cole (1887); Peabody, Sears (1896); Lynnfield, H. 8336 (1953); Nahant, S.S. Clark 29 (1916).

R. bulbosus, var. *dissectus* Babey - Peabody, Sears (1896).

THALICTRUM L. Meadow-Rue.

T. dioicum L. Early Meadow-Rue. Rich rocky woods; frequent. - Andover, Lloyd (1884); Boxford, Cushman (1907); Georgetown, Horner; Ipswich, Oakes; Topsfield, G.T. Dow; Middleton, Pease 27769 (1940); Danvers, Robinson (1879); Wenham Swamp, Sears (1884); Salem, Cole (1903); Swampscott, Cole (1889); Marblehead, Cole (1900).

T. revolutum DC. Purple Meadow-Rue. Dry open woods and meadows; occasional. - Methuen, Boott (1867); Andover, Pease 3782 (1904); North Andover, Root (1945); Ipswich, Oakes; Topsfield, L. Sweeney; Peabody, Sears (1886).

T. polygamum Muhl. Tall Meadow-Rue. Meadows and swamps; common. - Salisbury, H. 12211 (1956); Amesbury, H. 18224 (1958); Merrimac, H. 18356 (1958); Haverhill, H. 704 (1932); Lawrence, H. 11254 (1955); Andover, Pease 593 (1901); North Andover, Morss (1885); Boxford, M.E. Perley (1882); Groveland, H. 8793 (1953); Georgetown, Horner; Ispwich, H. 18182 (1958); Topsfield, M.E. Perley (1882); Middleton, H. 18087 (1958); Danvers, Robinson (1874); Wenham, Cole (1911); Beverly, Lowell (1847); Peabody, Mackintosh (1933).

T. polygamum, var. *hebecarpum* Fern. - Rockport, Smith 614 (1927); Manchester, Hubbard (1913).

ANEMONELLA Spach

A. thalictroides (L.) Spach. Rue-Anemone. Open woods; local. - Haverhill, H. 43 (1931); Andover, Pease 557 (1902); North Andover, Williams 1 (1900); Boxford, Frye (1881); Ipswich, Oakes; Topsfield, B.B. Towne; Danvers, Sears (1897); Salem, Cole (1904); Marblehead, Saunders (1868).

HEPATICA Mill. Hepatica.

H. americana (DC.) Ker. Dry woods; frequent. - Haverhill, H. 44 (1931); Andover, Pease 573 (1902); North Andover, Sears (1885); Boxford, Frye (1880); Newbury, Horner; Georgetown, Williams (1906); Ipswich, Oakes; Topsfield, G.T. Dow; Danvers, Frost (1873); Wenham, Sears (1896); Essex, Robinson (1875); [Rockport, A & W (1969)]; Salem, Cole (1887); Peabody, Mackintosh (1925); Swampscott, Robinson (1872).

ANEMONE L. Anemone.

A. cylindrica Gray. Thimbleweed. Dry open soil; occasional; - North Andover, Pease 3780 (1904); West Boxford, Horner; Boxford, H. 8107 (1952); Danvers, Oakes; Salem, Sears (1886).

A. virginiana L. Thimbleweed. Dry open woods; frequent. - Haverhill, H. 1834 (1935); Andover, Pease 554 (1900); North Andover, Morss; Boxford, M.E. Perley (1884); West Newbury, Robinson (1879); Georgetown, Horner; Topsfield, Mrs. J.W. Perkins; Danvers, Oakes; Wenham, Cole (1908); Salem, Mackintosh (1925); Swampscott, Sears (1888); Marblehead, Sears (1885).

A. canadensis L. Damp thickets; occasional. - Topsfield, A.M. Dow; Danvers, Sears (1883); Wenham, Sears (1902); Swampscott, Swan (1886).

A. quinquifolia L. Wood-Anemone. Open woods and clearings; common. - Haverhill, H. 45 (1931); Andover, Pease 555 (1901); North Andover, Morss (1885); Boxford, Frye (1881); Ipswich, Robinson (1872); Topsfield, H.R. Towne; Danvers, Sears (1885); Wenham, Sears (1899); Gloucester, Smith 944 (1928); Salem, Cole (1886); Saugus, Lund (1880).

A. nemerosa L. Introduced from Europe. Escape from cultivation; rare. - Danvers, Sears (1903), "persistant and spreading."

CLEMATIS L. Clematis.

C. virginiana L. Virgin's-bower. Devil's-darning-needle. Thickets and borders of woods; common. - Haverhill, H. 514 (1932); Andover, Pease (1901); North Andover, Morss (1885); Boxford, M.E. Perley (1881); Groveland, Sears (1885); Georgetown, Horner (1881); Ipswich, H. 18415 (1958); Ipswich, Plum Island, H. 12417 (1956); Topsfield, M. Clark; Middleton, H. 18085 (1958); Danvers, Sears (1897); Wenham, Robinson (1876); Salem, Robinson (1879); Peabody, Mackintosh (1933).

CALTHA L. Marsh-Marigold.

C. palustris L. "Cowslip." Swamps and wet meadows; common. - Haverhill, H. 304 (1932); North Andover, Morss (1885); Boxford, Frye (1881); West Newbury, H. 31891 (1968); Georgetown, Horner; Ipswich, Oakes; Topsfield, Wadleigh (1884); Danvers, Sears (1885); Wenham, Sears (1886); [Rockport, A & W (1969)] ; Salem, Cole (1887); Peabody, Cole (1879); Lynn, Forbes (1900).

PAEONIA L. Peony.

P. lactiflora Pall. Introduced from Asia. Escaped from cultivation; occasional on dumps. - Salisbury, H. 18994 (1958); Georgetown, H. 17677 (1958).

COPTIS Salisb. Goldthread.

C. groenlandica (Oeder) Fern. Canker-root. Mossy woods and swamps; common. - Amesbury, H. 18237 (1958); Haverhill, H. 411 (1932); Andover, Pease 571 (1902); North Andover, H. 17985 (1958); Boxford, H.E. Parkhurst (1882); Georgetown, Horner; Topsfield, W. Poor; Middleton, H. 18905 (1958); Danvers, Sears (1885); Essex, H. & Smith 2730 (1934); Gloucester, Magnolia, Sears (1896); Rockport, Anderson, Shafer & Smith (1931), [A & W (1969)] ; Manchester, Sears (1896); Beverly, Sears (1885); Salem, Cole (1887); Peabody, Mackintosh (1933); Lynnfield, Young (1877).

AQUILEGIA L. Columbine.

A. canadensis L. Rocky woods and slopes; common. - Haverhill, H. 359 (1932); Andover, Pease 559 (1902); North Andover, Zink (1943); Boxford, Frye (1880); Newbury, Schweinfurth (1950); Georgetown, R. Torrey; Ipswich, Oakes; Topsfield, G.T. Dow; Danvers, Frost (1874); Essex, Hubbard (1913); Beverly Farms, Shaw (1910); Salem, Robinson (1873); Peabody, Sears (1896); Marblehead, Robinson (1872).

A. canadensis, forma *phippenii* (J. Robins.) R. Hoffm. Named for its discoverer, George D. Phippen of Salem. - Reported by Robinson to have been found in Salem by Phippen in 1844, Robinson in 1875 and Waters in 1880.

A. canadensis, forma *flaviflora* (Tenney) Britt. - Salem, Robinson (1873).

A. vulgaris L. Garden Columbine. Introduced from Europe. Garden escape. - North Andover, Root (1957); Boxford, H. 7846 (1952); Danvers, Sears (1879).

A. caerulea James x *A. chrysantha* Gray. Garden escape. - Georgetown, H. 17676 (1958).

DELPHINIUM L. Larkspur.

D. ajacis L. Rocket-Larkspur. Adventive from Europe. Garden escape. - Rockport, Sanford (1945).

ACTAEA L. Baneberry.

A. rubra (Ait.) Willd. Red Baneberry. Woods and thickets; frequent. - Merrimac, H. 18346 (1958); Lawrence, H. 9849 (1955); Boxford, Frye (1881); Groveland, H. 8818 (1953); West Newbury, NEBC (1951); Topsfield, M.E. Perley (1883); Danvers, Sears (1884); Wenham, Cole (1911); [Rockport, A & W (1969)] ; Salem, Mackintosh (1933).

A. rubra, forma *neglecta* (Gillman) Robins. - Georgetown, Horner; Danvers, Sears (1884).

A. pachypoda Ell. White Baneberry. Rich woods and thickets; occasional. - Andover, Root (1950); Boxford, M.E. Perley (1883); Georgetown, Williams (1906); Salem, Cole (1890); Peabody, Mackintosh (1933); Saugus, Lund; Swampscott, Sears (1885).

XANTHORHIZA Marsh. Shrub-Yellowroot.

X. simplicissima Marsh. Introduced from the South. Garden escape. - Danvers, Miss M.W. Nichols (1890); Lynn, L.A. Wentworth (1919).

LARDIZABALACEAE (Lardizabala Family).

AKEBIA Dcne.

A. quinata Dcne. Introduced from Asia. Persistent in garden for over fifty years, Ipswich, Wade 19191 (1959).

BERBERIDACEAE (Barberry Family).

CAULOPHYLLUM Michx. Blue Cohosh.

C. thalictroides (L.) Michx. Papoose-root. Rich woods; rare. - Boxford, Kennedy (1906); Georgetown, Horner (1873, 1875).

BERBERIS L. Barberry.

B. vulgaris L. Common Barberry. Naturalized from Europe. Pastures and fencerows; common. - Merrimac, Bean & H. 7706 (1951); Haverhill, H. 46 (1931); Andover, Sutton 75 (1876); Boxford, M.E. Perley (1880); West Newbury, H. 31901 (1968); Newbury, Plum Island, H., McGregor & Schweinfurth 12743 (1957); Georgetown, Horner; Rowley, H. 12229 (1956); Rowley, Plum Island, H. 12062 (1956); Ipswich, H. 13914 (1957); Topsfield, H.R. Towne; Middleton, Bean & H. 7757 (1951); Wenham, Robinson (1874); Hamilton, H. 10810 (1955); Manchester, Hubbard (1913); Salem, Sears (1884); Peabody, Mackintosh (1925); Saugus, Bean & H. 9788 (1955); Swamspcott, Buck (1897); Marblehead, Cole (1902).

B. thunbergii DC. Japanese Barberry. Introduced from Asia. Spreading from cultivation; frequent. - Merrimac, Bean & H. 7711 (1951); Haverhill, H. 14159 (1957); West Boxford, Root & Zink (1945); Groveland, Bean & H. 8524 (1953); West Newbury, H. 10943 (1955); Georgetown, H. 17684 (1958); Ipswich, H. 13918 (1957); Topsfield, H. 11890 (1955); Middleton, H. 17690 (1958); Marblehead, Cole (1928).

MAGNOLIACEAE (Magnolia Family).

Magnolia L. Magnolia.

M. virginiana L. Laurel-Magnolia. Swamps in Gloucester, Magnolia; the most northern station of the species; now rare. - Gloucester, Magnolia, Frost (1872), H. 14166 (1957), [A & W (1969)].

M. tripetala L. Umbrella-tree. Introduced from South. Escape from cultivation. - Andover, Root (1952). Two trees 30 to 35 feet tall in swamp 1½ miles from nearest cultivated tree.

LIRIODENDRON L. Tulip-tree.

L. tulipifera L. Tulip-Poplar. Introduced from South. Probably present only as a cultivated tree. - Georgetown, Mrs. Bickford.

LAURACEAE (Laurel Family).

SASSAFRAS Nees Sassafras.

S. albidum (Nutt.) Nees, var. *molle* (Raf.) Fern. Red Sassafras. Woods and thickets; common but large trees now rare. - Haverhill, H. 47 (1931); North Andover, Morss (1885); Boxford, M.E. Perley (1884); Groveland, H. 11087 (1955); Newbury, Plum Island, H., McGregor & Schweinfurth 12739 (1957); Rowley, Bean & Stockman (1954); Rowley, Plum Island, H. 12551 (1956); Ipswich, Oakes; Topsfield, H.R. Towne; Middleton, Robinson (1880); Danvers, Robinson (1879); Gloucester, H. 10154 (1955); Rockport, Cushman (1908); Manchester, Hubbard (1913); Salem, Cole (1899); Peabody, Mackintosh (1933); Lynnfield, Robinson (1874); Saugus, Bean & H. 9800 (1955).

LINDERA Thunb. Wild Allspice.

L. benzoin (L.) Blume. Spicebush. Damp woods and brooksides; common. - Merrimac, H. 18353 (1958); Haverhill, H. 306 (1932); Methuen, Wheeler (1922); Andover, Pease 597 (1900); North Andover, Morss (1885); Boxford, M.E. Perley (1884); Topsfield, M. Clark; Danvers, Sears (1884); Gloucester, Annisquam, Smith 853 (1928); Rockport, Smith 4329 (1945); Salem, Cole (1924); Peabody, Sears (1887); Saugus, Lund (1880); Lynn, Bean (1933).

PAPAVERACEAE (Poppy Family).

SANGUINARIA L. Bloodroot.

S. canadensis L. Red Pucoon. Rich woods; frequent. - Haverhill, Bradford, Fernald 11347 (1914); Lawrence, H. 10078 (1955); North Andover, Morss (1885); Boxford, Frye (1881); Byfield, Horner; Georgetown, M.E. Perley (1883); Ipswich, Oakes; Topsfield, B.B. Towne; Salem, Sears (1885); Swampscott, Cole (1888).

CHELIDONIUM L. Celandine.

C. majus L. Naturalized from Europe. Damp soil about towns; common. - Merrimac, H. 10570 (1955); Haverhill, H. 411 (1932); Andover, Pease 601 (1901); North Andover, H. 18031 (1958); Boxford, M.P. Perley (1882); West Newbury, H. 31906 (1968); Rowley, H. 12241 (1956); Ipswich, Oakes; Topsfield, H.R. Towne; Danvers, Bean & H. 9980 (1955); Manchester, Hubbard (1913); Salem, Robinson (1879); [MNS, Snyder (1970)] ; Nahant, Bean & H. 9842 (1955).

PAPAVER L. Poppy.

P. somniferum L. Common Poppy. Introduced from Europe. Roadsides and waste places; occasional. - Boxford, H. 12196 (1956); Georgetown, Horner "Poppy sown itself for 40 years".

P. rhoeas L. Corn-Poppy. Adventive from Europe. Waste places; rare. - Saugus, H. 18637 (1958).

ADLUMIA Raf. Climbing Fumitory. Mountain Fringe

A. fungosa (Ait.) Greene. Probably an escape from cultivation. - Boxford, H. 13957 (1957); Georgetown, Horner; Rockport, Smith 11 (1921); Manchester, Hubbard (1913); [and R.J. Eaton (1917) "seems to be indigenous"] ; Salem, Robinson (1877).

DICENTRA Bernh. Bleeding-heart.

D. cucullaria (L.) Bernh. Dutchman's-breeches. Rich woods; rare. - Gloucester, Salt Island, Mrs. Babson (1875).

D. formosa (Andr.) Walp., ssp. *formosa* Turkey-corn. Escaped from cultivation. - Danvers, D.S. Fowler.

CORYDALIS Medic. Corydalis.

C. sempervirens (L.) Pers. Pale Corydalis. Rocky places and recent clearings; frequent. - Haverhill, H. 361 (1932); Boxford, M.E. Perley (1885); Byfield, Horner; Danvers, Frost (1872); Essex, Foye 9514 (1954); Gloucester, Lanesville, Hoyt; Manchester, Hubbard (1913); [Beverly

Farms, Snyder (1953)]; Peabody, Sears (1884); Lynnfield, Phippen (1857); Lynn, Sears (1896); Marblehead, Robinson (1872).

FUMARIA L. Fumitory.

F. officinalis L. Common Fumitory. Adventive from Europe. Cultivated and waste ground; rare. - Andover, Pease 3793 (1904); Boxford, M.P. Perley (1883); Ipswich, Oakes; Wenham, Lucy Davis (1875).

CAPPARIDACEAE (Caper Family).

CLEOME L.

C. serrulata Pursh. Stinking-Clover. Adventive from West. Waste places; rare. - Lawrence, Gutterson (1903).

C. spinosa Jacq. Spider-flower. Introduced from the Tropics. Garden escape; frequent. - Lawrence, H. 18750 (1958); Newburyport, Fernald & Moseley Clark 2571 (1932); Danvers, H. 18572 (1958); Saugus, H. 18614 (1958); Lynn, H. 11813 (1955); Marblehead, H. 18916 (1958).

POLANISIA Raf.

P. dodecandra (L.) DC., var. *trachysperma* (T. & G.) Ilfis, Boxford, H. 30761 (1966).

CRUCIFERAE (Mustard Family).

DRABA L.

D. reptans (Lem.) Fern. Dry sands and ledges; rare. - Salem, Great Pastures, Dr. Pickering (circa 1824), Robinson (1872, 1874, 1877).

D. verna L., var. *boerhaavii* Van Hall. Whitlow-grass. Naturalized from Europe. Roadsides and open places; occasional. - Haverhill, Ward Hill, H. 362 (1932); Boxford, Root & Zink (1945); Newburyport, Lewis 17624 (1958); Topsfield, Knowlton (1939); Danvers, Oakes; Sears (1898); Peabody, Robinson (1872) at station found by Dr. Andrew Nichols in 1826; Marblehead, Lewis 17625 (1958).

BERTEROA DC.

B. incana (L.) DC. Hoary Alyssum. Naturalized from Europe. Fields and roadsides; apparently becoming a common weed. - Amesbury, H. 12991 (1957); Methuen, H. 8958 (1953); Andover, H. 11067 (1955); Ipswich, H. 13592 (1957); Wenham, H. 9421 (1953); Hamilton, H. 12802 (1957); Gloucester, Magnolia, Bean (1940); Salem (1915).

LOBULARIA Desv. Sweet Alyssum.

L. maritima (L.) Desv. Introduced from Europe. Escape from cultivation. - Newburyport, Bean (1945); Georgetown, H. 17671 (1958); Danvers, H. 18569 (1958); Gloucester, H. 18868 (1958); Beverly, H. 11679 (1955); Salem, Robinson (1877); Marblehead, H. 11467 (1955).

ALYSSUM L.

A. alyssoides L. Naturalized from Europe. Roadsides and waste places; occasional. - Danvers, Sears (1880); Marblehead Neck, Sears (1883); Nahant, Mrs. Wilson (1898).

A. saxatile L. Introduced from Europe. Garden escape. - Topsfield, Mrs. F. Peabody.

THLASPI L. Penny-Cress.

T. arvense L. Mithridate-Mustard. Naturalized from Europe. Fields and waste places; frequent. - Haverhill, H. (1932); Lawrence, Gutterson (1903); Andover, Pease 25272 (1935); North Andover, Root (1945); West Boxford, Root & Zink (1945); Newburyport, H. 10360 (1955); Rowley, H. 8116 (1952); Topsfield, Bean & H. 9971 (1955); Gloucester, Magnolia, H. 18876 (1958); Beverly, H. 10181 (1955); Salem, Russell (1859).

LEPIDIUM L. Pepperwort.

L. perfoliatum L. Naturalized from Europe. Waste places; rare. - Rowley, H. 9538 (1954).

L. latifolium L. Naturalized from Europe. Tidal shores and waste ground; occasional. - Peabody, Mackintosh (1924); Saugus, H. 11865 (1955); Lynn, H. 11808 (1955).

L. campestre (L.) R.Br. Cow-Cress. Naturalized from Europe. Roadsides and waste places; frequent. - Haverhill, Ward Hill, H. 5693 (1933); Andover, Pease 2106 (1903); North Andover, Fernald (1909); Boxford, H. 7903 (1952); Groveland, H. 10596 (1955); Newburyport, H. 10378 (1955); Ipswich, H. 12009 (1956); Topsfield, H. 17641 (1958); Middleton, H. 9658 (1955); North Beverly, Robinson (1875); Nahant, Mrs. Wilson.

L. virginicum L. Poor-man's Pepper. Roadsides and waste places; common. - Haverhill, Ward Hill, H. 412 (1932); Lawrence, H. 9844 (1955); Andover, Pease 3731 (1904); North Andover, Root (1957); Boxford, M.P. Perley (1883); Groveland, Bean & H. 8527 (1953); Newburyport, Plum Island, H. 13041 (1957); Newbury, H. 10421 (1955); Newbury, Plum Island, H. 12036 (1956); Rowley, H. 10431 (1955); Ipswich, Oakes; Ipswich, Plum Island, H. 12090 (1956); Topsfield, H. 8075 (1952); Wenham, Robinson (1875); Essex, H. 10238 (1955); Gloucester, Jenks (1883); Manchester, Hubbard (1913); Beverly, H. 10170 (1955); Salem, Sears (1884); Peabody, Mackintosh (1933); Saugus, H. 18679 (1958); Marblehead, Bean & H. 9767 (1955).

L. ruderale L. Naturalized from Europe. Waste places; rare. - Gloucester, Eastern Point, Rand & B.L. Robinson (1896).

L. densiflorum Schrad. Probably adventive in our area. Roadsides and waste places; occasional. - Salisbury, H. 10501 (1955); North Andover, Morss (1878); Boxford, Forbes (1909); Topsfield, Bean & H. 9939 (1955); Hamilton, H. 10789 (1955); Rockport, Smith 608 (1927); Beverly, H. 10193 (1955); Saugus, Ripley 15537 (1915); Nahant, Forbes (1913).

CAPSELLA Medic. Shepherd's-purse.

C. bursa-pastoris (L.) Medic. Pickpocket. Naturalized from Europe. Weed of roadsides and cultivated ground; common. - Haverhill, H. 414 (1932); Andover, Pease 3526 (1904); North Andover, Morss (1885); Boxford, M.E. Perley (1881); Newburyport, H. 10366 (1955); Newbury, H. 10417 (1955); Rowley, H. 10442 (1955); Ipswich; Topsfield, H.R. Towne; Danvers, Sears (1904);

Wenham, Robinson (1875); Hamilton, Hubbard (1913); Essex, H. 10242 (1955); Gloucester, H. 12135 (1956); Beverly, H. 10167 (1955); Salem, Sears (1884); Marblehead, Bean & H. 9749 (1955).

CAMELINA Crantz False Flax.

C. sativa (L.) Crantz. Gold-of-Pleasure. Adventive from Europe. Roadsides and waste places; rare. - Danvers, Oakes; Salem, Derby Wharf, Robinson (1877).

C. microcarpa Andrz. Naturalized from Europe. Roadsides and waste places; rare. - Andover, Pease 2073 (1903).

CAKILE Hill. Sea-Rocket.

C. edentula (Bigel.) Hook. Common along the shores. - Andover, H. 17735 (1958); Newburyport, Plum Island, Cole (1898); Ipswich, Plum Island, Furber (1911); Gloucester, Prince (1883); Rockport, Smith 47 (1921); Salem, Sears (1884); Lynn, Robinson (1875); Marblehead, H. 888 (1933); Nahant, Merrill (1849).

RAPISTRUM

R. rugosum (L.) All. Adventive from Europe. Waste places. - Saugus, H. 18716 (1958).

RAPHANUS L. Radish.

R. raphanistrum L. Wild Radish. Naturalized from Europe. Waste places; common. - Salisbury, H. 10508 (1955); Amesbury, H. 10523 (1955); North Andover, Root (1945); Boxford, H. 8904 (1953); Newburyport, Robinson (1879); Georgetown, M.E. Perley (1882); Rowley, H. 18491 (1958); Ipswich Beach, M.E. Perley (1903); Topsfield, H.R. Towne; Danvers, Morse (1927); Hamilton, Hubbard (1913); Gloucester, H. 12134 (1956); Beverly, Lowell (1846); Salem, Robinson (1877); Peabody, H. 27065 (1964); Saugus, H. 10895 (1955); Lynn, H. 11848 (1955); Swampscott, Churchill (1891); Marblehead, H. 889 (1933).

R. raphanistrum, forma *sulphureus* (Babey) Hayek - Ipswich, Plum Island, H. 12095 (1956); Wenham, H. 9419 (1953).

R. raphanistrum, forma *candidus* (Opiz) G. Beck - Amesbury, H. 9550 (1954); West Newbury, Bean & H. 19468 (1959); Georgetown, H. 10514 (1955).

R. sativus L. Radish. Introduced from Europe. Fields and waste places; occasional. - West Andover, garden, Morse (1921); Boxford, M.P. Perley (1884); Rowley, H. 18503 (1958); Hamilton, Robinson (1875); Manchester, Farlow (1915).

BRASSICA L. Mustard. Turnip.

B. hirta Moench. White Mustard. Introduced from Europe. Waste places; occasional. - Boxford, H. 18809 (1958); Salem, Baker's Island, Sears (1890).

B. kaber (DC.) L. C. Wheeler, var. *pinnatifida* (Stokes) L. C. Wheeler. Crunch-weed. Naturalized from Eurasia. Cultivated and waste ground; frequent. - Haverhill, H. 415 (1932); Andover, Pease 615 (1902); North Andover, Pease 617 (1902); Boxford, M.P. Perley (1882); Rowley, Plum Island,

H. 12061 (1956); Danvers, Sears (1879); Essex, H. 10243 (1955); Saugus, H. 18612 (1958); Nahant, Kingman (1895).

B. juncea (L.) Coss. Chinese Mustard. Naturalized from Eurasia. Waste places; occasional. - North Andover, Root (1945); Boxford, M.E. Perley (1891); Gloucester, Magnolia, H. 18885 (1958); Salem, Sears (1904).

B. nigra (L.) Koch. Black Mustard. Waste places and cultivated ground; occasional. - Georgetown, Horner; Rowley, Plum Island, H. 12580 (1956); Ipswich, Oakes; Danvers, Kennedy (1907); Wenham, Robinson (1875); Essex Ponds, Lowell (1844); Gloucester, Magnolia, Clarke (1911), H. 12323 (1956).

B. rapa L. Bird's Rape. Naturalized from Eurasia. Cultivated ground and waste places; frequent. - Andover, Pease 1962 (1903); North Andover, Root (1945); Boxford, H. 9670 (1955); Newbury, H. 18459 (1958); Georgetown, H. 17668 (1958); Rowley, H. 18512 (1958); Ipswich, Grape Island, Sears (1888); Ipswich, Plum Island, H. 12093 (1956); Danvers, Sears (1891); Salem, Robinson (1877); Saugus, H. 18719 (1958); Nahant, Clark 21 (1916).

B. napus L. Introduced from Europe. Dumps and waste places; occasional. - Topsfield, H. 18300 (1958).

B. oleracea L. Cabbage. Introduced from Europe. Waste places; occasional. - Gloucester, Magnolia, H. 12378 (1956).

CONRINGIA Link Hare's-ear-Mustard.

C. orientalis (L.) Dumort. Naturalized from Europe. Waste places; occasional. - Haverhill, Ward Hill, H. 5697 (1933); Topsfield, H. 9589 (1954).

ALLIARIA B. Ehrh. Garlic-Mustard.

A. officinalis Andrz. Naturalized from Europe. Near habitations; rare. - Ipswich, H. & F.C. Wade 18784 (1958).

SISYMBRIUM L.

S. officinale (L.) Scop., var. *leiocarpum* DC. Hedge Mustard. Naturalized from Europe. Waste places; common. - Andover, H. 12298 (1956); North Andover, Root (1943); Boxford, M.E. Perley (1882); Newburyport, H. 10368 (1955); Rowley, H. 12242 (1956); Topsfield, H.R. Towne; Wenham, Robinson (1875); Gloucester, H. 12147 (1956); Rockport, F.E. & L.B. Smith 351 (1923); Beverly, Lowell (1847); Salem, Sears (1884); Saugus, H. 10902 (1955); Swampscott, H. 10654 (1955).

S. altissimum L. Tumble Mustard. Naturalized from Europe. Roadsides and waste places; common. - Haverhill, H. 416 (1932); South Lawrence, Pease 627 (1902); Andover, Pease 2000 (1903); North Andover, Root (1945); West Boxford, Root & Zink (1945); Georgetown, H. 17669 (1958); Rowley, H. 9539 (1954); Topsfield, Bean & H. 9940 (1955); Danvers, H. 9539 (1954); Gloucester, Collins (1915); Salem, Sears (1904); Marblehead, Bean & H. 9732 (1955).

ARABIDOPSIS Heynh. Mouse-ear-Cress.

A. thaliana (L.) Heynh. Naturalized from Europe. Dry fields; occasional. - West Boxford, Root (1945); Georgetown, H. 17670 (1958).

DESCURAINIA Webb & Berthelot Tansy-Mustard.

D. sophia (L.) Webb. Naturalized from Europe. Waste places; occasional. - Methuen, H. 12264 (1956); Andover, Lowell Junction, Hopkins & Pease 24058 (1934); West Boxford, Root (1945).

HESPERIS L. Rocket.

H. matrionalis L. Mother-of-the-evening. Naturalized from Europe. Spread from cultivation; common. - Merrimac, Bean & H. 7731 (1951); Haverhill, H. 417 (1932); Andover, Pease 3695 (1904); North Andover, Root (1945); Boxford, H. 12511 (1956); Newburyport, H. 10363 (1955); Georgetown, Horner; Topsfield, H. 18301 (1958); Danvers, Kennedy (1907); Beverly Farms, Hubbard (1913); Salem, Morse (1928); Nahant, Bean & H. 9841 (1955).

ERYSIMUM L. Treacle-Mustard.

E. cheiranthoides L. Wormseed-Mustard. Naturalized from Old World. Cultivated fields and waste places; occasional. - Haverhill, H. 10459 (1955); Andover, Pease 8042 (1905); North Andover, H. 14164 (1957); West Boxford, Root & Zink (1945); North Salem, Hall.

RORIPPA Scop. Yellow Cress.

R. sylvestris (L.) Bess. Naturalized from Europe. Meadows, shores and roadsides; occasional. - North Andover, H. 18024 (1958); Ipswich, H. 10231 (1955).

R. sessiliflora (Nutt.) Hitchc. Adventive from West. Cultivated land; rare. - Salem, Russell (1859) "a weed in garden and seen there for many years". Apparently the only collection from New England.

R. islandica (Oeder) Borbas. Wet shores and waste places; occasional. - Salisbury, A.P. Morse (1925); Gloucester (Magnolia) J.M. Hunnewell (1900); Danvers, Cushman (1905).

R. islandica, var. *Fernaldiana* Butt. & Abbe. Similar habitats; frequent. - North Andover, Root (1945); Plum Island, Sears (1888); Georgetown, Horner; Topsfield, H.R. Towne; Manchester, Hubbard (1913); Salem, Sears (1891); Peabody, Mackintosh (1933); Swampscott, Robinson (1877).

R. islandica, var. *hispida* (Desv.) Butt. & Abbe. Similar habitats; frequent. - Amesbury, H. 12995 (1957); Merrimac, H. 13685 (1957); Haverhill, H. 617 (1932); North Andover, H. 12920 (1957); Groveland, H. 11191 (1955); West Newbury, H. 9134 (1953); Georgetown, Horner; Ipswich, H. 18995 (1958); Topsfield, H. 8859 (1953); Wenham, Robinson (1875); Rockport, Smith 15 (1921); Peabody (1924).

NASTURTIUM R. Br. Watercress.

N. officinale R.Br. True Watercress. Introduced from Europe. Brooks and rills; occasional. - Newburyport, Bean (1940); Topsfield, H.R. Towne; Danversport, Sears (1886); Salem, Robinson (1879).

N. officinale, var. *microphyllum* (Boenn.) Thell. - Methuen, Root (1949).

ARMORACIA Gaertn., Mey. & Scherb.

A. lapathifolia Gilib. Horseradish. Naturalized from Europe. Escape from cultivation, moist ground; common. - Salisbury, Bean & H. 8394 (1953); Andover, Pease 2014 (1903); North Andover, H. 9892 (1955); Boxford, M.E. Perley (1884); Ipswich, Oakes; Topsfield, Charles Sweeney; West Manchester, Hubbard; Beverly, Robinson (1879); Salem, Sears (1884); Peabody, H. 11555 (1955); Marblehead, Cole (1933).

BARBAREA R.Br. Winter-Cress.

B. vulgaris R. Br. Yellow Rocket. Naturalized from Eurasia. Cultivated fields and waste places; common. - Haverhill, Ward Hill, Cushman (1906); Lawrence, H. 10089 (1955); Andover, Pease 614 (1902); North Andover, Root (1945); Boxford, M.E. Perley (1884); Newburyport, H. 10364 (1955); Georgetown, H. 17666 (1958); Rowley, H. 12248 (1956); Ipswich, Oakes; Topsfield, Bean & H. 9967 (1955); Middleton, Miss Norris (1915); Manchester, Robinson (1875); Beverly Farms, Shaw (1910); Salem, Robinson (1877); Saugus, Bean & H. 9799 (1955); Swampscott, H. 10667 (1955); Marblehead, Bean & H. 9740 (1955).

B. vulgaris, var. *arcuata* (Opiz). Fries - Merrimac, Bean & H. 8359 (1953); Haverhill, H. 418 (1932); Andover, Pease 612 (1902); North Andover, Root (1944); Groveland, Bean & H. 8561 (1953); Georgetown, H. 17667 (1958); Topsfield, H.R. Towne; Middleton, H. 9657 (1955); Wenham, Bean & H. 10012 (1955); Hamilton, H. 10798 (1955); West Manchester, Hubbard (1913); Peabody, Mackintosh (1933); Marblehead, Bean & H. 9760 (1955); Nahant, Clark 26 (1916).

B. verna (Mill.) Aschers. Belle-Isle Cress. Naturalized from Europe. Fields; rare. - Haverhill, H. 419 (1932).

CHORISPORA DC.

C. tenella (Willd.) DC. Adventive from Asia. Roadsides and fields; rare. - West Boxford, Root & Zink (1945).

LUNARIA L.

L. annua L. Honesty. Introduced from Europe. Garden escape; occasional. - Haverhill, Ward Hill, H. 7820 (1952); North Andover, Root 17628 (1958).

CARDAMINE L. Bitter Cress.

C. bulbosa (Schreb.) BSP. Spring-Cress. About springs or in meadows; occasional. - Haverhill, Bradford, Collins 866 (1910); Andover, Pease 7693 (1905); North Andover, Morss (1900); Boxford, Pease & St. John 11940 (1917); West Newbury, H. 31893 (1968); Topsfield, B.B. Towne; Middleton, Cole (1901); Danvers, Oakes.

C. pratensis L., var. *palustris* Wimm. & Grab. Lady's-smock. - Springs and swampy woods; rare. - Georgetown, Crane Meadows, Horner; West Newbury, H. 31880 (1968).

C. pensylvanica Muhl. Springs, rills and wet clearings, very variable; common. - Amesbury, H. 18221 (1958); Haverhill, H. 48 (1931); Lawrence, Merrill (1851); Andover, Robinson (1884); North Andover, Zink (1944); Boxford, H. 7889 (1952); West Newbury, H. 31908 (1968);

Georgetown, Horner; Ipswich, H. 13457 (1957); Topsfield, H. 13602 (1957); Middleton, H. 31912 (1968); Danvers, Robinson (1879); Rockport, Smith 4330 (1945); Beverly Farms, Hubbard (1913); Salem, Cole (1902); Marblehead, Cole (1931).

C. parviflora L., var. *arenicola* (Britt.) O.E. Schulz. Dry woods and sandy soils; rare. - Salem, Cole (1886); Peabody, Forbes (1912); Lynn, Bartlett, Eames & Fernald 217 (1908).

ARABIS L. Rock-Cress.

A. glabra (L.) Bernh. Tower-Mustard. Ledges, thickets and fields; rare. - West Boxford, Fernald (1909); Boxford, Forbes (1909).

A. drummondi Gray. Ledges, gravels and sand; occasional. - Haverhill, Bradford, Fernald 11346 (1914); Methuen, Bean & H. 7745 (1951); West Boxford, Zink (1945); West Newbury, Nubble Squid, Sears (1904); Newburyport, Moulton (1888); Newbury, Plum Island, H. 12012 (1956); Rowley, Plum Island, H., McGregor & Schweinfurth 12727 (1957); Beverly, Phillips (1878).

A. missouriensis Greene. Ledges or rocky woods; rare. - Salem, Ornes Point, Boott (1861).

A. canadensis L. Sicklepod. Rich woods and thickets; occasional. - Andover, Pease 610 (1900); Topsfield, Oakes; Danvers, Sears (1877).

RESEDACEAE (Mignonette Family)

RESEDA L. Mignonette.

R. lutea L. Adventive from Europe. Waste places; rare. - Rockport, Pease 645 (1898).

R. alba L. Adventive from Europe. Waste places; rare. - Topsfield, A. Porter.

SARRACENIACEAE (Pitcher-plant Family)

SARRACENIA L. Pitcher-plant.

S. purpurea L. Sidesaddle-flower. Sphagnous bogs; common. - Haverhill, H. 480 (1932); Methuen, H. 8922 (1953); North Andover, Morss (1872); Boxford, Frye (1880); Georgetown, Horner; Ipswich, Bullard (1881); Danvers, Oakes; Wenham, Morse (1925); Essex, Hubbard (1913); Gloucester, Prince (1883); Rockport, H. 19038 (1958); Manchester, Oakes; Lynnfield, Lund.

DROSERACEAE (Sundew Family).

DROSERA L. Sundew.

D. intermedia Hayne. Wet acid peat and sand; common. - Merrimac, H. 9209 (1953); Haverhill, H. 575 (1932); Andover, Root (1947); North Andover, Morss (1885); Boxford, W. Merrill, Jr. (1885); Groveland, Bean & H. 8477 (1953); Georgetown, H. 12791 (1957); Topsfield, H.R. Towne; Danvers, Robinson (1875); Chebacco, Robinson; Gloucester, Prince (1883); Rockport, Smith (1921); Manchester, H. 10725 (1955); Beverly, Lowell (1846); Salem, Cole (1910).

D. rotundifolia L. Round-leaved Sundew. Peaty or moist acid soils; common. - Salisbury, White 331 (1913); Haverhill, H. 574 (1932); Methuen, H. 8943 (1953); Andover, Root & Mr. & Mrs. J.B.C. Thomas (1950); North Andover, Morss (1885); Boxford, H. 12768 (1957); Newburyport, Horner (1873); Goergetown, Horner; Topsfield, H.R. Towne; Middleton, H. 18106 (1958); Wenham, Sears (1884); Chebacco, Robinson (1874); Essex Woods, Lowell (1846); Gloucester, Magnolia, J.M. Hunnewell (1899); Beverly, Prides Crossing, Hubbard (1913); Peabody, Sears (1886); Saugus, H. 13083 (1957).

CRASSULACEAE (Orpine Family).

Tillaea L. Pigmyweed.

T. aquatica L. Fresh to tidal shores; rare. - Newburyport, B.L. Robinson (1903).

SEDUM L. Stonecrop. Orpine.

S. acre L. Wallpepper. Naturalized from Europe. Rocks, walls and dry open places; occasional. - Topsfield, M.E. Perley (1882); Beverly, Lowell (1849); Salem, Sears (1884); Lynnfield, Bean (1916); Marblehead, Cole (1907).

S. sarmentosum Bunge. Introduced from eastern Asia. Escape from cultivation. - Marblehead, H. 18952 (1958).

S. rupestre L. Introduced from Europe. Escape from cultivation. - Rockport, Pigeon Cove, Robinson (1877).

S. ternatum Michx. Introduced from South. Escape from cultivation. - Boxford, M.P. Perley (1882); Danvers, Robinson (1877).

S. purpureum (L.) Link. Live-forever. Naturalized from Europe. Roadsides, open banks and waste places; frequent. - North Andover, Root (1945); Boxford, Howe (1885); Groveland, M.E. Perley (1885); Ipswich, Sears (1886); Topsfield, H.S. Porter; Danvers, Sears (1885); Wenham, Stone (1884); Gloucester, Magnolia, J.M. Hunnewell (1901); Salem, Sears (1885).

SEMPERVIVUM L. Houseleek.

S. tectorum L. Hens-and-chickens. Introduced from Europe. Escape from cultivation. - Swampscott, Robinson (1875).

SAXIFRAGACEAE (Saxifrage Family).

PENTHORUM L. Ditch-Stonecrop.

P. sedoides L. Wet low ground; common. - Salisbury, White 359 (1913); Merrimac, H. 9241 (1953); North Andover, Morss (1885); Boxford, M.P. Perley (1883); Groveland, H. 13962 (1957); Newburyport, Cole (1899); Georgetown, Horner; Ipswich, H. 18136 (1958); Topsfield, H. 8855 (1953); Danvers, Sears (1885); Essex, Fernald, Hunnewell & Long 9585 (1913); Manchester, J.M. Hunnewell (1901); Salem, Cole (1901); Peabody, Mackintosh (1933).

SAXIFRAGA L. Saxifrage.

S. virginiensis Michx. Early Saxifrage. Dry or wet rocks and open or shaded slopes, oak woods, upland marsh; common. - Haverhill, H. 1192 (1932); Andover, Pease 673 (1902); North Andover, Zink (1943); Boxford, M.E. Perley (1881); Georgetown, Horner; Ipswich, St. John 463 (1912) [R. Goodale (1966, '70)] ; Topsfield, Oakes; Danvers, Frost (1872); Wenham, Sears (1899); Beverly Farms, Shaw (1910); Salem, Cole (1888); Marblehead, Phippen (1857).

S. virginiensis, forma *glomerata* Fern. - Andover, Prospect Hill, Pease 671 (1902).

S. pensylvanica L. Swamp Saxifrage. Wet meadows and boggy thickets; occasional. - Haverhill, Bradford, Fernald 11350 (1914); Andover, Pease 670 (1901); North Andover, Robinson (1884); West Boxford, Root (1943); Groveland, Bean & H. 8468 (1953); West Newbury, H. 31894 (1968); Georgetown, Horner; Topsfield, H.R. Towne; Middleton, H. 31911 (1968); Danvers, Robinson (1879); Peabody, Mackintosh (1933).

TIARELLA L. False Miterwort.

T. cordifolia L. Rich woodlands; occasional. - Salisbury, Moulton (1888), Pope (1898); Newburyport, Moulton (1887); Topsfield, H.R. Towne; Danvers, Miss Mary Nichols (1882).

MITELLA L. Miterwort.

M. diphylla L. Coolwort. Loamy and rocky woods; occasional. - Salisbury, Moulton (1886, 1888); Amesbury, Beech Hill, Eaton; Newburyport, Moulton (1887).

M. nuda L. Cool woods or swamps; rare. - West Newbury, Indian Hill Run, Eaton (1897).

CHRYSOSPLENIUM L. Golden Saxifrage.

C. americanum Schwein. Water-mat. Rills and cold wet places; probably more common than collections indicate. - Amesbury, H. 18220 (1958); Merrimac, Bean & H. 8368 (1953); Haverhill, H. 422 (1932), Andover, Pease 3474 (1904); Boxford, H. 8446 (1953); Groveland, Bean & H. 8562 (1953); Georgetown, Horner (1886); Ipswich, Oakes; Danvers, Sears (1885); Salem, Cole (1908).

PARNASSIA L. Grass-of-Parnassus.

P. glauca Raf. Wet calcareous soils; rare and local. - Topsfield, Mrs. A. Towne; Danvers, Sears (1885), Mackintosh (1933); Lynnfield, Nichols (1839); Miss Olga E. Ahler & Mrs. Robert S. Russell (1917).

PHILADELPHUS L. Mock-Orange or "Syringa".

P. coronarius L. Introduced from Europe. Escape from cultivation. - Topsfield, H.R. Towne; Wenham, Robinson (1875).

P. sp. - Manchester, Hubbard (1913), escaped.

HYDRANGEA L. Hydrangea.

H. paniculata Sieb. Introduced from Asia. Escape from cultivation. - Newburyport, Carrs Island, Bean (1945), [MNS, Snyder & Leahy, 1971].

RIBES L. Currant. Gooseberry.

Probably more abundant before the Ribes eradication program to control the White Pine Blister Rust.

R. hirtellum Michx. Rocky or swampy woods; frequent. - Amesbury, H. 9528 (1954); West Boxford, Root & Zink (1945); West Newbury, H. 9533 (1954); Newbury, H. 8243 (1952); Newbury, Plum Island, H. 12004 (1956); Rowley, Plum Island, H., McGregor & Schweinfurth 12723 (1957); Ipswich, Oakes; Topsfield, Robinson (1884); Rockport, Smith 928 (1928); Salem, Sears (1886).

R. triste Pall. Red Currant. Cool woods; rare. - Andover, H. 17750 (1958); West Newbury, H. 31885 (1968).

R. sativum Syme. Garden-Currant. Introduced from Europe. Garden escape; occasional. - Andover, Knowlton (1903); North Andover, Root & Zink (1945); Boxford, Robinson (1877); Ipswich, H. 18155 (1958); Topsfield, M.E. Perley (1881); Danvers, Robinson (1875); Lynnfield, Bean (1916).

R. americanus Mill. Wild Black Currant. Rich thickets and slopes; occasional. - Andover, Pease 667 (1903); Boxford, Frye (1881), Zink (1945); Groveland, Horner; Newbury, Devil's Den, M.E. Perley (1886); Ipswich, Oakes; Topsfield, Sears (1878); Marblehead, Swan (1886).

R. nigrum L. Black Currant. Introduced from Europe. Escape from cultivation; occasional. - Merrimac, H. 10578 (1955); Boxford, M.E. Perley (1887).

R. odoratum Wendland. Buffalo-Currant. Introduced from the West. Escape from cultivation; occasional. - Andover, Robinson (1884); Salem, Robinson (1879).

HAMAMELIDACEAE (Witch-Hazel Family).

HAMAMELIS L. Witch-Hazel.

H. virginiana L. Dry or moist woods; common. - Amesbury, Bean & H. 7736 (1951); Haverhill, H. 50 (1931); Andover, Pease 1739 (1902); North Andover, H. 8667 (1953); Boxford, M.E. Perley (1880); Groveland, Bean & H. 8557 (1953); Topsfield, H.R. Towne; Wenham, Robinson (1874); Hamilton, Robinson (1876); Essex, H. 11985 (1955); Gloucester, Magnolia, H. 11988 (1955); Rockport, H. 10300 (1955); Manchester, Oakes; Peabody, H. 9668 (1955); Lynnfield, Cole (1887); Saugus, Bean & H. 9803 (1955); Marblehead, Cole (1933).

H. virginiana, var. *parvifolia* Nutt. - Georgetown, Horner.

PLATANACEAE (Plane-tree Family).

PLATANUS L. Sycamore. Buttonwood.

P. occidentalis L. Rich soils; occasional. - Haverhill, Ward Hill, H. 576 (1932); North Andover, H. 11030 (1955); Boxford, M.E. Perley (1885); Topsfield, H. 10131 (1955); Middleton, Sears (1887); Wenham, Robinson (1875); Swampscott, Buck (1896). The largest deciduous tree in eastern North America.

ROSACEAE (Rose Family).

PHYSOCARPUS Maxim. Ninebark.

P. opulifolius (L.) Maxim. Introduced from the South or West. Escape from cultivation; occasional. - North Andover, Root (1945); Georgetown, H. 13569 (1957); Hamilton, H. 10821 (1955).

P. opulifolius, var. *luteus* Zabel. - Georgetown, H. 17685b (1958).

SPIRAEA L. Spiraea.

S. latifolia (Ait.) Borkh. Meadow-sweet. Low grounds; common. - Salisbury, White 61 (1913); Amesbury, H. 10531 (1955); Haverhill, H. 515 (1932); Lawrence, H. 11220 (1955); North Andover, Morss (1885); Boxford, M.E. Perley (1884); West Newbury, H. 10962 (1955); Newbury, H. 18462 (1958); Georgetown, Horner; Rowley, H. 12253 (1956); Rowley, Plum Island, H. 12565 (1956); Ipswich, H. 9574 (1954); Topsfield, G.T. Dow; Danvers, Lowell (1846); Hamilton, H. 10816 (1955); Rockport, Smith 1007 (1928); Manchester, Hubbard (1913); Beverly, Robinson (1879); Salem, Cole (1908); Peabody, Mackintosh (1933); Saugus, H. 10852 (1955); Lynn, H. 10930 (1955); Swampscott, H. 10645 (1955).

S. tomentosa L. Hardhack. Sterile low grounds and pastures; common. - Haverhill, H. 516 (1932); Lawrence, H. 11225 (1955); Andover, Pease 751 (1901); North Andover, Morss (1885); Boxford, M.E. Perley (1883); West Newbury, H. 10963 (1955); Newbury, H. 18461 (1958); Georgetown, Horner; Rowley, Plum Island, H. 12549 (1956); Ipswich, Oakes; Ipswich, Plum Island, St. John & White 571 (1913); Topsfield, H.R. Towne; Wenham, Sears (1897); Rockport, Smith 711 (1927); Manchester, H. 11356 (1955); Beverly, Robinson (1879); South Salem, A.P. Morse (1922); Peabody, Mackintosh (1933); Saugus, H. 10851 (1955); Marblehead, H. 11442 (1955).

S. tomentosa, forma *albiflora* Macbr. - Ipswich, Oakes; Gloucester, Grover; Beverly, Lowell (1846); Swampscott, Buck (1897); Marblehead, Chapman.

SORBARIA A. Br. False Spiraea.

S. sorbifolia (L.) A. Br. Introduced from Asia. Escaping from cultivation to waste land and copses; occasional. - North Andover, Root (1945); Boxford, H. 12199 (1956); Ipswich, H. 10611 (1955); Topsfield, Bean & H. 9929 (1955); West Beverly, Stone (1879).

GILLENIA Moench Indian-Physic.

G. trifoliata (L.) Moench. Introduced from South. Escape from cultivation. Haverhill, Ward Hill, H. 481 (1932).

PYRUS L.

P. communis L. Common Pear. Introduced from Eurasia. Spread from cultivation; occasional. - Topsfield, Mrs. Jenkins; Gloucester, Dogtown Common, H. 12158 (1956); Salem, Sears (1885).

P. malus L. Apple. Introduced from Eurasia. Spread from cultivation; common. - Haverhill, Ward Hill, H. 364 (1932); North Andover, Robinson (1885); Ipswich, Plum Island, E., McGregor & Schweinfurth 12686 (1957); Topsfield, H.R. Towne; Danvers, Sears (1875); Salem, Cole (1924); Marblehead, Cole (1900).

P. arbutifolia (L.) L. f. Red Chokeberry. Swamps; rare. - Boxford, H. 17715 (1958); Lynn, Cedar Pond, Sears (1885, 1886), Robinson (1890).

P. floribunda Lindl. Purple Chokeberry. Low thickets, wet to dry clearings; common. - Haverhill, H. 618 (1932); Andover, Pease (1901); North Andover, Root & Zink (1945); Boxford, M.E. Perley (1885); Ipswich, Plum Island, H. 12101 (1956); Topsfield, H.R. Towne; Gloucester, H. 12165 (1956); Rockport, Fernald, Hunnewell & Long 9613 (1913); West Manchester, Hubbard (1913); Salem, Cole (1945); Saugus, Bean & H. 9782 (1955); Lynn, Sears (1885); Swampscott, Sears (1885); Marblehead, Cole (1943).

P. melanocarpa (Michx.) Willd. Black Chokeberry. Drier thickets; occasional. - Haverhill, H. 686 (1932); Andover, Ballardvale, Zink (1945); Boxford, H. 7834 (1952); Georgetown, Horner; Rowley, H. 8263 (1952); Topsfield, Oakes; Rockport, Williams (1898); Swampscott, Sears (1885); Marblehead Neck, H. 7828 (1933).

P. americana (Marsh.) DC. American Mountain-Ash. Woods; rare. - Topsfield, E. Morrisey & B. Woodberry; Beverly, Lowell (1848); Beverly Farms, Robinson & Sears (1885); Salem, Cole (1902).

P. aucuparia (L.) Gaertn. Rowan. Introduced from Europe. Spread from cultivation; occasional. - Merrimac, Bean & H. 7712 (1951); Haverhill, H. 551 (1932); Andover, Pease 1945 (1903); North Andover, Root (1945); Boxford, M.E. Perley (1885); Groveland, H. 10986 (1955); Newbury, Williams (1898); Topsfield; Danvers, Frost; Salem, Sears (1885).

CYDONIA

C. vulgaris Pers. Quince. Introduced from Asia, cult. since ancient times. Spread from cultivation. - Topsfield, C. Jackman.

AMELANCHIER Medic. Shadbush.

A. sanguinea (Pursh) DC. Open woods; rare. - Ipswich, Plum Island, H. 12446 (1956), H., McGregor & Schweinfurth 12684 (1957).

A. stolonifera Wieg. Dry sterile habitate; frequent. - Haverhill, H. 423 (1932); South Lawrence, Morss (1885); Andover, Pease 683 (1902); Boxford, Kennedy (1910); Groveland, Bean & H. 8511 (1953); Newbury, Plum Island, H. 12003 (1956); Ipswich, Plum Island, H., McGregor & Schweinfurth 12698 (1957); Topsfield, Mackintosh (1925); Gloucester, Magnolia, Robinson (1887); Rockport, Forbes (1903); Manchester, Hubbard (1913); Nahant, S.S. Clark 5 (1916).

A. canadensis (L.) Medic. Low grounds and thickets; common. - Amesbury, H. 10520 (1955); Andover, H. 17887 (1958); North Andover, Root & Zink (1945); Boxford, M.E. Perley (1881); West Newbury, H. 31886 (1968); Rowley, Plum Island, H. 12050 (1956); Ipswich, Oakes; Topsfield, H.R. Towne; Danvers, Sears (1885); Wenham, Robinson (1884); East Gloucester, Sears (1891); Beverly, Lowell (1849); Salem, Mackintosh (1925); Peabody, Robinson (1872); Marblehead, Bean & H. 9731 (1955).

A. canadensis X *laevis* - Andover, Pease 679 (1902).

A. arborea (Michx. f.) Fern. Rich woods; rare. - Andover, Pease 678 (1902); Gloucester, Cole (1905); [A. Robinson (1971)] ; Salem, Cole (1902).

A. laevis Wieg. Dry to moist thickets; occasional. - Lawrence, Merrill (1852); Boxford, H. 7824 (1952); West Newbury, Robinson (1879); Newbury, Plum Island, H., McGregor & Schweinfurth 12736 (1957); Ipswich, Oakes; Topsfield, P. Conley; Lynnfield, Bean (1919); Swampscott, Robinson (1875).

CRATAEGUS L. Hawthorn.

C. monogyna Jacq. Introduced from Eurasia. Spreading from cultivation; occasional. - Haverhill, H. 687 (1932); Andover, Pease 689 (1901); Newburyport, Bean (1940); Topsfield, M. Hobson; Rockport, H. 19014 (1958); Salem, Robinson (1879).

C. foetida Ashe. Thickets and rocky hillsides; rare. - Rockport, Smith 931 (1926).

C. crus-galli L. Thickets and open ground; rare. - Newbury, H. 18470 (1958).

C. chrysocarpa Ashe. Thickets and rocky ground; frequent. - Boxford, H. 9665 (1955); Ipswich, H. 12007 (1956); Topsfield, Robinson (1875); Danvers, Robinson (1879); Hamilton, H. 12500 (1956); Gloucester, Dogtown Common, H. 12182 (1956); Manchester, Robinson (1876); Salem, Robinson (1879); Peabody, Sears (1885).

C. chrysocarpa, var. *phoenica* Palmer - Salisbury, Bean & Stockman (1953); Andover, Root (1950); Newburyport, Bean (1940).

C. dodgei Ashe. Thickets and borders of woods; rare. - North Andover, Root (1950); West Boxford, Root (1950).

C. macrosperma Ashe. Woods and thickets; occasional. - Andover, Root (1950); North Andover, Root (1949); Boxford, M.E. Perley (1888); Gloucester, Dogtown Common, H. 12171 (1956); Rockport, Forbes 1998 (1902, 1903).

C. macrosperma, var. *demissa* (Sarg.) Egglest. - North Andover, Root (1950).

C. pedicellata Sarg. Thickets and banks of streams; occasional. - Andover, Root (1950); North Andover, Root (1950); Ipswich, Sears (1887); Rockport, Sears (1885).

C. anomala Sarg. ? Rocky banks and open woods; rare. - North Andover, Root (1950).

C. submollis Sarg. Wooded hillsides; occasional. - Andover, Pease 3615 (1904); North Andover, Root (1950).

C. succulenta Link. Thickets and borders of woods; occasional. - Ipswich, Oakes; Sears (1880); South Salem, Sears (1887).

C. succulenta, var. *macrantha* (Lodd.) Egglest. - West Newbury, Eaton (1896); Ipswich, Oakes.

FRAGARIA L. Strawberry.

F. virginiana Duchesne. Fields and open slopes; common. - Haverhill, H. 424 (1932); Andover, Pease 694 (1901); North Andover, Zink (1943); Boxford, Frye (1881); West Newbury, H. 31892 (1968); Newbury, H. 10410 (1955); Rowley, H. 10437 (1955); Rowley, Plum Island, H., McGregor & Schweinfurth 12713 (1957); Ipswich, G.M. Allen (1912); Ipswich, Plum Island, H., McGregor &

Schweinfurth 12674 (1957); Topsfield, B.B. Towne; Danvers, M.W. Nichols (1891) fruit white; Rockport, Anderson, Shafer & Smith (1931); West Manchester, Hubbard (1913); Beverly Farms, Shaw (1910); Salem, Cole (1899); Swampscott, M.K.R. (1910).

X. *F. ananassa* Duchesne. Cultivated Strawberry. Escape from cultivation. - Georgetown, H. 13573 (1957); Beverly, H. 18900 (1958); Saugus, H. 18633 (1958).

F. vesca L. Woodland Strawberry. Rocky woods and openings; rare. - Haverhill, H. 51 (1931); Boxford, C.E. Faxon, Robinson & Sears (1877), Alcott (1882).

WALDSTEINIA Willd.

W. fragarioides (Michx.) Tratt. Barren Strawberry. Introduced ? - Danvers, Haswell & Woods, Miss M.W. Nichols (1869, 1891).

POTENTILLA L. Cinquefoil.

P. fruticosa L. Shrubby Cinquefoil. Wet or dry open ground; occasional but locally abundant. - Boxford, M.E. Perley (1881); Georgetown, Horner; Rowley, H. 10440 (1955); Ipswich, Oakes; Topsfield. H.R. Towne; Danvers, Sears (1880); Essex, Knowlton (1907); Lynnfield, Bean (1909).

P. tridentata Ait. Three-toothed Cinquefoil. Occasional near the seashore. - Newbury, Schweinfurth (1950); Ipswich, Plum Island, St. John & White 572 (1913); H., McGregor & Schweinfurth 12693 (1957); Topsfield, H.S. Porter; Hamilton, Wentworth (1903); Gloucester, Russell (1861), H. 10324 (1955); Rockport, Smith 610 (1927); Salem Neck, Sears (1886); Marblehead, Lindsay (1883); Bean & H. 9729 (1955).

P. palustris (L.) Scop. Marsh-Five-Finger. Wet meadows and swales; rare. - Topsfield, Pierce; Wenham, Cole (1903); Lynnfield, Young (1879); Essex County, Oakes.

P. arguta Pursh. Tall Cinquefoil. Rocky or alluvial soils; occasional. - Andover, Pease 708 (1900); Georgetown, Horner (1881); Rowley, Nelson's Island, Sears (1888); Rowley, Plum Island, H. 12056 (1956); Salem, Oakes; Salem, Misery Island, Cole (1898); Lynnfield, Horner (1877).

P. argentea L. Silvery Cinquefoil. Naturalized from Europe. Dry open ground; common. - Salisbury, White 72 (1913); Haverhill, H. 619 (1932); Methuen, H. 9616 (1954); Andover, Sutton 189 (1874); Boxford, M.P. Perley (1883); Newbury, H. 8235 (1952); Georgetown, Horner; Ipswich, H. 9578 (1954); Ipswich, Plum Island, St. John & White 588 (1913); Topsfield, S.J. Towne; Middleton, Bean & H. 7781 (1951); Danvers, H. 7968 (1952); Wenham, Robinson (1875); Gloucester, Eastern Point, Prince (1883); West Manchester, Hubbard (1913); Beverly, H. 10191 (1955); Salem, Cole (1910); Peabody, Mackintosh (1933); Marblehead, Bean & H. 9753 (1955).

P. recta L. Naturalized from Europe. Dry fields and roadsides; first appeared 1886 and now a common weed. - Salisbury, H. 12223 (1956); Amesbury, H. 10528 (1955); Andover, H. 7958 (1952); North Andover, Pease 28053 (1940); Rowley, H. & Wetherall 9560 (1954); Ipswich, H. 9583 (1954); Topsfield, A. Dow; Danvers, Sears (1899); Wenham, H. 10004 (1955); Gloucester, Annisquam, E.S. Morse (1899); Manchester, H. 10717 (1955); Beverly, H. 10207 (1955); Salem, Robinson (1886); Peabody, Mackintosh (1933); Swampscott, H. 10651 (1955); Marblehead, Bean & H. 9755 (1955).

P. norvegica L. Thickets, roadsides and waste places; common. - Salisbury, White 29 (1913); Haverhill, H. 733 (1932); North Andover, Root & Zink (1945); Boxford, M.E. Perley (1885);

Groveland, Bean & H. 8515 (1953); Newburyport, Plum Island, H. 13049 (1957); Georgetown, Horner; Ipswich, Plum Island, H., McGregor & Schweinfurth 12659 (1957); Topsfield, H.R. Towne; Wenham, Robinson (1875); Rockport, L.B. & R.C. Smith (1934); Manchester, Hubbard (1913); Beverly, Lowell (1846); Salem, H. 9703 (1955); Peabody, Mackintosh (1933); Lynn, Cushman 2984 (1908).

P. reptans L. Adventive from Europe. Waste places; rare. - Haverhill, Bradford, H. 18812 (1958).

P. canadensis L. Dry open soil; common. - Salisbury, Cushing (1819); Haverhill, H. 308 (1932); Andover, Pease 709 (1901); North Andover, Zink (1943); Boxford, M.E. Perley (1882); Georgetown, Horner; Ipswich, Oakes; Ipswich, Plum Island, H. 12105 (1956); Topsfield, H.R. Towne; Wenham, Robinson (1874); West Manchester, Hubbard (1913); Beverly, H. 10168 (1955); Salem, Cole (1902); Peabody, Mackintosh (1925); Marblehead, H. 9684 (1955).

P. simplex Michx. Old-field Cinquefoil. Dry or moist fields and thickets; frequent. - Haverhill, Ward Hill, Cushman 2926 (1908); Lawrence, Morss (1900); Andover, Pease 3649 (1904); Boxford, H.E. Parkhurst (1882); Newburyport, E.N. Cole (1861); Newbury, H. 10415 (1955); Ipswich, Oakes; Topsfield, H.R. Towne; Essex, C.E. Brown (1912); Rockport, H. 10269 (1955); Beverly Farms, Shaw (1910); Saugus, Bean & H. 9792 (1955); [MNS, Snyder (1973)].

P. simplex, var. *calvescens* Fern. - Middleton, M.E. Perley (1884); Wenham, Robinson (1875); Gloucester, Eastern Point, Rand & B.L. Robinson (1896); West Manchester, Hubbard (1913).

P. egedei Normsk., var. *groenlandica* (Tratt.) Polunin. Seacoast; common. - Salisbury, Cushing (1819); Newbury, W.S. Horner; Newbury, Plum Island, H. 12024 (1956); Rowley, Pingree Island, Horner; Ipswich, Oakes; Ipswich, Plum Island, H. 12117 (1956); East Gloucester, Swan (1881); Rockport, Smith 605 (1927); West Manchester, Hubbard (1913); South Salem, Robinson (1872); Peabody, Mackintosh (1925); Marblehead, H. 9674 (1955).

FILIPENDULA Mill.

F. rubra (Hill) Robins. Queen-of-the-Prairie. Introduced from the West. Escaping from cultivation; occasional. - Georgetown, H. 10932 (1955); Wenham, Robinson (1875); West Manchester, Sears (1890).

F. hexapetala Gilib. Dropwort. Introduced from Europe. Escape from cultivation; occasional. - Boxford, H. 12746 (1957); Marblehead, Bean & H. 9724 (1955).

GEUM L. Avens.

G. canadense Jacq. Thickets and borders of woods; occasional. - West Boxford, Root (1945); West Newbury, Bean (1940); South Georgetown, Horner; Ipswich, Oakes; Danvers, Lowell (1846); Rockport, Pigeon Cove, Robinson (1877).

G. canadense, var. *camporum* (Rydb.) Fern. Common. - Salisbury, H. 12227 (1956); Merrimac, H. 10569 (1955); South Lawrence, Morss (1884); Andover, Pease 701 (1901); Boxford, M.E. Perley (1884); Groveland, H. 10971 (1955); West Newbury, H. 10940 (1955); Newburyport, White 194 (1913); Rowley, H. 12233 (1955); Ipswich, H. 18162 (1958); Topsfield, H.R. Towne; Danvers, Sears (1885); Gloucester, Williams (1897); West Manchester, Hubbard (1913); Salem, Robinson (1877); Peabody, Mackintosh (1933); Marblehead, Cole (1927).

G. laciniatum Murr. Damp thickets and meadows; occasional. - Lawrence, H. 10097 (1955); Groveland, Bean & H. 8550 (1953); Danvers, Sears (1885).

G. aleppicum Jacq., var. *strictum* (Ait.) Fern. Thickets and meadows; occasional. - Ipswich, Oakes; Topsfield, H.S. Porter, Middleton, Manning (1884); Wenham, Robinson (1875).

G. urbanum L. Adventive from Europe. Waste places; rare. - South Salem, Sears (1885).

G. rivale L. Water-Avens. Wet meadows and bogs; frequent. - Haverhill, Palfrey (1890); Andover, Pease 704 (1901); North Andover, Morss (1884); West Boxford, Horner; Groveland, Bean & H. 8484 (1953); Georgetown, Horner; Topsfield, B.B. Towne; Danvers, Oakes; Swampscott, Swan (1886).

RUBUS L. Bramble.

R. pubescens Raf. Damp slopes and low thickets; occasional. - Amesbury, H. 18272 (1958); Methuen, H. 18377 (1958); Andover, Pease 741 (1901); Boxford, Cushman 764 (1907); Georgetown, Horner; Ipswich, H. 18132 (1958); Topsfield, Mrs. Herrick; Middleton, H. 18078 (1958); Danvers, Sears (1877); Wenham, Sears (1899); Marblehead, Cole.

R. odoratus L. Purple-flowering Raspberry. Thickets and borders of woods; occasional. - Merrimac, Bean & H. 8591 (1953); Haverhill, H. 1837 (1935); North Andover, Root (1945); Boxford, M.E. Perley (1883); Topsfield, Mrs. G.W. Towne; Wenham, Robinson (1875).

R. idaeus L., var. *strigosus* (Michx.) Maxim. Raspberry. Thickets and clearings; common. - Haverhill, H. 620 (1932); Andover, Pease 2016 (1903); Boxford, M.E. Perley (1884); Ipswich, H. 18427 (1958); Ipswich, Plum Island, H., McGregor & Schweinfurth 12681 (1957); Topsfield, H.R. Towne; Danvers, Sears (1885); Wenham, Robinson (1875); Manchester, Hubbard (1913).

R. idaeus, var. *canadensis* Richards - Andover, H. 17767 (1958); Rowley, H. 10430 (1955); Rowley, Plum Island, H. 12452 (1956); Salem, Little Misery Island, H. 13287 (1957).

X *R. neglectus* Peck. A series of hybrids of *R. idaeus,* var. *strigosus;* usually sterile. - Andover, Pease 2125 (1903).

R. occidentalis L. Black Raspberry. Rich thickets and borders of woods; common. - Merrimac, H. 10579 (1955); Lawrence, H. 9846 (1955); Andover, Sutton 104 (1874); North Andover, H. 18009 (1958); Boxford, M.E. Perley (1885); Groveland, Bean & H. 8555 (1953); West Newbury, Bean (1951); Newbury, Cushing (1819); Ipswich, H. 18166 (1958); Topsfield, Mrs. H. Gould; Danvers, Sears (1885); Wenham, Robinson (1875).

R. flagellaris Willd. Dry fields and borders of thickets; frequent. - Amesbury, Rich (1906); Merrimac, H. 18331 (1958); Methuen, H. 18394 (1958); North Andover, H. 17945 (1958); Boxford, M.E. Perley (1882); Newburyport, Williams (1898); Georgetown, Horner; Ipswich, H. 13909 (1957); Topsfield, H.H. Roberts; Wenham, Robinson (1875); Essex, Hubbard (1913); Gloucester, Dogtown Common, H. 12157 (1956); Marblehead, Cole (1931).

R. tetricus Bailey. Dry open soil; occasional. - Salisbury, White 15 (1913); Andover, Pease 738 (1902); Ipswich, Plum Island, St. John & White 575 (1913), H., McGregor & Schweinfurth 12696 (1957).

R. baileyanus Britt. Dry to moist sandy open woods; occasional. - Haverhill, H. 735 (1932); Manchester, Hubbard (1913).

R. curtipes Bailey. Open habitats; rare. - Andover, Pease 2084 (1903).

R. hispidus L. Moist or dry open soil; occasional. - Amesbury, H. 18240 (1958); Rockport, Forbes 1902 (1903).

R. hispidus, var. *obovalis* (Michx.) Fern. Common. - Salisbury, White 89 (1913); Andover, Pease 743 (1901); North Andover, Morss (1885); Boxford, M.P. Perley (1884); Groveland, H. 12861 (1957); Ipswich, H. 18135 (1958); Topsfield, H.R. Towne; Middleton, H. 18093 (1958); Danvers, Lowell (1846); Wenham, Robinson (1875); Hamilton, Hubbard (1913); Beverly, Lowell (1846).

R. setosus Bigel. Swales and damp thickets; occasional. - Danvers, Sears (1877); Wenham, Robinson (1875); Salem, Sears (1885).

R. vermontanus Blanch. Dry to moist thickets; occasional. - Wenham, Robinson (1884); Lynnfield, Kennedy (1911).

R. allegheniensis Porter. Sow-teat Blackberry. Dry clearings and thickets; common. - Andover, Pease 745 (1902); Boxford, Frye (1881); West Newbury, NEBC. (1951); Goergetown, M.E. Perley (1884); Ipswich, Plum Island, H., McGregor & Schweinfurth 12690 (1957); Gloucester, Dogtown Common, H. 12159 (1956); Manchester, Hubbard (1913); Beverly, Robinson (1879); Saugus, Bean & H. 9818 (1955); Marblehead, Cole (1931).

R. allegheniensis, var. *gravesii* Fern. - Boxford, R.J. Eaton 5234 (1962).

R. floricomus Blanch. Thickets and clearings; Occasional. - Haverhill, H. 734 (1932); Andover, Pease 736 (1902); North Andover, Morss (1884).

R. pensilvanicus Poir. Borders of woods; rare. - Rowley, Bean & H. 8426 (1953); Lynn, Cushman 2987 (1908).

R. recurvans Blanch. Dry to moist thickets; rare. - Ipswich, Plum Island, H. 12107 (1956).

DALIBARDA Kalm False Violet.

D. repens L. Robin-Run-Away. Woods; rare. - Middleton, Sears (1883); Manchester, Robinson (1875).

AGRIMONIA L. Agrimony.

A. gryposepala Wallr. Thickets and borders of woods; occasional. - Amesbury, H. 18212 (1958); North Andover, Root (1945); Boxford, M.E. Perley (1883); Newbury, Palmer 46289 (1942); Ipswich, Oakes; Topsfield, M. Todd; Danvers, Sears (1904); Chebacco Woods, Robinson (1874); Marblehead, Cole (1899).

A. striata Michx. Thickets and borders of woods; frequent. - Haverhill, H. 52 (1931); Andover, Pease 677 (1902); West Newbury, H. 10954 (1955); Newbury, H. 12200 (1956); Georgetown, Williams (1906); Ipswich, Knowlton (1909); Danvers, Sears (1904).

SANGUISORBA L. Burnet.

S. canadensis L. Canadian Burnet. Peaty or boggy soil; most common near coast. - Salisbury, White 322 (1913); Amesbury, Bean (1950); Methuen, H. 8928 (1953); Andover, Root (1945);

Boxford, Williams (1906); West Newbury, Cole (1899); Newburyport, Bean (1925); Newbury, Churchill (1897); Newbury, Plum Island, H. 18604 (1958); Georgetown, Horner (1884); Rowley, H. 8248 (1952); Ipswich, Oakes; Topsfield, Sears (1880); Wenham, Fernald, Hunnewell & Long 9687 (1913); Essex, Bean (1938); Rockport, Williams (1898); Manchester, Lothrop (1927); Beverly, Lowell (1849).

ROSA L. Rose.

R. multiflora Thunb. Introduced from eastern Asia. Escape from cultivation; occasional. - Merrimac, H. 18365 (1958); Newburyport, H. 10200 (1955); Georgetown, H. 17683 (1958); Beverly, H. 10345 (1955).

R. gallica L. French Rose. Introduced from Europe. Escape from cultivation; occasional. - Andover, Pease 2185 (1903); Marblehead, Sears (1885).

R. eglanteria L. Sweet-Brier. Introduced from Europe. Thickets and roadsides; occasional. - Andover, Pease 2127 (1903); Boxford, M.E. Perley (1885); Topsfield, N.R. Towne; Middleton, Sears (1887); Danvers, Sears (1885); Wenham, Robinson (1875); Manchester, Lynnfield, Cummin.

R. micrantha Sm. Introduced from Europe. Thickets and roadsides; occasional. - Andover, Pease 2128 (1903); Georgetown, Horner (1882); Rockport, Sears (1890).

R. canina L. Dog-Rose. Introduced from Europe. Dry banks and thickets; occasional. - Andover, Pease 3779 (1904); Ipswich, H. & F.C. Wade 18782 (1958).

R. nitida Willd. Bogs, wet thickets and pond margins; frequent. - Haverhill, H. 621 (1932); Boxford, M.E. Perley (1885); Newbury, H. 18455 (1958); Ipswich, Oakes; Danvers, Sears (1886); Wenham, Sears (1886); Gloucester, Sears (1890); Rockport, Sears (1890); Salem, Bakers Island, Robinson (1890); Peabody, Sears (1885); Lynnfield, Bean (1919).

R. virginiana Mill. Damp to dry thickets, swamps and shores; common. - Salisbury, White 62 (1913); Amesbury, H. 18192 (1958); Merrimac, H. 18334 (1958); Andover, Pease 732 (1902); North Andover, Morss (1885); Boxford, M.E. Perlev (1885); West Newbury, Bean & Pease (1940); Newbury, White 427 (1913); Georgetown, Horner; Rowley, Plum Island, H. 12451 (1956); Ipswich, H. 13894 (1957); Middleton, H. 18099 (1958); Danvers, Sears (1885); Wenham, Robinson (1875); Gloucester, Magnolia, Jenks (1882); Rockport, H. 10304 (1955); Manchester; Salem, Bakers Island, Robinson (1890); West Peabody, Mackintosh (1933); Marblehead Neck, Sears (1885).

R. palustris Marsh. Swamps, wet thickets and shores; frequent. - Methuen, H. 9628 (1954); Lawrence, H. 10094 (1956); Andover, Pease 730 (1902); West Newbury, H. 9131 (1953); Ipswich, Plum Island, H., McGregor & Schweinfurth 12695 (1957); Middleton, Bean & H. 7765 (1952); Danvers, Sears (1885); Wenham, Robinson (1875); Rockport, Sears (1890); West Peabody, Sears (1885); Swampscott, Sears (1885); Marblehead, Sears (1885).

R. carolina L. Dry sandy, rocky or open habitats; frequent. - Haverhill, H. 622 (1932); Andover, Pease 2174 (1903); Rowley, Plum Island, H., McGregor & Schweinfurth 12711 (1957); Ipswich, Plum Island, H., McGregor & Schweinfurth 12691 (1957); Topsfield, H. 18308 (1958); Wenham, Robinson (1875); Gloucester, Eastern Point, Prince (1883); Rockport, Forbes 1971 (1902); Beverly Farms, Hubbard (1913); Salem, Cole (1902); Saugus, M.K. Crosby (1861); Marblehead, Sears (1885).

R. rugosa Thunb. Introduced from eastern Asia. Seashore thickets and dune sands; occasional. - Andover, H. 17916 (1958); Newburyport, White 499 (1913); Newburyport, Plum Island, H. 17768 (1958); Ipswich, H. 12641 (1957); Ipswich, Plum Island, H., McGregor & Schweinfurth 12664 (1957); Gloucester, H. 12343 (1956); Marblehead, Cole (1931).

R. cinnamomea L. Cinnamon Rose. Introduced from Europe. Roadsides and fencerows; occasional. - Amesbury, Robinson (1879).

PRUNUS L.

P. cerasifera Ehrh. Cherry Plum. Introduced from Asia. Escape from cultivation. - Boxford, H. 19067 (1959).

P. domestica L. Plum. Introduced from Asia. Escape from cultivation. - Boxford, H. 9655 (1955).

P. maritima Marsh. Beach-Plum. Sandy soil along or near coast; common. - Amesbury, Robinson (1879); Merrimac, H. 18187 (1958); Haverhill, H. 482 (1932); Andover, Pease 2660 (1902); North Andover, Morss (1885); Boxford, M.E. Perley (1884); Newbury, Plum Island, H. 12021 (1956); Ipswich, Plum Island, St. John & White (1913); Topsfield, B.B. Towne; Middleton, H. 9659 (1955); Beverly, Robinson (1879); Salem, A.P. Morse (1924); Peabody, Mackintosh (1925); Nahant, Keyes (1886).

P. nigra Ait. Canada Plum. Thickets and stream-banks; rare. - Boxford, M.P. Perley (1884); East Boxford, Robinson (1891); Essex County, Oakes.

P. persica (L.) Batsch. Peach. Introduced from Asia. Spread from cultivation; occasional. - Boxford, H. 9654 (1955); Georgetown, H. 13595 (1957); Gloucester, H. 12344 (1956); Salem, Cole (1928); Saugus, H. 18269 (1958); Marblehead, Cole (1928); [MNS Synder, 1971].

P. susquehanae Willd. Sand-Cherry. Sandy or other acid habitats; occasional. - South Lawrence, Morss (1886); Andover, Lowell Junction, Pease 2450 (1903); West Newbury, H. 31904 (1968); Peabody, Sears (1885).

P. pensylvanica L. f. Bird-Cherry. Dry woods and openings; frequent. - Haverhill, H. 53 (1931); Metheun, Root 17629 (1958); Andover, Pease 722 (1901); North Andover, Morss (1886); Boxford, Frye (1881); West Newbury, H. 31895 (1968); Chebacco, Robinson (1879); Wenham, Cole (1924); Essex, Buck (1897); Gloucester, Brier Neck, Smith 936 (1928); Manchester, Oakes; Beverly Farms, Shaw (1910); Peabody, Mackintosh (1915); Lynn, Young (1880); Marblehead, Cole (1900); [MNS, Snyder, 1971].

P. avium L. Sweet Cherry. Introduced from Eurasia. Spread from cultivation; occasional. - Andover, Sutton 24 (1874); Salem, Cole (1900); Marblehead, Cole (1928).

P. cerasus L. Sour Cherry. Introduced from Asia. Roadside thickets; occasional. - between Topsfield and Beverly, C.E. Faxon & Robinson (1891).

P. mahaleb L. Perfumed Cherry. Introduced from Eurasia. Spread from cultivation; occasional. - South Salem, Sears (1887).

P. serotina Ehrh. Black Cherry. Dry woods and fencerows; common. - Haverhill, H. 688b (1932); North Andover, Morss (1900); Boxford, M.E. Perley (1885); Newbury, Plum Island, H. 12014

(1956); Rowley, H. 12236 (1956); Rowley, Plum Island, H. 12052 (1956); Ipswich, H. 13922 (1957); Ipswich, Plum Island, H. 12110 (1956); Topsfield, H.H. Roberts; Danvers, Sears (1885); Wenham, Robinson (1879); Gloucester, Dogtown Common, H. 12177 (1956); Manchester, Williams (1895); Marblehead, H. 9700 (1955).

P. serotina X *P. virginiana* - Newburyport, McGregor (1939).

P. virginiana L. Choke-Cherry. Thickets, shores and borders of woods; common. - Haverhill, H. 688a (1932); Methuen, H. 11000 (1955); Andover, Robinson (1884); North Andover, Root (1943); Boxford, M.E. Perley (1883); Groveland, Bean & H. 8526 (1953); West Newbury, H. 31903 (1968); Newburyport, E.M. Cole (1861); Newbury, H. 18464 (1958); Newbury, Plum Island, H. 131 (1930); Rowley, Plum Island, H., McGregor & Schweinfurth 12725 (1957); Ipswich, Buck (1896); Topsfield, H.R. Towne; Middleton, H. 7878 (1952); Danvers, Robinson (1879); Hamilton, H. 10804 (1955); Salem, Robinson (1879); Marblehead, H. 9699 (1955).

LEGUMINOSAE (Pulse Family).

GLEDITSIA L. Honey-Locust.

G. triacanthos L. Honey-shuck. Introduced from South. Spread from cultivation; occasional. - Methuen, H. 11709 (1955); Lawrence, Root (1949); [W. Newbury, Snyder (1950's)] ; Newburyport, Cole (1901); Saugus, Young (1879); Lynn, H. 18822 (1958).

G. triacanthos, forma *inermis* (Pursh). Schneid. - Topsfield, H. 11369 (1955); Saugus, S. 18693 (1958); Lynn, H. 11806 (1955).

CASSIA L. Senna.

C. hebecarpa Fern. Wild Senna. Alluvial soil; rare. - West Boxford, Horner (1882); Georgetown, Tenney (1872).

C. tora L. Sicklepod. Adventive from South. Waste places; rare. - Gloucester, Magnolia, H. 12369 (1956).

C. fasciculata Michx. Partridge-Pea. Adventive from West; rare. - Rowley, Miss B. Dodge (1885).

C. nictitans L. Wild Sensitive-Plant. Sandy soil; occasional in Merrimack Valley. - Salisbury, Horner (1885); Amesbury, Eaton (1896); Merrimac, Cole (1900); Haverhill, Ward Hill, Bean (1916); Groveland, Russell (1859); West Newbury, H. 1269 (1934); Newburyport, Deer Island, Robinson (1879).

BAPTISIA Vent. False Indigo.

B. tinctoria (L.) R. Br. Dry open woods and clearings; common. - Haverhill, Ward Hill, Cushman 2937 (1908); North Andover, Morss (1879); Boxford, M.E. Perley (1883); Groveland, Williams (1906); Newburyport, White 458 (1913); Georgetown, Horner (1885); Rowley, H. 12244 (1956); Topsfield, H.R. Towne; Danvers, Frost; Wenham, Sears (1897); Essex, Fernald, Hunnewell & Long 9715 (1913); Beverly, Lowell (1853); Peabody, Mackintosh (1933); Lynnfield, Cole (1898); Saugus, H. 10857 (1955); Lynn, H. 10917 (1955).

THERMOPSIS R. Br.

T. mollis (Michx.) M.A. Curtis. Introduced from South; rare. - Newburyport, H.F. Harris 19068 (1959). Reported from Beverly, 1945, but no specimen seen.

CROTALARIA L. Rattlebox.

C. sagittalis L. Dry sandy or gravelly soil; rare. - West Andover, Pease 759 (1901); North Andover, Miss A.M. Greene.

GENISTA L. Woad-Waxen.

G. tinctoria L. Dyer's Greenweed. Naturalized from Europe. Dry sterile soil; common in Salem, Peabody and Danvers and occasional elsewhere. - Andover, Lowell Junction, H. 17939 (1958); Boxford, H. 10968 (1955); Georgetown, W.S. Horner; Ipswich, Prince (1896); Topsfield, B.B. Towne; Middleton, Hopkins & Pease 24031 (1933); Danvers, Oakes; Hamilton, H. 10841 (1955); Manchester, Hubbard (1913); Beverly, Swan (1881); Salem, Russell (1863); Peabody, Carlton (1893); Lynnfield, Bean (1919); Lynn, Cushman 3001 (1908); Swampscott, Morss (1899); Marblehead, H. 11917 (1955); Nahant, Huggins.

ULEX L. Gorse.

U. europaeus L. Introduced from Europe. Sands and rocky headlands; rare. - Nahant, Rossbach 963 (1928).

LUPINUS L. Lupine.

L. perennis L. Wild Lupine. Dry open woods, clearings and openings; occasional in northern portion of County. - Andover, Sutton 73 (1874); West Boxford, Horner; Georgetown, Bullard (1881); Ipswich, Oakes; Middleton, Frost.

TRIFOLIUM L. Clover.

T. arvense L. Rabbit-foot-Clover. Naturalized from Europe. Dry roadsides and fields; common. - Merrimac, H. 9230 (1953); Haverhill, H. 514 (1932); Methuen, H. 9617 (1954); Lawrence, H. 11231 (1955); North Andover, Morss (1885); Boxford, Frye (1880); West Newbury, H. 10955 (1955); Newbury, Plum Island, White 145 (1913); Georgetown, Horner; Ipswich, Furber (1911); Ipswich, Plum Island, H. 12418 (1956); Topsfield, B. Dow; Hamilton, H. 10818 (1955); Gloucester, Eastern Point, Prince (1883); Rockport, Smith 717 (1927); Manchester, H. 10714 (1955); Beverly, H. 11672 (1955); Salem, Robinson (1879); Salem, Little Misery Island, H. 13299 (1957); Peabody, Mackintosh (1933); Lynnfield, Kennedy (1862); Saugus, H. 10855 (1955); Marblehead, H. 11444 (1955); Nahant, Merrill (1849).

T. arvense, forma *albiflorum* Sylven - Lawrence, H. 12293 (1956).

T. incarnatum L., var. *elatius* Gibelli & Belli. Crimson Clover. Introduced from Europe. Waste ground; occasional. - Andover, Mrs. S.E. Abbott (1898); Georgetown, Horner (1886); Essex, A.H. Clark (1925).

T. pratense L. Red Clover. Naturalized from Europe. Fields and roadsides; common. - Salisbury, White 25 (1913); Amesbury, H. 10565 (1955); Haverhill, H. 428 (1932); North Andover, Morss (1885); Boxford, Frye (1880); Newburyport, H. 10371 (1955); Newbury, Plum Island, H. 12028

(1956); Georgetown, Horner; Ipswich, Plum Island, H. 12096 (1956); Topsfield, H.R. Towne; Danvers, Frost; Wenham, Robinson (1875); Hamilton, H. 10791 (1955); Gloucester, Dogtown Common, H. 12149 (1956); Rockport, Smith 710 (1927); West Manchester, Hubbard (1913); Beverly, H. 10172 (1955); Salem, H. 10629 (1955); Peabody, Mackintosh (1933); Saugus, Bean & H. 9823 (1955); Lynn, H. 10918 (1955); Swampscott, H. 10672 (1955); Marblehead, H. 9686 (1955).

T. pratense, forma *leucochraceum* Aschers & Prantl. - Wenham, Robinson (1875); Salem, A.P. Morse (1925); Saugus, Robinson (1874).

T. pratense, var. *sativum* (Mill.) Schreb. - Newburyport, Robinson (1875); Topsfield, H. 18050 (1958).

T. medium L. Zigzag Clover. Naturalized from Europe. Fields and roadsides; occasional. - Danvers, Oakes; Salem, Russell (1863), Sears (1901); Marblehead, Chapman (1884), white flowered.

T. repens L. White Clover. Naturalized from Europe. Grasslands and roadsides; common. - Salisbury, H. 10470 (1955); Haverhill, H. 429 (1932); Methuen, H. 11725 (1955); North Andover, Morss (1884); Boxford, M.E. Perley (1881); Rowley, Plum Island, H. 12058 (1956); Ipswich, Plum Island, H. 12114 (1956); Topsfield, H.R. Towne; Danvers, Lowell (1847); Wenham, Robinson (1875); Rockport, H. 10270 (1955); Manchester, Robinson (1875); Beverly, H. 10160 (1955); Salem, McGrane (1875); Peabody, Mackintosh (1933); Saugus, H. 18611 (1958); Marblehead, H. 9687 (1955).

T. hybridum L. Alsike Clover. Introduced from Europe. Fields and roadsides; common. - Salisbury, H. 10485 (1955); Haverhill, H. 430 (1932); North Andover, Morss (1885); Groveland, M.E. Perley (1885); Newburyport, H. 10377 (1955); Newbury, H. 10388 (1955); Georgetown, Horner (1875); Rowley, H. 7937 (1952); Ipswich, Plum Island, H. 12115 (1956); Topsfield, H.R. Towne; Wenham, Bean & H. 10013 (1955); Essex, Williams (1896); Gloucester, Lanesville, Collins (1915); Manchester, Hubbard (1913); Beverly, Sears (1885); Swampscott, H. 10674 (1955); Marblehead, H. 9685 (1955).

T. agrarium L. Yellow Clover. Naturalized from Europe. Roadsides, dry fields and waste places; common. - Salisbury, H. 10484 (1955); Amesbury, H. 10563 (1955); Methuen, H. 9614 (1954); Andover, Bullard (1880); North Andover, Morss (1885); Boxford, M.P. Perley (1884); Groveland, W.S. Horner (1886); West Newbury, H. 10331 (1955); Newburyport, H. 10349 (1955); Rowley, H. 7938 (1952); Ipswich, Bean & H. 9978 (1955); Topsfield, M. Conley; Wenham, Robinson (1875); Hamilton, Sears (1896); Essex, H. 9819 (1955); Rockport, Smith 1001 (1928); Manchester, Hubbard (1913); Beverly, H. 10211 (1955); Salem, Cole (1888); Saugus, Young (1879); Lynn, A. W. & C. A. Cheever (1900).

T. procumbens L. Low Hop-Clover. Naturalized from Europe. Roadsides and waste places; frequent. - North Andover, Pease 788 (1901); Boxford, M.E. Perley (1885); Georgetown, Horner; Ipswich, Oakes; Topsfield, Mrs. Herrick; Danvers, Russell (1859); Wenham, Lowell (1847); Chebacco, Robinson (1875); Gloucester, Jenks (1881); Rockport, Williams (1898); Salem, Cole (1899); Saugus, Young (1879).

T. macraei Hook. & Arn. Introduced from California. Rare adventive. - Georgetown, Parker River Mills, Horner (1886).

T. tomentosum L. Introduced from Rare adventive. - Georgetown, Mill Yard, (Horner (1886).

MELILOTUS Mill. Sweet Clover.

M. officinalis (L.) Lam. Yellow Melilot. Naturalized from Europe. Waste or cultivated ground; common. - Haverhill, H. 519 (1932); Lawrence, Pease 3753 (1904); North Andover, Root (1945); Newburyport, H. 10379 (1955); Newbury, Plum Island, H. 13985 (1957); Topsfield, M.E. Perley (1883); Manchester, W.S. Horner; Salem, Robinson (1879); Peabody, Mackintosh (1933); Swamscott, Cole (1902).

M. alba Desr. White Melilot. Naturalized from Europe. Rich soil and roadsides; common. - Haverhill, H. 520 (1932); Methuen, H. 11760 (1955); Lawrence, H. 11213 (1955); North Andover, Root (1945); Boxford, M.E. Perley (1885); Newburyport, White 160 (1913); Newbury, H. 10390 (1955); Georgetown, Horner (1880); Rowley, Plum Island, H. 12575 (1956); Topsfield, H. 11396 (1955); Hamilton, H. 10846 (1955); Rockport, Gates & Smith 995 (1928); Beverly, Lowell (1846); Salem, Robinson (1879); Peabody, Mackintosh (1933); Swampscott, H. 10675 (1955); Marblehead, H. 11481 (1955).

MEDICAGO L. Medick.

M. sativa L. Alfalfa. Naturalized from Old World. Roadsides and old fields; occasional. - Andover, Pease 773 (1902); North Andover, Root (1945); Newburyport, Cole (1902); Topsfield, Horner (1885); Danvers, Sears (1879).

M. falcata L. Adventive from Europe. Roadsides and waste places; rare. - Middleton, Manning (1884, 1886).

M. lupulina L. Black Medick. Naturalized from Europe. Roadsides and waste places; common. - Haverhill, H. 431 (1932); Andover, Sutton 87 (1847); Boxford, M.E. Perley (1885); Newburyport, H. 10348 (1955); Georgetown, Horner; Rowley, Plum Island, H. 12059 (1956); Ipswich, H. & Wade 18781 (1958); Topsfield, H.R. Towne; Danvers, H. 7981 (1952); Gloucester, Swan (1881); West Manchester, Hubbard (1913); Beverly, Lowell (1849); Salem, Robinson (1877); Saugus, H. 10899 (1955); Marblehead, Russell (1858).

M. lupulina, var. *glandulosa* Neilr. - Georgetown, Horner; Ipswich, Plum Island, H. 12086 (1956); Beverly, Barlow (1888); Swamspcott, H. 10671 (1955).

M. arabica (L.) Huds. Spotted Medick. Adventive from Old World. Waste places; rare. - North Lawrence, Pease 776 (1902); Georgetown, Parker River Mills, Horner (1885).

M. hispida Gaertn. Bur-Clover. Adventive from Europe. Waste places; occasional. - Lawrence, Pease 775 (1902); North Andover, Pease 2663 (1903); Georgetown, Parker River Mills, Horner (1885).

LOTUS L. Birdsfoot-Trefoil.

L. corniculatus L. Introduced from Europe. Roadsides, a recent arrival becoming established in County. - Rowley, F.C. Wade 26810 (1964); Ipswich, Wade 26812 (1964); Hamilton, Wade 26811 (1964).

AMORPHA L.

A. fruticosa L. False Indigo. Introduced from the South. Escape from cultivation; occasional. - Haverhill, H. 14157 (1957); Rockport, Prybot 31783 (1967); Beverly, Mackintosh (1933); Salem, Cole (1930); Lynn, Cushman 2993 (1908); Marblehead, H. 18914 (1958).

DALEA Juss.

D. alopecuroides Willd. Foxtail-Dalea. Adventive from West. Roadsides and waste places; rare. - Rowley, Plum Island, Snyder (1950); Lynn, H. 11817 (1955).

TEPHROSIA Pers. Hoary Pea.

T. virginiana (L.) Pers. Goat's-Rue. Dry sandy woods and openings; rare. - Andover, West Parish, Pease 784 (1902); Topsfield, Davenport.

SESBANIA Scop.

S. exaltata (Raf.) Cory. Adventive from South. - Lynn, H. 18832 (1958).

ROBINIA L. Locust.

R. pseudo-acacia L. Black Locust. Naturalized from South. Woods and thickets; common. - Haverhill, Ward Hill, Cushman 2941 (1908); North Andover, Bean (1916); Boxford, Frye (1880); Newbury, Williams 1 (1899); Georgetown, Horner (1881); Ipswich, H. 13398 (1957); Topsfield, H.R. Towne; Danvers, Lowell (1853); Wenham, Robinson (1875); Gloucester, Churchill (1887); Rockport, H. 19050 (1958); Beverly, Robinson (1879); [MNS, Snyder (1971)]; Salem, Sears (1885); Peabody, H. 27066 (1964); Lynnfield, H. 14140 (1957); Saugus, H. 18715 (1958).

R. viscosa Vent. Clammy Locust. Introduced from South. Escape from cultivation; occasional. - Gerogetown, M.E. Perley (1881); Topsfield, Bean & H. 9964 (1955); Wenham, Robinson (1875); Gloucester, Cole (1903).

R. hispida L. Bristly Locust. Introduced from South. Escape from cultivation; occasional. - Boxford, H. 9921 (1955); Georgetown, H. 9718 (1955); Topsfield, H.R. Towne.

COLUTEA L. Bladder-Senna.

C. arborescens L. Introduced from Europe. Escape from cultivation; occasional. - Andover, H. 17917 (1958); Boxford, M.E. Perley, H. 8622 (1953); Danvers, Fowler.

GLYCYRRHIZA L. Licorice.

G. lepidota (Nutt.) Pursh. Wild Licorice. Adventive from West. Waste places; rare. - Methuen, H. 11742 (1955); Lawrence, Pease 2160 (1903).

DESMODIUM Desv. Tick-Trefoil.

D. nudiflorum (L.) DC. Rich or dry woods; frequent. - Haverhill, H. 552 (1932); Andover, Pease 763 (1895); North Andover, Root (1945); Boxford, M.E. Perley (1883); Georgetown, Horner (1884); Middleton, G. Towne; Danvers, Oakes; Gloucester, Magnolia, J.M. Hunnewell (1901); Lynnfield, Young (1879); Lynn, Phippen (1857).

D. glutinosum (Muhl.) Wood. Dry or rocky woods; occasional. - Haverhill, Robinson (1879); North Andover, Cole (1898); Georgetown, Horner (1881); Middleton, Mrs. G. Towne; Danvers, Sears (1887).

D. rotundifolium DC. Dry woods; occasional. - Andover, Pease 762 (1902); Boxford, R.J. Eaton & Richardson 5482 (1962); Danvers, Lowell (1847), Sears (1888); Essex County, Oakes.

D. canescens (L.) DC. Dry sandy woods; rare. - Ipswich, Oakes.

D. rigidum (Ell.) DC. Dry sandy woods; rare. - North Andover, Miss A.M. Greene; Georgetown, Horner.

D. marilandicum (L.) DC. Sandy pinelands; rare. - North Andover, Cole (1898); Essex County, Oakes.

D. canadense (L.) DC. Open woods and roadsides; common. - Amesbury, H. 10551 (1955); Merrimac, Cole (1900); Haverhill, Bradford, Morss (1885); Methuen, H. 8907 (1953); Lawrence, H. 11234 (1955); Andover, Pease 764 (1901); North Andover, H. 11040 (1955); Boxford, M.E. Perley (1883); Groveland, Horner (1880); Ipswich, Oakes; Topsfield, B.B. Towne; Danvers; Lowell (1845); Gloucester, Magnolia, J.M. Hunnewell (1901); Beverly, Lowell (1846); Peabody, Mackintosh (1933); Marblehead, H. 11446 (1955).

D. cuspidatum (Muhl.) Loud. Rich woods; rare. - Danvers, Cole (1906); Wenham Swamp, Sears (1887).

D. paniculatum (L.) DC. Borders of dry woods; rare. - North Andover, Cole (1898); Middleton, G. Towne; Danvers, Oakes, Sears (1887).

D. perplexum Schub. Sandy woods; rare. - North Andover, Bean (1947); Georgetown, Horner; Wenham Swamp, Sears (1887).

LESPEDEZA Michx. Bush-Clover.

L. procumbens Michx. Dry sandy or rocky woods and clearings; rare. - Gloucester, Magnolia, Lobster Cove, Sears (1887).

L. violacea (L.) Pers. Dry woods and thickets; rare. - West Andover, Morss (1885); Danvers, Sears (1887).

L. nuttallii Darl. Dry woods and openings; rare. - Boxford, L.B. Twitchell (1883).

L. intermedia (S. Wats.) Britt. Dry open woods and thickets; frequent. - Lawrence, H. 11218 (1955); Andover, Pease 1514 (1902); Boxford, Williams (1906); Georgetown, Horner (1874); Danvers, Lowell (1847); Wenham, Sears (1880); Gloucester, Magnolia, Deane (1901); Lynnfield, Cole (1898); Saugus, H. 10853 (1955).

L. capitata Michx. Dry open soil; frequent. - Newbury, H. 18448 (1958); Ipswich, Plum Island, Cole; Topsfield, O. Floyd & B. Woodberry; Hamilton, H. 10823 (1955); Lynnfield, Rich (1892); Saugus, H. 10854 (1955).

L. capitata, var. *vulgaris* T. & G. Common. - Merrimac, Cole; Haverhill, Ward Hill, Bean (1916); Lawrence, H. 11212 (1955); North Andover, H. 8996 (1953); Boxford, H. 8170 (1952); Groveland, Robinson (1879); Newbury, Plum Island, H. 12481 (1956); Byfield, Pease 25761 (1936); Georgetown, H. 11161 (1955); Topsfield, H.R. Towne; Peabody, Sears (1886); Lynnfield, H. 9344 (1953).

L. capitata, var. *velutina* (Bickn.) Fern. Frequent. - Haverhill, Ward Hill, Bean (1916); Andover, Pease 1515 (1902); Groveland, M.E. Perley (1885); Danvers, Lowell (1845); Lynnfield, Young (1880); Lynn, H. 10911 (1955); Marblehead, Cole (1930).

L. hirta (L.) Hornem. Dry soil; frequent. - Andover, Pease 1516 (1902); North Andover, Root (1945); Boxford, M.E. Perley (1885); Groveland, H. 11162 (1955); Georgetown, H. 11160 (1955); Danvers, Lowell (1847); West Peabody, Bean (1949); Lynnfield, Bean (1949).

L. hirta, var. *dissimulans* Fern. Occasional. - Haverhill, Bradford, Morss (1885); Groveland, Horner; West Peabody, Bean (1949); Lynnfield, Bean (1949).

VICIA L. Vetch.

V. sativa L. Spring-Vetch. Introduced from Europe. Spread from cultivation; occasional. - North Andover, Root (1945); Ipswich, Sears (1890); Topsfield, Horner; Lynn, Phippen (1859).

V. angustifolia Reichard, var. *segetalis* (Thuill.) W.D.J. Koch. Naturalized from Europe. Roadsides and waste places; occasional. - Andover, Pease 797 (1901); North Andover, Morss (1885); Wenham, Robinson (1877); Rockport, Forbes 2213 (1902); Salem, Sears (1884); "On ye beach at Nahant," Manasseh Cutler.

V. sepium L. Naturalized from Europe. Roadsides; rare. - Topsfield, Bean & H. 9950 (1955), 9951 (1955) flowers white.

V. tetrasperma (L.) Moench. Naturalized from Europe. Waste places and roadsides; occasional. - Methuen, H. 9621 (1954); North Andover, Root (1945); Newburyport, H. & Wetherall 9567 (1954).

V. tetrasperma, var. *tenuissima* Druce - Middleton, Bean (1950); Marblehead, Lindsey (1883).

V. hirsuta (L.) S.F. Gray. Naturalized from Europe. Roadsides and waste places; rare. - Newburyport, Moulton (1886).

V. cracca L. Tufted Vetch. Naturalized from Europe. Fields and thickets; common. - Amesbury, H. 10530 (1955); Haverhill, H. 484 (1932); Methuen, H. 9608 (1954); Lawrence, H. 12295 (1956); North Andover, Root (1943); West Newbury, Robinson (1879); Newburyport, H. 10382 (1955); Newbury, H. 10395 (1955); Rowley, Plum Island, H. 12048 (1956); Topsfield, H. 18292 (1958); Middleton, Bean (1950); Danversport, Sears (1886); Wenham, Bean & H. 10011 (1955); Hamilton, H. 10807 (1955); Rockport, Smith 633 (1927); Manchester, Hubbard (1913); Beverly, H. 10190 (1955); Peabody, Mackintosh (1933); Saugus, Bean & H. 9827 (1955); Swampscott, H. 10692 (1955); Marblehead, Ives (1926).

V. villosa Roth. Naturalized from Europe. Fields and roadsides; occasional. - Haverhill, Ward Hill, H. 8463 (1953); North Andover, H. 9910 (1955); Rowley, Plum Island, H. 12048 (1956); Danvers, Hathorne, St. John 517 (1912); Marblehead, Ives (1926).

LATHYRUS L. Vetchling.

L. japonicus Willd., var. *glaber* (Ser.) Fern. Beach-Pea. Gravelly seashores; common. - Salisbury, H. 10478 (1955); Newburyport, Plum Island, H. 12017 (1956); Rowley, Plum Island, H. 12572 (1956); Ipswich, M.E. Perley (1903); Ipswich, Plum Island, H., McGregor & Schweinfurth 12682

(1957); Gloucester, Eastern Point, Rand & B.L. Robinson (1896); West Manchester, Hubbard (1913); Salem Willows, Manning (1882); Marblehead, Frye (1881); Nahant, Hoar (1857).

L. japonicus, var. *pellitus* Fern. - Newburyport, Plum Island, Cole (1886); Gloucester, H. 10318 (1955); Rockport, Smith 651 (1927); Salem, Manning (1882); Marblehead, Cole (1888); Nahant, Merrill (1849).

L. palustris L., var. *pilosus* (Cham.) Ledeb. Vetchling. Damp sand, gravels and shores; occasional. - Salisbury, White 70 (1913); Newburyport, E.N. Cole (1861); Ipswich, Oakes; Topsfield, Horner; Salem, Sears (1883); Marblehead, Lindsay (1889).

L. latifolius L. Perennial Pea. Introduced from Europe. Escape from cultivation; occasional. - Amesbury, H. 12992 (1957); Newburyport, H. 10935 (1955); Salem, M.E. Perley (1882); Marblehead, Cole (1947).

PISUM L.

P. sativum L. Garden Pea. Introduced from Europe. Waste places; occasional waif. - Newbury, H. 20705 (1959); Ipswich, H. 13853 (1957); Danvers, Sears (1885); Salem, Derby Wharf, (1877).

APIOS Medic. Groundnut.

A. americana Medic. Rich thickets; common. - Salisbury, White 336 (1913); Amesbury, H. 18958 (1958); Haverhill, H. 1840 (1935); West Andover, Morss (1885); Boxford, M.E. Perley (1885); Newburyport, Cole (1886); Georgetown, Horner; Ipswich, Oakes; Topsfield, H.R. Towne; Danvers, Frost (1875); Hamilton, Hunnewell 4618 (1916); Rockport, Gates & Smith 2397 (1928) [A & W (1969)] ; Beverly, Loring (1876); Salem, Phippen (1838-39); Peabody, Mackintosh (1933); Saugus, Cliftondale, Lund (1882).

PHASEOLUS L. Kidney-Bean.

P. vulgaris L. Introduced from South America via Europe. Escape from cultivation. - Haverhill, H. 18813 (1958); Lawrence, H. 11775 (1955); Boxford, H. 18807 (1958); Newbury, H. 20704 (1959); Danvers, H. 20763 (1959).

P. vulgaris, var. *humilis* Alef. Bush-Bean. - Haverhill, H. 12624 (1932); Boxford, H. 11564 (1955); Ipswich, H. 13854 (1957); Essex, H. 11313 (1955); Marblehead, H. 11475 (1955).

STROPHOSTYLES Ell. Wild Bean.

S. helvoela (L.) Ell. Damp thickets and shores; occasional. - Haverhill, Bradford, Morss (1885); Andover, Lowell Junction, Pease 783 (1901); Groveland, Paine (1885); Newburyport, Deer Island, Robinson & Sears (1878); Beverly, Lowell (1849); Salem, Phippen (1856); Lynn; Swampscott, Swan (1886); Nahant Beach, Williams (1894).

AMPHICARPA Ell. Hog-Bean.

A. bracteata (L.) Fern. Damp woodlands; frequent. - Haverhill, H. 676 (1932); West Andover, Morss (1895); Boxford, M.E. Perley (1881); West Newbury, H. 11134 (1955); Georgetown, Horner; Ipswich, Oakes; Topsfield, H.S. Porter; Danvers, Sears (1875); Gloucester, Magnolia, J.M. Hunnewell (1899); Beverly, Lowell (1846); Peabody, Mackintosh (1933); Marblehead, Cole (1930).

A. bracteata (L.) Fern. var. *bracteata* Groveland, Seymour 4863 (1936).

LINACEAE (Flax Family)

LINUM L. Flax.

L. usitatissimum L. Common Flax. Introduced from Europe. Waste places; occasional. - Boxford, H. 8616 (1953); Newburyport, Robinson (1879); Topsfield, Foye 18775 (1958); Danvers, H. 20764 (1959); Salem, Phippen (1857), Sears (1904).

L. virginianum L. Open woods and clearings; rare. - Andover, West Parish, Pease 2617 (1903); Byfield, Eaton (1896).

L. striatum Walt. Damp sands, peats and low woods; rare. - Georgetown, Horner (1885); Essex Woods, Lowell (1853); Robinson & Sears (1880); Peabody, Sears (1887).

OXALIDACEAE (Wood-Sorrel Family).

OXALIS L. Wood-Sorrel.

O. montana Raf. Common Wood-Sorrel. Damp woods; rare. - Salisbury, Raynal Dodge (1890).

O. violacea L. Violet Wood-Sorrel. Woods and shaded places; rare. - Ipswich, Oakes.

O. intermedia A. Richard. Introduced from the West Indies. Rare waif. - Newburyport, McGregor (1937, 1940), around greenhouses, subhardy.

O. corniculata L. Creeping Lady's-Sorrel. A cosmopolitan weed. Waste places; chiefly about greenhouses; occasional. - Andover, Pease 812 (1902); North Andover, H. 14165 (1957); Newburyport, Bean (1949), in greenhouses; Ipswich, H. 14011 (1957); Salem, Robinson (1896).

O. stricta L. Dry open soil; occasional. - Haverhill, H. 690 (1932); North Andover, Root (1945); Boxford, H. 9022 (1953); Topsfield, H. 18051 (1958); Manchester, Hubbard (1913); Lynnfield, H. 8337 (1953); Marblehead, H. 11439 (1955).

O. europea Jord. Cultivated ground and roadsides; frequent. - Salisbury, White 32 (1913); Merrimac, H. 9266 (1953); Andover, Sutton (1874); Ipswich, Plum Island, H., McGregor & Schweinfurth 12653 (1957); Topsfield, H.R. Towne; Hamilton, H. 12499 (1956); Manchester, Hubbard (1913); Lynn, floating in Flax Pond, H. 13177 (1957).

O. europea, forma *pilosella* Wieg. - Saugus, H. 18632 (1958).

O. europea, forma *cymosa* (Small) Wieg. - Haverhill, Ward Hill, H. 432 (1932); Andover, Pease 815 (1902); North Andover, Morss (1885); Boxford, H. 8769 (1953); Topsfield, H. 11368 (1955); Wenham, Robinson (1875); Hamilton, H. 10820 (1955); Salem, Robinson (1875); Peabody, Mackintosh (1933); Lynnfield, H. 14128 (1957); Saugus, H. 10870 (1955); Lynn, Cushman 2991 (1908).

O. europea, forma *villicaulis* Wieg. - Andover, Pease 813 (1902); Boxford, M.E. Perley (1881); Georgetown, H. 13586 (1957); Gloucester, Magnolia, H. 12368 (1956); Salem, Cole (1910); Marblehead, H. 11470 (1955).

TROPAEOLACEAE (Tropaeolum Family).

TROPAEOLUM L. Nasturtium.

T. majus L. Introduced from South America. Rare waif on dumps. - Lynn, H. 18844 (1958).

GERANIACEAE (Geranium Family).

GERANIUM L. Cranesbill.

G. maculatum L. Wild Geranium. Woods and thickets; common. - Haverhill, Bradford, Cushman 2948 (1908); Lawrence, H. 10083 (1955); Andover, Pease 2024 (1903); North Andover, Morss (1884); Boxford, Frye (1881); West Newbury, H. 31888 (1968); Ipswich, Oakes; Topsfield, H.R. Towne; Rockport, Smith 2630 (1932) [A & W (1967)] ; Beverly, Robinson (1879); Salem, Cole (1902); Saugus, Ripley 15529 (1915); Marblehead, Robinson (1872).

G. pratense L. Naturalized from Europe. Fields and roadsides; rare. - Topsfield, Bean & H. 9954 (1955); Swampscott, L.A. Wentworth (1903).

G. nepalense Sweet, var. *thunbergii* (Sieb. & Zucc.) Kudo. Adventive from eastern Asia. Weed in garden; rare. - Newburyport, Stockman (1953).

G. carolinianum L., var. *confertiflorum* Fern. Dry rocky or sandy soil; occasional. - Haverhill, Ward Hill, H. 9889 (1955); North Andover, H. 18035 (1958); Danvers, Sears (1878); Salem, Robinson (1879), Sears (1896).

G. pusillum L. Naturalized from Europe. Roadsides and waste places; rare. - Andover, Pease 805 (1901); Topsfield, G.T. Dow; Rockport, Smith 228 (1922).

G. robertianum L. Herb-Robert. Rocky woods and gravelly shores; frequent. - Andover, Follansbe; Rowley, Plum Island, H., McGregor & Schweinfurth 12717 (1957); Hamilton, Hunnewell (1916); Gloucester, Salt Island, A.P.Morse (1927); Rockport, Forbes (1902); Manchester, Williams (1896); Beverly, Lowell (1846); Salem, Sears (1884); Lynn, Phippen (1857); Marblehead, Cole (1886).

ERODIUM L'Her. Storksbill.

E. cicutarium (L.) L'her. Alfilaria. Naturalized from Europe. Waste places; occasional. - Methuen, H. 11727 (1955); Lawrence, H. 18765 (1958); North Andover, Robinson & Sears (1877); West Boxford, Root (1945); Georgetown, Horner; Salem, 201; Peabody, H. 18596 (1958).

E. moschatum (L.) L'Her. Adventive from Europe. Waste, often from woolen mills; rare. - Lawrence, Pease 801 (1902); Peabody, H. 18597 (1958).

E. ciconium (L.) L'Her. Adventive from Europe. Waste places; rare. - Lawrence, H. 18747 (1958).

RUTACEAE (Rue Family).

XANTHOXYLUM Gmel. Prickly Ash.

X. americanum Mill. Toothache-tree. Introduced from West; rare. - Boxford, Sears (1875); Georgetown, Robinson (1875); M.E. Perley (1882).

[*X. clava-hercules* L. Hercules-club, MNS, Snyder (1950's).]

PHELLODENDRON Rupr. Cork-tree.

P. sachalinense (Fr. Schmidt.) Sarg. Introduced from Asia. Escape from cultivation; rare. - Boxford, Root (1951).

PTELEA L. Hop-tree.

P. trifoliata (L.) L'Her. Introduced from West. Escape from cultivation; rare. - Andover, Ballardvale, Root (1951); Lynn, A.P. Morse (1919).

CITRUS

C. sp. Introduced from Asia. Seedlings occasional on dumps. - Haverhill, Bradford, H. 677 (1932); Newbury, H. 20703 (1959); Danvers, H. 18579 (1958); Peabody, H. 18593 (1958); Saugus, H. 18720 (1958).

SIMAROUBIACEAE (Quassia Family).

AILANTHUS Desf. Tree-of-Heaven.

A. altissima (Mill.) Swingle. Copal-tree. Naturalized from Asia. Escaped from cultivation; occasional. - Lawrence, Root (1949); Andover, Bullard (1880); Gloucester, Annisquam, Ripley 15918 (1915); Beverly, Robinson (1879); Salem, Sears (1884); Saugus, H. 18634 (1958); Lynn, H. 11804 (1955).

POLYGALACEAE (Milkwort Family).

POLYGALA L. Milkwort.

P. paucifolia Willd. Fringed Polygala. Woods; occasional. - Haverhill, Bradford, Morss; Methuen, Root 17626 (1958); Andover, Pease 3813 (1904); North Andover, Morss (1884); Manchester, Oakes.

P. polygama Walt., var. *obtusata* Chodat. Dry open soil; occasional. - Haverhill, Ward Hill, H. 13123 (1957); North Andover, Gutterson (1902); Boxford, M.E. & M.P. Perley (1882); Georgetown, Horner; Topsfield, A.M. Dow; Essex Woods, Lowell (1847); Gloucester, Lanesville Hoyt; West Manchester, Hubbard (1913); Peabody, Sears (1896).

P. sanguinea L. Sterile, moist, open soils; common. - Salisbury, White 293 (1913); Haverhill, H. 54 (1931); Metheun, H. 10996 (1955); North Andover, Greene (1886); Boxford, M.E. Perley (1883); Newbury, Williams (1899); Georgetown, Horner; Ipswich, Allen (1912); Topsfield, F.

Gould; Hamilton, Phippen (1857); Essex, Bean (1938); Gloucester, Prince (1883); [Rockport, A & W (1969)] ; Beverly, Cole (1899); Salem, Sears (1883); Peabody, Mackintosh (1933); Lynnfield, Boott (1854); Saugus, H. 10887 (1955); Swampscott, Cole (1889); Marblehead Neck, Mann (1849).

P. cruciata L., var. *aquilonia* Fern. & Schub. Damp peat and sands; occasional. - Andover, Gutterson (1907); Newbury, White 428 (1913); Manchester, Oakes; Beverly, Robinson (1874).

P. verticillata L. Sterile open soils; occasional. - Newburyport, Williams (1908); Danvers, Mackintosh (1924); Essex, Hunnewell (1914); Salem, Cole (1887).

P. verticillata, var. *isocycla* Fern. - Haverhill, H. 1832 (1935); North Andover, Greene (1886); Georgetown, Horner (1881); Danvers, Lowell (1847); Gloucester, Boott (1856); Salem, Russell (1859).

P. verticillata, var. *ambigua* (Nutt.) Wood - Newburyport, Bean (1945).

EUPHORBIA (Spurge Family).

MERCURIALIS L. Mercury.

M. annua L. Boys-and-Girls. Adventive from Europe. Waste places and ballast flats: rare. - Newburyport, Moulton (1886).

ACALYPHA L. Three-seeded Mercury.

A. rhomboidea Raf. Roadsides and fields; frequent. - Haverhill, H. 12515 (1956); Methuen, H. 11017 (1955); Boxford, H. 8643 (1953); West Newbury, H. 11105 (1955); Newburyport, H. 11280 (1955); Topsfield, H.R. Towne; Danvers, H. 9033 (1953); Beverly, H. 11676 (1955); Salem, A.P. Morse (1925); Peabody, Mackintosh (1933).

A. virginica L. Dry soils; occasional. - Andover, Pease 2800 (1903); Boxford, H. 9010 (1953); Georgetown, Horner; Rowley, H. 12252 (1956); Ipswich, H. & F.C. Wade 18797 (1958); Hamilton, H. 12485 (1956); Salem, Cole (1908).

A. gracilens Gray. Dry sterile soils; occasional. - Andover, Pease 9865 (1906); North Andover, Pease 2667 (1903); Boxford, M.E. Perley (1885); South Groveland, H. 13967 (1957); Danvers, H. 11626 (1955); Chebacco, Robinson (1874); Gloucester, H. 18875 (1958).

EUPHORBIA L. Spurge.

E. cyparissias L. Cypress Spurge. Naturalized from Europe. Roadsides and old fields; common. - Haverhill, H. 434 (1932); Andover, Sutton 66 (1874); North Andover, H. 7880 (1952); Boxford, M.E. Perley (1882); Ipswich, H. 9577 (1954); Topsfield, H.R. Towne; Danvers, Lowell (1848); Wenham, Robinson (1875); Rockport, Cushman 3351 (1908); Beverly, Robinson (1879); Peabody, Mackintosh (1929).

E. esula L. Naturalized from Europe. Roadsides and waste places; formerly common only about Newburyport but now widespread in the County. - Salisbury, Collins (1890); Amesbury, Knowlton (1904); Merrimac, Bean & H. 8563 (1953); Haverhill, H. 1199 (1932); Andover, H. 8045 (1952); North Andover, H. 9895 (1955); Boxford, Bean & H. 8467 (1953); West Newbury, A.W. & C.A. Cheever & Sanford (1933); Newburyport, Robinson (1875); Newbury, White 243 (1913); Newbury,

Plum Island, H. 12011 (1956); Rowley, Sears (1888); Ipswich, Oakes; Topsfield, H. 17638 (1958); Danvers, H. 17639 (1958); Wenham, H. 12133 (1956); Beverly, Mackintosh (1915); Peabody, Mackintosh (1933); Saugus, H. 7983 (1952); Nahant, Bean & H. 9835 (1955).

E. peplus L. Naturalized from Europe. Weed of cultivated and waste ground; rare. - Salem, Robinson (1901).

E. marginata Pursh. Introduced from West. Escape from cultivation; occasional. - Methuen, H. 12272 (1956); Newburyport, Bean (1945); Peabody, H. 11518 (1955); Saugus, H. 18624 (1958); Lynn, H. 27052 (1964); Marblehead, H. 18932 (1958).

E. supina Raf. Milk-Purslane. Dry open soil; common. - Haverhill, Ward Hill, H. 521 (1932); Methuen, H. 11724 (1955); North Andover, H. 11033 (1955); Boxford, M.E. Perley (1885); Groveland, H. 11178 (1955); Newburyport, H. 20738 (1959); Newbury, H. 10408 (1955); Georgetown, Horner; Topsfield, H.R. Towne; Danvers, H. 11623 (1955); Wenham, Robinson (1875); Gloucester, Magnolia, H. 12364 (1956); Beverly, Montserrat-Prides, Hubbard (1913); Salem, Forbes 2315 (1903); Peabody, H. 11538 (1955); Saugus, H. 18642 (1958).

E. polygonifolia L. Seaside Spurge. Coastal sands; common. - Salisbury, White 296 (1913); Newbury, Plum Island, Pease 3079 (1903); Rowley, Plum Island, H. 12571 (1956); Ipswich, Oakes; Ipswich, Plum Island, H. 12440 (1956); Rockport, Robinson (1875); Beverly, Lowell (1847).

E. maculata L. Dry open soil; rare. - Haverhill, (Dr. Nichols) Robinson (1875); Newburyport, Cole (1899).

CALLITRICACEAE (Water-Starwort Family).

Callitriche L. Water-Starwort.

C. deflexa A. Br., var. *austini* (Engelm.) Hegelm. Damp earth; rare. - Essex, Chebacco Lake, H. 13339 (1957).

C. stagnalis Scop. Naturalized from Europe. Running or quiet waters; rare. - Amesbury, Tuxbury Pond, H. 12997 (1957).

C. palustris L. Quiet waters and wet shores; occasional. - Merrimac, Bean & H. 8566 (1953); North Andover, H. 14087 (1957); Groveland, H. 13972 (1957).

C. heterophylla Pursh. Quiet waters and mud; common. - Amesbury, H. 13020 (1957); Metheun, Boott (1860); North Andover, Pease 2654A (1903); Boxford, Robinson (1875); West Newbury, H. 11128 (1955); Georgetown, H. 12952 (1957); Rowley, Sears; Ipswich, H. 13466 (1957); Hamilton, H. 12827 (1957); Essex, Sears (1887); Gloucester, H. 12145 (1956); Rockport, Smith 4333 (1945); Lynnfield, H. 14151 (1957).

ANACARDIACEAE (Cashew Family).

COTINUS Duham. Smoke-tree.

C. coggygria Scop. Introduced from Eurasia. Escaped from cultivation; occasional. - Boxford, M.E. Perley (1884); Beverly Farms, Hubbard (1913).

RHUS L. Sumac.

R. typhina L. Staghorn Sumac. Dry, rocky or gravelly soil; common. - Amesbury, H. 10548 (1955); Haverhill, H. 522 (1932); Groveland, Paine (1885); Newbury, White 240 (1913); Rowley, H. 10425 (1955); Rowley, Plum Island, H. 12550 (1956); Ipswich, Plum Island, H. 12432 (1956); Chebacco, Robinson (1875); Gloucester, Annisquam, Ripley 15864 (1915); Rockport, Cushman 3352 (1908); Salem, Lowell (1847); Salem, Little Misery Island, H. 13290 (1957); Saugus, Bean & H. 9810 (1955); Swampscott, Robinson (1875); Marblehead, Cole (1937).

R. glabra L. Smooth Sumac. Dry soil; common. - Amesbury, Robinson (1875); Haverhill, H. 523 (1932); Methuen, H. 9605 (1954); Andover, Pease 4235 (1904); North Andover, Morss (1885); Boxford, M.E. Perley (1880); Newburyport, H. 10341 (1955); Newbury, White 239 (1913); Georgetown, Horner; Rowley, H. 10451 (1955); Topsfield, G.T. Dow; Middleton, H. 8695 (1953); Danvers, Sears (1885); Hamilton, H. 10796 (1955); Salem, Lowell (1847); Saugus, Bean & H. 9829 (1955); Marblehead, Cole (1943).

R. copallina L., var. *latifolia* Engler. Dwarf Sumac. Dry woods and openings; common. - Salisbury, White 372 (1913); Haverhill, H. 577 (1932); West Andover, Morss (1885); Boxford, M.E. Perley (1885); West Newbury, NEBC. (1951); Newburyport, H. 10342 (1955); Newbury, H. 18447 (1958); Newbury, Plum Island, H. 11293 (1955); Georgetown, Horner; Rowley, H. 8289 (1952); Rowley, Plum Island, H. 12066 (1956); Topsfield, 199; Gloucester, Annisquam, Ripley 15873 (1915); Rockport, Pigeon Cove, Williams (1897); Manchester, Williams (1895); Beverly, Lowell (1846); Lynn, Russell (1859).

R. vernix L. Poison Sumac. Wooded swamps; all too common but seldom collected. - West Boxford, Zink (1944); Middleton, Sears (1887); [Gloucester, A & W (1969)]; Beverly, Montserrat-Prides, Hubbard (1913); Peabody, Sears (1884); Swampscott, Buck (1896); Marblehead, Cole (1926).

R. radicans L. Poison Ivy. Everywhere; common but collected only by dedicated botanists. - Plum Island, St. John 852 (1913).

R. radicans, var. *vulgaris* (Michx.) DC. - Andover, Pease 836 (1902); Topsfield, H.R. Towne; Salem, Robinson (1880); Peabody, Sears (1884); Marblehead, Cole (1931).

R. radicans, var. *rydbergii* (Small) Rehd. - Haverhill, Bradford, Churchill (1914); Rockport, Rouleau 4300 (1945).

AQUIFOLIACEAE (Holly Family).

ILEX L. Holly.

I. opaca Ait. American Holly. Moist woodlands; very rare. - Rockport, Lufkin (three leaves). Said by Robinson to have come from Dogtown Common from trees of considerable size.

I. verticillata (L.) Gray. Black Alder. Swamps and damp thickets; common. - Salisbury, White 80 (1913); Merrimac, H. 13688 (1957); Haverhill, H. 524 (1932); Andover, Pease 3785 (1904); North Andover, Root (1945); Boxford, Frye (1880); West Newbury, White 518 (1913); Newbury, Plum Island, H., McGregor & Schweinfurth 12737 (1957); Georgetown, Horner (1875); Rowley, Plum Island, H., McGregor & Schweinfurth 12730 (1957); Ipswich, Ware (1911); Ipswich, Plum Island, St. John & White 544 (1913); Topsfield, H.R. Towne; Middleton, Bean (1950); Wenham, Robinson

(1875); Hamilton, Knowlton (1907); Chebacco, Robinson (1879); Gloucester, Magnolia, J.M. Hunnewell (1898); Manchester, Hubbard (1913); Beverly, Lowell (1849); Salem, Sears (1885); Peabody, Sears (1884); Saugus, Ripley 15518 (1915); [MNS, Snyder (1950-1971)] ; Lynn, Forbes (1911).

I. verticillata, forma *chrysocarpa* Robins. - West Boxford, Cole (1884); Newburyport, White 478 (1913).

I. verticillata, var. *padifolia* (Willd.) T. & G. - Amesbury, H. 9239 (1953); Merrimac, H. 10521 (1955); Boxford, M.E. Perley (1885); Ipswich, H. 18163 (1958).

I. laevigata (Pursh) Gray. Smooth Winterberry. Wooded swamps; occasional. - Georgetown, Horner (1876); Peabody, Sears (1885); Lynn, Sears (1885); Essex County, Oakes.

I. glabra (L.) Gray. Inkberry. Low peaty soil; rare, apparently only in Gloucester. - Gloucester, Bass Rocks, Pease 838 (1902); Gloucester, Magnolia, Robinson & Sears (1877); Bean & H. 7799 (1951).

NEMOPANTHUS Raf. Mountain-Holly.

N. mucronata (L.) Trel. Damp woods and thickets; frequent. - Haverhill, H. 435 (1932); Andover, Sutton 76; North Andover, H. 17971 (1958); Groveland, Bean & H. 8501 (1953); Ipswich, Oakes; Hamilton, Hubbard (1913); Essex Woods, Lowell (1857); Beverly, Robinson (1877); Peabody, Sears (1885); [MNS, Snyder (1950s)] ; Lynn, Farlow & Robinson (1891).

CELASTRACEAE (Staff-Tree Family).

EUONYMUS L. Spindle-Tree.

E. fortunei (Turcz.) Hand-Maz. Introduced from Asia [MNS, Snyder (1971)] .

E. alatus (Thunb.) Sieb. Introduced from eastern Asia. Escape from cultivation; rare. - Boxford, H. 19053 (1958).

CELASTRUS L.

C. scandens L. Climbing Bittersweet. Thickets; frequent. - Salisbury, White 294 (1913); Merrimac, Bean & H. 7714 (1951); Haverhill, H. 651 (1932); Andover, Sutton 115 (1874); North Andover, M.E. Perley (1885); Boxford, Mackintosh (1924); Rowley, Plum Island, H. 12055 (1956); Ipswich, Plum Island, H., McGregor & Schweinfurth 12697 (1957); Topsfield, B.B. Towne; Danvers, Sears (1887); Salem, Robinson (1879); Peabody, Mackintosh (1933); Lynnfield, Bean (1916); Lynn, Cushman 2990 (1908); Marblehead, H. 11480 (1955).

ACERACEAE (Maple Family).

ACER L. Maple.

A. pseudo-platanus L. Sycamore Maple. Introduced from Europe. Frequently planted and occasionally escaping. - Danvers, Robinson (1875); Lynn, H. 18827 (1958); Nahant, A.P. Morse (1919).

A. ginnala Maxim. Introduced from Asia. Escape from cultivation; occasional. - Topsfield, H. 10103 (1955).

A. spicatum Lam. Mountain-Maple. Cool woods; rare. - Salisbury, Moulton; Haverhill, Ward Hill Neck, Root (1951).

A. pensylvanicum L. Striped Maple. Rich cool woods; local. - Haverhill, Ward Hill, Root (1951); Andover, Knowlton (1903); Boxford, Horner; Gloucester, Magnolia, Williams (1896); Rockport, Pool (1875); Manchester, Oakes.

A. platanoides L. Norway Maple. Introduced from Europe. A common shade tree; occasionally escaping. - Metheun, Root (1929); North Andover, Root (1957); Newbury, H. 10414 (1955); Topsfield, G.T. Dow; Salem, Robinson (1875); Marblehead, Sears (1887).

A. saccharum Marsh. Often cultivated but some apparently wild. - North Andover, Root (1949); Boxford, Sears (1875); Topsfield, H.R. Towne; Wenham, Robinson (1877); Saugus, H. 18688 (1958).

A. rubrum L. Red Maple. Swamps or uplands; common. - Amesbury, H. 10540 (1955); Haverhill, H. 57 (1931); Lawrence, H. 9881 (1955); North Andover, Root (1957); Boxford, M.E. Perley (1880); West Newbury, H. 31905 (1968); Newbury, Plum Island, White 321 (1913); Georgetown, H. 10705 (1955); Rowley, H. 10444 (1955); Rowley, Plum Island, H., McGregor & Schweinfurth 12733 (1957); Ipswich, Oakes; Ipswich, Plum Island, H. 12102 (1956); Topsfield, Robinson & Sears (1875); Hamilton, Hubbard (1913); Essex, Robinson (1875); Rockport, H. 10284 (1955); Manchester, Robinson (1875); Salem, Sears (1884); Peabody, Mackintosh (1925); Saugus, Bean & H. 9780 (1955); Swampscott, H. 10653 (1955); Marblehead, Cole (1930).

A. rubrum, var. *trilobum* K. Koch. - Salisbury, Bean & H. 8379 (1953); Middleton, H. 8692 (1953); Saugus, Bean & H. 9789 (1955).

A. saccharinum L. Silver Maple. Often cultivated and frequent as a wild tree on river-banks and bottomlands. - Haverhill, H. 56 (1931); Methuen, Root (1949); South Lawrence, Morss (1885); North Andover, Root (1957); Groveland, M.E. Perley (1885); West Newbury, H. 11114 (1955); Georgetown, Horner (1875); Ipswich, Oakes; Topsfield, Oakes; Middleton, Oakes; Danvers, Sears (1887).

A. negundo L. Box-Elder. Introduced from South. Naturalized; occasional. - Haverhill, Bradford, Fernald 11373 (1914); Methuen, Root (1949); Lawrence, H. 18754 (1958); North Andover, H. 9003 (1955); Groveland, H. 8815 (1953); Georgetown, H. 17687 (1958); Gloucester, Magnolia, Deane (1901); Peabody, Mackintosh (1933); Saugus, Bean & H. 9783 (1955).

HIPPOCASTANACEAE (Buckeye Family).

AESCULUS L. Horse-Chestnut.

A. hippocastanum L. Horse-Chestnut. Introduced from Europe Occasionally seeding from cultivated trees but it is impossible to tell if most of cited specimens were from spontaneous or cultivated trees. - West Newbury, H. 31887 (1968) spontaneous; Georgetown, H. 17674 (1958), spontaneous; Rowley, H. 18515 (1958); Topsfield, B.B. Towne; Danvers, Frost (1874); Salem, Robinson (1877); Saugus, H. 18704 (1958).

BALSAMINACEAE (Touch-Me-Not Family).

IMPATIENS L. Jewelweed.

I. capensis Meerb. Spotted Touch-Me-Not. Wet ground; common. - Amesbury, H. 18225 (1958); Haverhill, H. 525 (1932); North Andover, Morss (1885); Boxford, H.E. Parkhurst (1882); Groveland, H. 11077 (1955); Georgetown, Horner; Rowley, H. 13490 (1957); Topsfield, H.R. Towne; Wenham, H. 13743 (1957); Hamilton, Phippen (1857); Gloucester, Magnolia, J.M. Hunnewell (1898); Rockport, L.B. & R.C. Smith 2707 (1933); Manchester, Hubbard (1913); Beverly, Lowell (1846); Salem, Sears (1884); Peabody, Mackintosh (1933); Marblehead, H. 899 (1933).

I. balsamina L. Balsam. Introduced from Asia. Escape from cultivation. - Saugus, H. 18609 (1958).

RHAMNACEAE (Buckthorn Family).

RHAMNUS L. Buckthorn.

R. cathartica L. Common Buckthorn. Introduced from Europe. Open woods and fence-rows; frequent. - Andover, Pease 1937 (1903); West Boxford, Root (1949); West Newbury, H. 18056 (1958); Newbury, H. 8225 (1952); Topsfield, A.M. Dow; Salem, Robinson (1877); Peabody, Mackintosh (1933); Marblehead, Cole (1947).

R. frangula L. Alder-Buckthorn. Naturalized from Europe. Fence-rows and thickets; frequent and becoming an obnoxious weedy shrub. - North Andover, Root (1949); Boxford, H. 8615 (1953); Ipswich, H. 18424 (1958); Topsfield, H. 8051 (1952); Lynnfield, Snyder (1950).

CEANOTHUS L. Redroot.

C. americanus L. New Jersey Tea. Dry open woods; frequent. - Andover, Pease 24032 (1933); North Andover, Bean (1916); Boxford, M.E. Perley (1884); Groveland, M.E. Perley (1885); Georgetown, Horner; Ipswich, Oakes; Topsfield, H. 9639 (1954); Middleton, Bean & H. 7774 (1951); Danvers, Sears (1884); Salem, Phippen (circa 1838); Peabody, Mackintosh (1933).

C. ovatus Desf. Sandy or rocky slopes; rare. - Andover, Gowing (1888), Pease (1904); Fernald (1911).

VITACEAE (Vine Family).

PARTHENOCISSUS Planch. Virginia Creeper.

P. quinquefolia (L.) Planch. Woodbine. Woods and rocky banks; frequent. - Amesbury, H. 18250 (1958); Merrimac, H. 18341 (1958); Haverhill, H. 553 (1932); Andover, Pease 2446 (1903); North Andover, H. 17986 (1958); Boxford, Frye (1880); Newburyport, Bean (1945); Rowley, Plum Island, H. 12555 (1956); Ipswich, Morong (1875); Topsfield, H.R. Towne; Danvers, H. 18565 (1958); Wenham, Robinson (1875); Gloucester, Lanesville, Robinson (1877); Rockport, L.B. & R.C. Smith 4338 (1945); Beverly, Lowell (1847); Salem, Sears (1884); Saugus, H. 18622 (1958); Swampscott, H. 10680 (1955); Marblehead, H. 18939 (1958).

P. inserta (Kerner) K. Fritsch. Thickets; probably more frequent than collections indicate. - Andover, Kennedy (1908); Boxford, H. 25138 (1962); Newburyport, Knowlton, (1945); Salem, Cole (1930).

VITIS L. Grape.

V. labrusca L. Fox-Grape. Wet or dry thickets; common. - Amesbury, H. 18194 (1958); Haverhill, H. 691 (1932); Andover, Sutton 116 (1874); Boxford, M.E. Perley (1882); Ipswich, H. 13474 (1957); Topsfield, H.R. Towne; Danvers, Robinson (1875); Wenham, Sears (1887); Chebacco, Robinson (1875); West Gloucester, Robinson & Sears (1877); Beverly, Lowell (1847); Lynnfield, Bean (1919); Saugus, Ripley 15520 (1915); Marblehead, Cole (1930).

V. labrusca, forma *alba* (Prince) Fern. - Danvers, Sears (1877); Peabody, Sears (1887).

V. aestivalis Michx., var. *argentifolia* (Munson) Fern. Summer-Grape. Dry woods and thickets; occasional. - Andover, Pease 857 (1902); West Boxford, Sears (1888); Ipswich, Oakes; Middleton, Sears (1887); Gloucester, Lanesville, Robinson & Sears (1877); Peabody, Sears (1887); Lynn, Cushman 2979 (1908).

V. riparia Michx. Frost-Grape. River-banks and rich thickets; common. - Merrimac, H. 18319 (1958); Boxford, Robinson & Sears (1877); Newburyport, Cole (1901); Rowley, H. 12235 (1956); Rowley, Plum Island, H. 12552 (1956); Ipswich, Russell (1860); Ipswich, Plum Island, H., McGregor & Schweinfurth 12654 (1957); Middleton, Sears (1887); Essex, Robinson (1875); Gloucester, Robinson & Sears (1877); Beverly, Kidder (1884).

V. novae-angliae Fern. New England Grape. Rich thickets; rare. - Georgetown, Williams (1907).

V. vinifera L. Introduced from Europe. Seedlings on dumps. - Lawrence, H. 18767 (1958); Boxford, H. 12505 (1956); Rowley, H. 18489 (1958); Danvers, H. 18578 (1958); Gloucester, Magnolia, H. 12359 (1956); Beverly, H. 18890 (1958); Peabody, H. 18595 (1958); Saugus, H. 18624 (1958); Lynn, H. 18833 (1958).

TILIACEAE (Linden Family).

TILIA L. Linden. Basswood.

T. americana L. Rich woods; common. - Amesbury, Robinson (1879); Merrimac, Bean & H. 8338 (1953); Haverhill, H. 900 (1933); Lawrence, H. 10091 (1955); Andover, Pease 861 (1901); North Andover, Root (1949); Boxford, Robinson (1877); Groveland, Bean & H. 8471 (1953); West Newbury, H. 10326 (1955); Rowley, H. 10449 (1955); Ipswich, Oakes; Topsfield, M. Spofford; Danvers, Robinson (1877); Salem, Robinson (1877); Peabody, Mackintosh (1933); Swampscott, Cole (1889).

T. europaea L. European Linden. Introduced from Europe. Spread from cultivation. - Andover, Pease 2187 (1903); Boxford, M.E. Perley (1884); Georgetown, Horner; Danvers, Mrs. C. Whipple; Salem, Cole (1930); Salem, Little Misery Island H. 13300 (1957).

T. neglecta Spach. Rich woods; rare. - Newburyport, Bean (1945).

MALVACEAE (Mallow Family).

MALVA L. Mallow.

M. sylvestris L. High Mallow. Introduced from Europe. Escape from cultivation. - Topsfield, M.R. Peabody; Salem, Cole (1932); Marblehead, H. 20961 (1959).

M. neglecta Wallr. Cheeses. Naturalized from Europe. A common weed of barnyards and waste places. - Haverhill, H. 652 (1932); North Andover, Root (1945); Boxford, M.E. Perley (1882); Newburyport, H. 20728 (1959); Georgetown, Bullard (1881); Rowley, H. 11965 (1955); Ipswich, H. 11605 (1955); Topsfield, M. Clark; Wenham, Robinson (1875); Essex, H. 18857 (1958); Rockport, Forbes 2416 (1903); Salem, Sears (1884); Salem, Little Gooseberry Island, H. 13278 (1957); Saugus, H. 18651 (1958); Marblehead, H. 18953 (1958).

M. verticillata L. Adventive from Europe. Gardens and waste places; occasional. - Essex, Mackintosh (1924); Salem, Robinson (1880).

M. parviflora L. Naturalized from Europe. Waste places; occasional. - Peabody, H. 18599 (1958).

M. moschata L. Musk-Mallow. Naturalized from Europe. Escape from gardens. - Haverhill, Bradford, Mrs. C.P.U.; Andover, T.E.N. Eaton (1877); Boxford, M.E. Perley (1882); Essex, Robinson & Sears (1877).

M. moschata, forma *laciniata* (Desr.) Hayek - Topsfield, H. 18280 (1958).

ALTHAEA L.

A. rosea Cav. Hollyhock. Introduced from Eurasia. Seedlings occasional on dumps. - Metheun, H. 11748 (1955); Lawrence, H. 11766 (1955); North Andover, H. 18015 (1958); Newbury, H. 20722 (1959); Rowley, H. 11953 (1955); Ipswich, H. 11600 (1955); Beverly, H. 18894 (1958); Peabody, H. 11554 (1955); Saugus, H. 18680 (1958); Lynn, H. 11824 (1955); Marblehead, H. 11468 (1955).

SIDA L.

S. spinosa L. Prickly Mallow. Naturalized from the Tropics. Waste places; rare. - Peabody, H. 18594 (1958).

ABUTILON Mill. Indian Mallow.

A. theophrasti Medic. Velvet-leaf. Naturalized from India. Waste places; occasional. - West Newbury, H. 12309 (1956); Georgetown, Horner; Ipswich, Morong (1872); Topsfield, H. 18303 (1958); [Wenham, Dodge (1953)]; Danvers, H. 11651 (1955); Beverly, Lowell (1853); Salem, Perkins (1877) [Cole (1910)]; Peabody, H. 27074 (1964); Saugus, H. 18653 (1958).

HIBISCUS L. Rose-Mallow.

H. palustris L. Swamp-Rose-Mallow. Saline, brackish or fresh marshes; rare. - Salisbury, Horner (1882).

H. trionum L. Flower-of-an-Hour. Introduced from Europe. Waste places; occasional. - Andover, Pease 858 (1900); Wenham, Mrs. M.W. Kimball (1877).

GUTTIFERAE (St. John's-wort Family).

Hypericum L. St. John's-wort.

H. perforatum L. Common St. John's-wort. Naturalized from Europe. Dry pastures and roadsides; common. - Salisbury, H. 10472 (1955); Amesbury, H. 10562 (1955); Haverhill, H. 692 (1932); Metheun, H. 9619 (1954); Lawrence, H. 11210 (1955); Andover, Zink (1943); North Andover, Morss (1884); Boxford, M.E. Perley (1884); Newburyport, Plum Island, H. 13054 (1957); Newbury, Plum Island, H., McGregor & Schweinfurth 12378 (1957); Georgetown, Horner; Rowley, H. 12243 (1956); Rowley, Plum Island, H., McGregor & Schweinfurth 12728 (1957); Ipswich, H. 9576 (1954); Ipswich, Plum Island, H. 12419 (1956); Topsfield, H.R. Towne; Wenham, Robinson (1875); Hamilton, H. 10795 (1955); Rockport, Cushman 3361 (1908); Beverly, Robinson (1879); Salem, Forbes 2454 (1903); Saugus, H. 10859 (1955); Lynn, H. 10920 (1955); Swampscott, H. 10676 (1955).

H. punctatum Lam. Thickets and borders of woods; occasional. - Merrimac, H. 18357 (1958); North Andover, Root (1945); Boxford, H. 18537 (1958); Georgetown, Horner (1874); Ipswich, Oakes; Harris 18157 (1958).

H. spathulatum (Spach) Steud. Shrubby St. John's-wort. Introduced from the West. Escape from cultivation; rare. - Boxford, H. 8870 (1953).

H. ellipticum Hook. Damp shores and wet places; occasional. - Haverhill, Bradford, Fernald 11388 (1914); South Lawrence, Morss (1885); Andover, Pease 3811 (1904); North Andover, Root (1945); Boxford, M.E. Perley (1882); Georgetown, Horner; Ipswich Beach, Hubbard (1913); [Rockport, A & W (1969)] ; Salem, Sears (1884).

H. boreale (Britt.) Bickn. Damp sand and shallow water; common. - Salisbury, White 325 (1913); Amesbury, H. 13018 (1957); Merrimac, H. 9213 (1953); Haverhill, H. 58 (1931); Metheun, H. 10999 (1955); North Andover, H. 13085 (1957); Groveland, H. 28368 (1965); West Newbury, H. 9135 (1953); Rowley, H. 8128 (1952); Ipswich, Forbes (1913); Topsfield, H. 8856 (1953); Middleton, H. 9425 (1953); Essex, H. 13354 (1957); Gloucester, H. 13824 (1957); Rockport, Fernald, Hunnewell & Long 9896 (1913); Manchester, H. 11341 (1955); Peabody, H. 9312 (1953); Saugus, H. 13032 (1957).

H. boreale, forma *callitrichoides* Fassett. Submersed; occasional. - Haverhill, Bradford, H. 13135 (1957); North Andover, H. 13079 (1957); Boxford, H. 8741 (1953); Ipswich, H. 13262 (1957); Essex, H. 13374 (1957).

H. mutilum L., var. *parviflorum* (Willd.) Fern. Low ground; common. - Salisbury, White 355 (1913); Andover, Pease 3104 (1903); North Andover, H. 8669 (1953); Boxford, H. 9007 (1953); Georgetown, Horner; Ipswich, H. 13248 (1957); Topsfield, Williams (1903); Wenham, H. 13752 (1957); Gloucester, Prince (1883); Rockport, Smith 2718 (1933); Manchester, H. 11342 (1955); Peabody, Mackintosh (1933).

H. majus (Gray) Britt. Wet or dry open soil; frequent. - Haverhill, Ward Hill, H. 1200 (1932); Andover, H. 14088 (1957); North Andover, H. 11057 (1955); Boxford, H.E. Perley (1885); Groveland, H. 11073 (1955); Georgetown, H. 12974 (1957); Ipswich, Plum Island, H. 12415 (1956); Topsfield, H. 18277 (1958); Manchester, H. 10769 (1955).

H. canadense L. Wet or dry ground; common. - Haverhill, H. 61 (1931); Metheun, H. 8926 (1953); Andover, Swan (1882); North Andover, Morss (1885); Boxford, Williams (1899); West

Newbury, H. 10964 (1955); Georgetown, Horner; Rowley, H. 11954 (1955); Ipswich, H. 18148 (1958); Wenham, H. 13753 (1957); Essex, H. 13335 (1957); Manchester, Hubbard (1913); Salem, Cole (1886); Peabody, Cole (1910).

H. dissimulatum Bickn. Peaty or wet sandy open soil; occasional. - North Andover, H. 8670 (1953); Newburyport, H. 11285 (1955); Manchester, H. 11325 (1955).

H. gentianoides (L.) BSP. Orange-grass. Sandy, sun-baked soil; common. - Merrimac, H. 9231 (1953); Haverhill, H. 62 (1931); Andover, J. Blake 8296; North Andover, Morss (1884); Boxford, M.E. Perley (1883); Groveland, H. 10976 (1955); Newburyport, Bean (1945); Newbury, H. 18469 (1958); Newbury, Plum Island, White 310 (1913); Georgetown, Horner; Rowley, H. 11951 (1955); Rowley, Plum Island, H. 12446 (1956); Ipswich, Plum Island, H. 12445 (1956); Topsfield, Mrs. A. Dodge; Danvers, H. 11650 (1955); Wenham, Lowell (1847); Hamilton, Phippen (1857); Gloucester, Annisquam, Ripley (1915); Rockport, Pigeon Cove, Prince (1883); Peabody, Sears (1884); Marblehead, H. 11477 (1955).

H. virginicum L. Marsh-St. John's-wort. Wet sands, bogs and swamps; common. - Merrimac, H. 9220 (1953); Haverhill, Bradford, H. 13138 (1957); Methuen, H. 8911 (1953); Lawrence, Robinson (1879); Andover, Pease 4274 (1904); North Andover, Greene (1882); Boxford, M.E. Perley (1883); Groveland, H. 12869 (1957); Georgetown, Horner; Rowley, H. 13503 (1957); Ipswich, H. 13903 (1957); Ipswich, Plum Island, H., McGregor & Schweinfurth 12704 (1957); Topsfield, M. Clarke; Wenham, H. 13755 (1957); Hamilton, H. 12825 (1957); Essex, H. 13356 (1957); Gloucester, Annisquam, Ripley 15890a (1919); Rockport, Moore & Smith 1139 (1930); Salem, Cole (1908); Peabody, H. 9314 (1953); Lynnfield, Phippen (1839); Saugus, H. 13030 (1957); Marblehead, Kennedy (1887).

H. virginicum, var. *fraseri* (Spach) Fern. - Salisbury, White 377 (1913); Amesbury, H. 9194 (1953); Haverhill, H. 554 (1932).

ELATINACEAE (Waterwort Family).

ELATINE L. Waterwort.

E. minima (Nutt.) Fisch. & Mey. Margins and shores of ponds; common but easily overlooked. - Amesbury, H. 13017 (1957); Merrimac, H. 13733 (1957); Haverhill, Bradford, H. 13142 (1957); Andover, E. & C.E. Faxon (1882); North Andover, H. 14070 (1957); Boxford, H. 13787 (1957); South Groveland, H. 13979 (1957); Wenham, Fernald, Hunnewell & Long 9935 (1913); Hamilton, H. 9398 (1953); Essex, Knowlton (1908); Gloucester, H. 13849 (1957); Lynnfield, H. 14117 (1957); Lynn, Young (1880).

E. americana (Pursh) Arn. Muddy shores and margins of ponds; occasional. - Merrimac, H. 9210 (1953); Boxford, H. 13532 (1957); Danvers, Lowell (1849); Wenham, Sears (1886).

CISTACEAE (Rockrose Family).

HELIANTHEMUM Mill. Rockrose.

H. canadense (L.) Michx. Frostweed. Dry open woods and clearings; common. - Salisbury, H. 18986 (1958); Merrimac, Bean & H. 7708 (1951); Haverhill, Bradford, Bean (1915); Lawrence, H. 9876 (1955); Andover, Pease 868 (1902); North Andover, Root (1944); Boxford, Frye (1881);

Rowley, Bean & H. 8422 (1953); Rowley, Plum Island, H. 12068 (1956); Topsfield, Mrs. S. Conley; Danvers, Sears (1885); Wenham, Cole (1900); Hamilton, Hubbard (1913); Beverly Farms, Hubbard (1913); Salem, Sears (1885); Saugus, Bean & H. 9791 (1955).

H. bicknellii Fern. Frostweed. Dry open woods and clearings; common. - Merrimac, H. 9248 (1953); Haverhill, Ward Hill, Bean (1916); Andover, Pease 3760 (1904); North Andover, H. 11046 (1955); Boxford, Horner; Groveland, H. 8828 (1953); Danvers, Sears (1885); Gloucester, Lanesville, Collins (1915); Salem, Sears (1885); Lynn, H. 10914 (1955); Swampscott, H. 10641 (1955).

HUDSONIA L. Hudsonia.

H. ericoides L. Golden-Heather. Dry sands and acidic rocks; rare. - Ipswich, St. John 11890 (1916).

H. tomentosa Nutt. Poverty-grass. Sandy areas, chiefly near coast; common. - Amesbury, Eaton (1898); Newburyport, Plum Island, Cole (1901); Newbury, Plum Island, Pease 2647 (1903); Rowley, Plum Island, H. 12042 (1956); Ipswich, Oakes (1839); Ipswich, Plum Island, H. 12123 (1956); Topsfield, H. 8079 (1952); Gloucester, Jenks; Nahant, Russell (1859).

LECHEA L. Pinweed.

L. villosa Ell. Dry open woods and clearings; common. - Lawrence, H. 11236 (1955); Andover, Pease 871 (1900); North Andover, Bean (1945); Boxford, Robinson (1877); Groveland, H. 8823 (1953); Georgetown, Horner; Ipswich, H. 18431 (1958); Topsfield, Robinson (1877); Danvers, Lowell (1849); Hamilton, H. 10809 (1955); Gloucester, Mrs. Wheeler (1880); Beverly, Lowell (1847); Peabody, Sears (1885).

L. minor L. Sandy woods and clearings; rare. - Lynnfield, Bean (1932).

L. tenuifolia Michx. Dry open woods; frequent. - Boxford, Robinson (1877); Danvers, Oakes; Beverly, Lowell (1847); Peabody, Sears (1885); Saugus, H. 10850 (1955); Lynn, Russell (1859).

L. maritima Leggett. Sandy soil; common. - Merrimac, H. 9249 (1953); Haverhill, Bradford, Fernald 11380 (1914); Andover, Pease 1777 (1902); Groveland, H. 11176 (1955); Newbury, Plum Island, Pease 3080 (1903); Rowley, Plum Island, H. 12581 (1956); Ipswich, Oakes; Ipswich, Plum Island, H. 12346 (1956); Topsfield, Mrs. A. Towne; Gloucester, Annisquam, Mrs. Wheeler (1880); Lynnfield, Bean (1936).

L. intermedia Leggett. Dry sterile soil; common. - Boxford, H. 8177 (1952); Georgetown, Horner; Rowley, H. 11950 (1955); Ipswich, Robinson (1877); Middleton, Bean & H. 7766 (1951); Wenham, Robinson (1877); Hamilton, Hunnewell (1916); Essex, Fernald, Hunnewell & Long 9962 (1913); Gloucester, Magnolia, Deane (1901); Rockport, Smith 2712 (1933); Beverly, H. 11673 (1955); Lynnfield, Bean (1932).

L. intermedia, var. *juniperina* (Bickn.) Robins. - North Andover, Bean (1916); Newburyport, Bean (1945); Rockport, H. 19029 (1958).

VIOLACEAE (Violet Family)

VIOLA L. Violet.

V. pedata L., var. *lineariloba* DC. Birdfoot-Violet. Dry sunny openings; common. - Haverhill, Ward Hill, H. 437 (1932); South Lawrence, Williams (1900); Andover, J. Blake (1880); Boxford, Horner (1886); West Newbury, W.S. Horner (1879); Newburyport, E.M. Cole (1861); Topsfield, G.T. Dow; Middleton, H. 31917 (1968); Hamilton, Lund (1882); Beverly, Robinson (1872); Peabody, Sears (1885); Lynn, Phippen (1858).

V. pedata, var. *lineariloba,* forma *alba* (Thurb.) Britt. - Andover, Root (1951); Ipswich.

V. cucullata Ait. Wet meadows and swamps; common. - Amesbury, H. 18261 (1958); Haverhill, Bradford, Collins (1910); Methuen, Root 17630 (1958); Andover, Pease 2021 (1903); Boxford, Frye (1880); West Newbury, H. 31890 (1968); Topsfield, H.R. Towne; Danvers, Sears (1885); Essex, Fernald, Hunnewell & Long 9977 (1913); Rockport, Smith 940 (1928); West Manchester, Hubbard (1913); Salem, Robinson (1873); Saugus, Bean (1940); Marblehead, H. 18951 (1958).

V. cucullata X *V. fimbriatula* - Amesbury, H. 18257 (1958); Hamilton, Hubbard (1913).

V. papilionacea Pursh. Damp woods and meadows; occasional. - Andover, Pease 910 (1902); North Andover, Morss (1885); Newburyport, B.L. Robinson (1903); Middleton, H. 31913 (1968); Rockport, Sears (1885).

V. papileonacea X. *V. fimbriatula* - Newburyport, L. L. Robinson (1903).

V. sororia Willd. Meadows and low woods; occasional. - Haverhill, Bradford, Fernald 11379 (1914); Georgetown, Horner; Manchester, H. & Smith 2727 (1934); Saugus, Bean (1937).

V. septentrionalis Greene Open woods and clearings; frequent. - Haverhill, H. 64 (1931); Methuen, Churchill (1909); North Andover, Root (1943); Boxford, Kennedy (1910); Georgetown, Horner; West Manchester, Hubbard (1913); Peabody, Mackintosh (1925).

V. septentrionalis X. *V. fimbriatula* Haverhill, Bradford, Fernald (1914); Lynn, Bartlett, Eames & Fernald (1908).

V. fimbriatula Sm. Dry sterile woods and clearings; common. - Salisbury, H. 18976 (1958); Amesbury, H. 18205 (1958); Haverhill, H. 65 (1931); Andover, Pease 896 (1902); North Andover, Zink (1943); Boxford, M.E. Perley (1882); Georgetown, Horner; Ipswich, H. 18418 (1958); Ipswich, Plum Island, H., McGregor & Schweinfurth 12669 (1957); Danvers, H. 9987 (1955); Gloucester, Rand & B.L. Robinson (1896); Rockport, Sears (1885); West Manchester, Hubbard (1913); Beverly, Robinson (1884); Salem, Robinson (1873); Peabody, Mackintosh (1925); Nahant, S.S. Clark (1916).

V. sagittata Ait. Damp to dry open woods; occasional. - Newburyport, B.L. Robinson (1903); Topsfield, B.B. Towne; Middleton, Bean & H. 7758 (1951).

V. triloba Schwein. Rich woods; occasional. - Andover, Pease 911 (1902) "spontaneous in garden"; Boxford, Robinson (1890); West Boxford, Root & Zink (1945).

V. palmata L. Rich deciduous woods; rare. - Boxford, M.E. Perley.

V. pallens (Banks) Brainerd. Wet woods and thickets; common. - Haverhill, Ward Hill, Cushman 2980 (1908); Lawrence, Merrill (1853); Andover, Pease 892 (1902); North Andover, Zink (1943); Boxford, Cushman 766 (1907); West Newbury, H. 31910 (1968); Newburyport, E.N. Cole (1861); Georgetown, Horner (1875); Ipswich, Oakes; Topsfield, Mackintosh (1925); Danvers, Sears (1885); Essex, H. & Smith 2731 (1931); Salem, Sears (1885); Marblehead, Cole (1888).

V. incognita Brainerd, var. *forbesii* Brainerd. Wet to dryish woods and thickets; occasional. - Salisbury, Eaton (1899); Haverhill, H. 71 (1931); Andover, Pease 2393 (1903); Essex Woods, Lowell (1853).

V. blanda Willd. Rich woods; occasional. - Andover, Sutton 30 (1874); Groveland, Horner; Ipswich, Topsfield, B.B. Towne; Middleton, Bean & H. 7767 (1951); Rockport, Smith 934 (1928).

V. lanceolata L. Lance-leaved Violet. Damp open ground; common. - Haverhill, H. 68 (1931); Methuen, H. 18402 (1958); Lawrence, Sears (1885); Andover, Pease 879 (1900); North Andover, Morss (1885); Boxford, Frye (1880); Groveland, H. 9093 (1953); Newbury, Bean & H. 8406 (1953); Georgetown, Horner; Rowley, H. 7921 (1952); Ipswich, G.M. Allen (1912); Ipswich, Plum Island, Mackintosh (1925); Topsfield, B.B. Towne; Essex, Fernald, Hunnewell & Long 10006 (1913); Rockport, Smith 729 (1927); West Manchester, Hubbard (1913); Beverly, Lowell (1849); Salem, Sears (1884).

V. primulifolia L., var. *acuta* (Bigel.) T. & G. Open shores, meadows and thin woods; occasional. - Andover, Pease 2003 (1903); Newburyport, B.L. Robinson (1903); Newbury, Cushing (1819); Topsfield, B.B. Towne; Rockport, Sears (1885); Beverly Farms, Shaw (1910); Salem, Sears (1884); Marblehead Neck, Sears (1885).

V. rotundifolia Michx. Round-leaved Yellow Violet. Rich woods; occasional. - Salisbury, Moulton (1888); Amesbury, Eaton (1899); Newburyport, Moulton (1887); Gloucester, Babson (1866).

V. odorata L. English Violet. Naturalized from Europe. Spread from cultivation. - Lawrence, H. 18763 (1958); Topsfield, A.F. Towne; Salem, Robinson (1876); Peabody, Mackintosh (1933).

V. pubescens Ait. Downy Yellow Violet. Rich deciduous woods; frequent. - Haverhill, Bradford, A.W. & C.A. Cheever (1913); Andover, Pease 4881 (1904); North Andover, Cole (1901); Boxford, Frye (1880); Georgetown, Horner; Ipswich, Morong (1875); Danvers, Lowell (1848); Saugus, Svenson (1917).

V. pensylvanica Michx., var. *leiocarpa* (Fern. & Wieg.) Fern. Smooth Yellow Violet. Damp woods; rare. - Andover, Pease 912 (1902); Boxford, M.E. Perley (1882); Topsfield, B.B. Towne.

V. conspersa Reichenb. American Dog-Violet. Meadows and damp woods; common. - Haverhill, H. 73 (1931); Andover, Robinson (1872); North Andover, Morss (1885); Boxford, H.E. Parkhurst (1882); Georgetown, Horner; Topsfield, B.B. Towne; Middleton, H. 18064 (1958); Danvers, Sears (1885); Saugus, Lund; Swampscott, Cole (1889).

V. tricolor L. Johnny-jump-up. Introduced from Europe. Fields and waste places; occasional. - Methuen, H. 7995 (1952); Andover, Pease (1903); Boxford, H. 9647 (1954); Ipswich, H. 13869 (1957); Topsfield, H.R. Towne; Beverly, H. 18905 (1958); Salem, Robinson (1875); Marblehead, H. 11438 (1955).

THYMELAEACEAE (Mezereum Family).

DIRCA L. Leatherwood.

D. palustris L. Wicopy. Rich woods; rare. - Salisbury, Moulton & Sears (1887); Newburyport, Moulton (1889).

DAPHNE L. Mezereum.

D. mezereum L. Introduced from Europe. Escape from cultivation; occasional. - Topsfield, G.F. Dow (1916); Danvers, H.C. Stevens (1927); Salem, Waters (1884).

D. mezereum, var. *alba* West. - Topsfield, F. Wade 19065 (1959) spontaneous.

LYTHRACEAE (Loosestrife Family).

ROTALA L.

R. ramosior (L.) Koehne. Tooth-cup. Sandy shores; rare. - Danvers, Oakes; Saugus, Fernald & Svenson 980 (1928).

DECODON J. F. Gmel. Swamp Loosestrife.

D. verticillatus (L.) Ell. Water-Willow. Swamps and shallow pools; common. - Andover, T.E.N. Eaton (1879); Boxford, H. 8836 (1953); Groveland, H. 12888 (1957); West Newbury, Robinson (1879); Topsfield, H.R. Towne; Middleton, Bean (1950); Wenham, H. 13102 (1957); Hamilton, H. 9412 (1953); Essex ponds, Lowell (1848); East Gloucester, Sheffield (1926); Rockport, Smith 2717 (1933); Swampscott, Cole (1898).

D. verticillatus, var. *laevigatus* T. & G. - Georgetown, Horner (1881).

LYTHRUM L. Loosestrife.

L. hyssopifolia L. Marshes and sterile soil near the coast; occasional. - Boxford, M.E. Perley (1885); West Newbury, H. 18058 (1958); Newbury, White 108 (1913); Newbury, Plum Island, Pease 2836 (1903); Georgetown, Horner (1881); Ipswich Beach, Hubbard (1913); Ipswich, Plum Island, H., McGregor & Schweinfurth 12709 (1957); Gloucester, Fernald, Hunnewell & Long 10029 (1913); Rockport, Robinson (1875); Manchester, Lowell (1847); South Salem, Sears (1884); Nahant.

L. salicaria L. Spiked Loosestrife. Naturalized from Europe. Wet places; a handsome but very agressive weed. First reported in Gloucester, 10 August 1864, Proc. Essex, Inst. 4 liii (1866). - Salisbury, White 264 (1913); Amesbury, Bean (1949); Merrimac, H. 13706 (1957); Haverhill, H. 485 (1932); Methuen, H. 8948 (1953); Lawrence, H. 11243 (1955); North Andover, Root (1945); Boxford, H. 11969 (1955); Groveland, H. 9064 (1953); West Newbury, Pease 28532 (1940); Newburyport, White 465 (1913); Georgetown, Horner (1882); Rowley, H. 12238 (1956); Rowley, Plum Island, H. 12455 (1956); Ipswich, Sears (1904); Ipswich, Plum Island, H. 12412 (1956); Topsfield, G.T. Dow; Middleton, H. 8712 (1953); Danvers, H. 18571 (1958); Hamilton, H. 10781 (1955); West Peabody, Horner; Lynnfield, H. 14126 (1957); Marblehead, A.P. Morse (1924).

L. salicaria, var. *tomentosum* (Mill.) DC. - North Andover, H. 18728 (1958); Georgetown, Horner (1875); Rowley, H. 18513 (1958); Topsfield, H.A. Porter (1898); West Danvers, Horner; Saugus, H. 18627 (1958).

CUPHEA P. Br.

C. procumbens Gav. Introduced from southern Mexico. - Andover, Gutterson (1901), *Rhodora* 4 247 (1902).

NYSSACEAE (Sour Gum Family).

NYSSA L. Tupelo.

N. sylvatica Marsh. Black Gum. Low acid woods, swamps and shores; frequent. - Salisbury, H. 18987 (1958); Amesbury, H. 13010 (1957); Andover, Pease 982 (1902); North Andover, Root (1957); Boxford, H. 7694 (1951); Georgetown, H. 10704 (1955); Rowley, Plum Island, H. 12039 (1956); Wenham, Robinson (1875); Hamilton, H. 9402 (1953); Chebacco, Robinson (1880); Gloucester, H. 12180 (1956); Beverly, Sears (1886); Peabody, Sears (1888); Saugus, Bean & H. 9808 (1955); Marblehead, Sears (1895).

MELASTOMATACEAE (Melastoma Family).

RHEXIA L. Meadow-Beauty.

R. virginica L. Peats and wet sands; occasional. - Andover, Pease 4272 (1904); Boxford, M.P. Perley (1882); Georgetown, Horner (1881); Ipswich, Oakes; Topsfield, G.T. Dow; Danvers, Horner; Wenham, Sears (1897); Essex ponds, Lowell (1846); Beverly, Cole (1899); Peabody, Mackintosh (1933); Lynnfield, Miss M.T. Saunders (1870); Saugus, Lund (1882).

ONAGRACEAE (Evening-Primrose Family).

LUDWIGIA L. False Loosestrife.

L. alternifolia L. Seedbox. Swamps; rare. - Amesbury, Robinson (1879); Lynn, Tracy.

L. palustris (L.) Ell., var. *americana* (DC.) Fern. & Grisc. Water-Purslane. Wet open places; common. - Salisbury, White 368 (1913); Amesbury, H. 9176 (1953); Merrimac, H. 9245 (1953); Haverhill, H. 74 (1931); Methuen, H. 8932 (1953); Andover, H. 14102 (1957); North Andover, H. 9508 (1954); Boxford, H. 8194 (1952); Groveland, H. 9108 (1953); West Newbury, White 527 (1913); Georgetown, H. 12965 (1957); Rowley, H. 8123 (1952); Ipswich, Oakes; Topsfield, H. 8857 (1953); Middleton, H. 9428 (1953); Danvers, Lowell (1848); Wenham, H. 13750 (1957); Essex, H. 13353 (1957); Rockport, Fernald, Hunnewell & Long 10041 (1913); Manchester, H. 10754 (1955); Beverly, Robinson (1876); Salem, Robinson (1874); Peabody, H. 9335 (1953); Lynnfield, H. 14148 (1957); Saugus, H. 10871 (1955).

L. palustris, var. *americana,* forma *elongata* Fassett. Shallow water; frequent. - Boxford, H. 13220 (1957); Georgetown, H. 12961 (1957); Rowley, H. 13497 (1957); Ipswich, H. 13246 (1957); Topsfield, H. 13616 (1957); Middleton, H. 12898 (1957); Peabody, H. 13319 (1957); Lynn, H. 13181 (1957).

EPILOBIUM L. Willow-herb.

E. angustifolium L. Fireweed. Recent clearings and burns; common. - Haverhill, H. 737 (1932); South Lawrence, Morss (1884); Andover, Pease 930 (1902); Boxford, Alcott (1883); West Newbury, H. 11298 (1955); Georgetown, Horner; West Manchester, Hubbard (1913) [Gloucester, A & W (1969)]; Beverly, H. 18903 (1958); Salem, Cole (1899).

E. hirsutum L. Naturalized from Europe. Roadside thickets and waste places; frequent. - Lawrence, H. 18770 (1958); Andover, Bean (1936); Boxford, H. 8145 (1952); West Newbury, H. 11111 (1955); Ipswich, Plum Island, H. 12406 (1956); Topsfield, H. 11367 (1955); Danvers, H. 11645 (1955); Gloucester, Magnolia, H. 12336 (1956); Beverly, H. 18909 (1958); Salem, Robinson (1875); Peabody, Mackintosh (1933); Saugus, H. 18469 (1958); Swampscott, H. 10659 (1955); Marblehead, H. 11437 (1955).

E. strictum Muhl. Bogs, mossy thickets and meadows; occasional. - Haverhill, Horner; Newburyport, Sears (1886); Danvers, Lowell (1848); Wenham, Hunnewell 4625 (1916).

E. leptophyllum Raf. Low ground; occasional. - Metheun, H. 8941 (1953); Boxford, H. 8715 (1953); Ipswich, Forbes (1913); Peabody, C.E. & E. Faxon (1891).

E. palustre L. Swamps and shores; occasional. - Boxford, Robinson (1876); Wenham Swamp, Robinson (1875); Peabody, Sears (1886).

E. coloratum Biehler. Low ground; frequent. - North Andover, Morss (1885); Boxford, Robinson (1876); Georgetown, Horner; Topsfield, G.T. Dow; Danvers, Sears (1887); Manchester, Williams (1895); Salem, Lowell (1847).

E. glandulosum Lehm., var. *adenocaulon* (Haussk.) Fern. Damp thickets; common. - Merrimac, H. 9242 (1953); Haverhill, H. 693 (1932); Boxford, Twitchell (1883); Groveland, H. 11086 (1955); West Newbury, H. 11132 (1955); Ipswich, H. 14031 (1957); Ipswich, Plum Island, H. 11297 (1955); Topsfield, H. 8863 (1953); Essex, H. 13334 (1957); Gloucester, Magnolia, J.M. Hunnewell (1902); Manchester, H. 11324 (1955); Beverly, Lowell (1846); South Salem, Sears (1884); Peabody, Mackintosh (1933); Marblehead, H. 7827 (1933).

OENOTHERA L. Evening-Primrose.

O. depressa Greene. Gloucester, Magnolia, J.M. Hunnewell (1898), fide Munz 1961.

O. biennis L. Dry open soil; common. - Methuen, H. 11761 (1955); North Andover, Morss (1885); Boxford, Frye (1880); Groveland, H. 10972 (1955); Newburyport, Bean (1940); Ipswich, Plum Island, H. 12395 (1956); Topsfield, H.R. Towne; Danvers, H. 11661 (1955); Gloucester, Magnolia, H. 12376 (1956); Salem, Sears (1884); Swampscott, H. 10661 (1955).

O. grandiflora Ait. Naturalized from Europe. Spread from cultivation; occasional. - Topsfield, H.R. Towne; Salem, Robinson (1879).

O. cruciata Nutt. Dry open soil; rare. - Haverhill, Ward Hill, Bean (1916).

O. parviflora L. Dry openings; common. - North Andover, H. 11037 (1955); Newburyport, Plum Island, H. 13038 (1957); Newbury, H. 20709 (1959); Rowley, H. 18504 (1958); Rowley, Plum Island, H. 12471 (1956); Topsfield, H.R. Towne; Hamilton, H. 10828 (1955); Gloucester, Eastern

Point, Prince (1893); Rockport, Fernald, Hunnewell & Long 10063 (1913); Manchester, H. 11322 (1955); Beverly, Lowell (1853); Salem, Little Misery Island, H. 13288 (1957).

O. laciniata Hill. Naturalized from South. Sandy open ground; occasional, - Andover, Pease 9857 (1906); Topsfield, H.R. Towne.

O. perennis L. Dry to moist open ground; common. - Methuen, H. 10997 (1955); Andover, Pease 933 (1902); North Andover, Morss (1885); Boxford, M.P. Perley (1884); Newbury, Palmer 46294 (1942); Rowley, H. 7920 (1952); Ipswich, H. 9580 (1954); Ipswich, Plum Island, H., McGregor & Schweinfurth 12665 (1957); Topsfield, H.R. Towne; South Danvers, Sears (1885); Hamilton, Phippen (1857); Gloucester, Eastern Point, Prince (1883); Rockport, Cushman 3371 (1908); Manchester, Hubbard (1913); Beverly, Lowell (1846); Salem, Sears (1885); Peabody, Sears; Marblehead, Ives (1927).

CLARKIA Pursh

C. pulchella Pursh. Introduced from the West. Escape from cultivation. - Topsfield, Mrs. C. Towne.

CIRCAEA L. Enchanter's Nightshade.

C. quadrisulcata (Maxim.) Franch. & Sav., var. *canadensis* (L.) Hara. Rich woods; common. - Amesbury, H. 18218 (1958); Merrimac, H. 9282 (1953); Haverhill, H. 75 (1931); Andover, Pease 2103 (1902); North Andover, Root (1945); Boxford, M.E. Perley (1883); Groveland, H. 8830 (1953); West Newbury, Robinson (1879); Newbury, White 202 (1913); Georgetown, Williams (1899); Ipswich, Lowell; Topsfield, G.F. Dow; Middleton, Bean (1950); Danvers, Frost; Beverly, Lowell (1846).

C. alpina L. Cool moist woods; occasional. - Boxford, M.E. Perley (1883); Topsfield, H.S. Porter; Essex Woods, Lowell; Beverly Farms, Hubbard (1913).

HALORAGACEAE (Water- Milfoil Family).

MYRIOPHYLLUM L. Water-Milfoil.

M. alterniflorum DC. Lakes and ponds; rare. - Methuen, Boott (1863, 1865); Boxford, H. 13553 (1957).

M. exalbescens Fern. Ponds and quiet streams; occasional. - Metheun, Boott (1865); Andover, (1882); Wenham, Robinson (1875); Hamilton, Sears; Peabody, H. 9336 (1953).

M. verticillatum L., var. *pectinatum* Wallr. Shallow waters; occasional. - Georgetown, H. 12953 (1957); Peabody, H. 12766 (1957).

M. humile (Raf.) Morong. Sandy or muddy shores, frequent. - Amesbury, H. 9157 (1953); Merrimac, H. 9202 (1953); Boxford, H. 13189 (1957); South Groveland, H. 13977 (1957); Middleton, H. 9429 (1953); Danvers, Lowell (1848); Gloucester, Knowlton (1929); Lynnfield, H. 14152 (1957); Saugus, H. 13036 (1957); Lynn, Oakes.

M. humile, forma *capillaceum* (Torr.) Fern. Shallow waters, occasional. - Haverhill, H. 28388 (1965); Boxford, H. 13217 (1957); Georgetown, Sears (1876); Danvers, Sears (1876); Gloucester, H. 13813 (1957); Saugus, H. 13035 (1957).

M. humile, forma *natans* (DC.) Fern. Shallow water; occasional. - Gloucester, H. 13815 (1957); Rockport, Sears (1886).

M. tenellum Bigel. Shallow margins of ponds and pools; frequent. - Haverhill, H. 14154 (1957); North Andover, H. 14059 (1957); Boxford, H. 13545 (1957); South Groveland, H. 13976 (1957); Ipswich, H. 13270 (1957); Wenham, Sears (1872); Chebacco, Robinson (1875); Gloucester, Sears (1888); Beverly, Sears (1886); Lynn, H. 13169 (1957).

PROSERPINACA L. Mermaid-weed.

P. palustris L. Shallow water and shores; common. - Haverhill, H. 486 (1932); North Andover, H. 9509 (1954); Boxford, H. 8188 (1952); Groveland, H. 8800 (1953); West Newbury, H. 9122 (1953); Newbury, H. 8236 (1952); Georgetown, H. 12955 (1957); Rowley, H. 13516 (1957); Topsfield, H. 13599 (1957); Danvers, Sears (1885); Wenham, Fernald, Hunnewell & Long 10076 (1913); Essex ponds, Lowell (1846); Manchester, H. 10753 (1955); Salem, Cole (1899); Peabody, H. 9303 (1953); Lynn, Tracy (1859).

ARALIACEAE (Ginseng Family).

ARALIA L.

A. racemosa L. Spikenard. Rich woods and thickets; occasional. - North Andover, Russell (1859); Boxford, M.E. Perley (1881); Georgetown, Horner; Topsfield, H.R. Towne.

A. hispida Vent. Bristly Spikenard. Rocky and sandy open woods and clearings; common. - Salisbury, White 90 (1913); Andover, Pease 942 (1900); Boxford, Alcott (1883); Groveland, Bean & H. 8520 (1953); Rowley, H. 10435 (1955); Topsfield, Mrs. S. Conley; Wenham, Robinson (1875); Essex Woods, Lowell (1846); Gloucester, Magnolia, J.M. Hunnewell (1898); Rockport, Cushman 3353 (1908); Beverly Farms, Hubbard (1913); Saugus, Bean & H. 9779 (1955); Lynn, Cushman 2977 (1908).

A. nudicaulis L. Wild Sarsaparilla. Woodlands; frequent. - Haverhill, H. 438 (1932); Andover, Pease 941 (1901); North Andover, Root (1943); Boxford, M.E. Perley (1882); Ipswich, Oakes; Topsfield, Mrs. L.T. Brown (1912); Danvers, Phippen (1857); Hamilton, Cole (1887); Rockport, Anderson, Shafer & Smith (1931); West Manchester, Hubbard (1913); Peabody, Mackintosh (1925); Marblehead, Cole (1887).

PANAX L. Ginseng.

P. trifolius L. Ground-nut. Rich woods; frequent. - Haverhill, H. 440 (1932); Andover, Sutton 35 (1874); Boxford, Fernald (1909); Georgetown, Horner (1881); Ipswich, Oakes; Topsfield, B.B. Towne; Danvers, Robinson (1872); Essex, Robinson (1872); Gloucester, Cole (1905); Rockport, Smith 943 (1928); Peabody, Cole (1899).

UMBELLIFERAE (Parsley Family).

HYDROCOTYLE L. Water-pennywort.

H. umbellata L. Pond shores and low ground; rare. - Metheun, C.E. Perkins (1880); Essex, Lowell (1846); H, 13329 (1957).

H. americana L. Meadows and damp woods; frequent. - Salisbury, White 330 (1913); Amesbury, H. 18226 (1958); Merrimac, Bean & H. 8573 (1953); Andover, H. 17745 (1958); Boxford, M.E. Perley (1881); Georgetown, Horner (1882); Ipswich, H. 18184 (1958); Middleton, H. 18079 (1958); Wenham, H. 13751 (1957); Beverly, Lowell (1847); Lynnfield, Boott (1854).

SANICULA L. Black Snakeroot.

S. marilandica L. Meadows and open woods; frequent. - Andover, Pease 967 (1902); North Andover, Root & Zink (1945); Boxford, H. 8353 (1953); Georgetown, Horner (1873); Ipswich, St. John 11964 (1916); Topsfield, A. Dodge, Middleton, H. 8688 (1953); Danvers, Sears (1885); Hamilton, H. 10773 (1955); Essex, Hubbard (1913); Gloucester, Magnolia, Williams (1896).

S. gregaria Bickn. Rich woods and thickets; occasional. - Andover, Pease 971 (1902); Newbury, Sears (1888); Rowley, Sears (1888).

SCANDIX L. Shepherd's Needle.

S. pecten-veneris L. Native of Europe. Rare waif in birdseed. - Boxford, H. 8606 (1953).

OSMORHIZA Raf. Sweet Cicely.

O. claytoni C.B. Clarke. Sweet Jarvil. Woods; occasional. - Georgetown, H.E. Parkhurst (1882); Danvers, Sears (1887); Swampscott, Robinson (1877).

O. longistylis (Torr.) DC. Anise-root. Rich woods and thickets; frequent. - Haverhill, H. (1933); Andover, Sutton 93 (1874); Georgetown, Horner; Ipswich; Topsfield, C.T. Dow; Middleton, Pease 27709 (1940); Danvers, Sears (1977); Hamilton, Mackintosh (1915); Manchester, Churchill (1882); Marblehead, Sears (1885).

CONIUM L. Poison Hemlock.

C. maculatum L. Naturalized from Europe. Waste places; occasional. - Metheun, H. 9812 (1953); West Boxford, Horner; Ipswich, Oakes; Hamilton, Robinson (1875).

APIUM

A. graveolens L. Celery. Introduced from Europe. Escape from cultivation. - Saugus, H. 18707 (1958); Lynn, H. 18818 (1958).

ZIZIA W. D. J. Koch

Z. aurea (L.) W.D.J. Koch Golden Alexanders. Meadows, shores and damp thickets; common. - Amesbury, Eaton; Haverhill, Bradford, Cushman 2958 (1908); Metheun, Robinson (1875); Andover, Pease 974 (1901); North Andover, Robinson (1884); Boxford, Sears (1902); Groveland, Horner; West Newbury, A.W. & C.A. Cheever & Sanford (1933); Georgetown, Horner; Ipswich, Oakes; Topsfield, A.M. Dow; Beverly, Lowell (1849).

CICUTA L. Water Hemlock.

C. bulbifera L. Swamps and wet thickets; common. - Merrimac, H. 13692 (1957); Haverhill, H. 76 (1931); North Andover, H. 14079 (1957); Boxford, M.E. Perley (1885); Groveland, H. 9080

(1953); Georgetown, Horner; Ipswich, H. 18416 (1958); Danvers, Sears (1884); Wenham, H. 13744 (1957); Essex, Fernald, Hunnewell & Long 10093 (1913); Saugus, Robinson (1874).

C. maculata L. Meadows, swales and low thickets; common. - Salisbury, White 78 (1913); Amesbury, Bean (1950); Merrimac, H. 18317 (1958); Haverhill, H. 624 (1932); Lawrence, H. 12285 (1956); North Andover, Morss (1884); Boxford, M.E. Perley (1883); Groveland, H. 10597 (1955); Newburyport, White 182 (1913); Georgetown, Horner; Ipswich, Oakes; Topsfield, H.R. Towne; Middleton, H. 8687 (1953); Danvers, Lowell (1846); Hamilton, Robinson (1875); Beverly, Lowell (1847); Salem, Phippen; Peabody, Mackintosh (1933); Saugus, H. 10909 (1955); Swampscott, Cole (1889); Marblehead, Cole (1886).

CRYPTOTAENIA DC. Honewort.

C. canadensis (L.) DC. Rich woods and thickets; frequent. - Merrimac, Bean & H. 8361 (1953); Andover, Pease 3764 (1904); Boxford, Robinson (1876); Ipswich, Oakes; Topsfield, M.E. Perley (1882); Danvers, Sears (1885); Beverly, Lowell (1847).

CARUM L. Caraway.

C. carvi L. Naturalized from Europe. Neglected fields; occasional. - Andover, Pease 2124 (1903); Boxford, M.E. Perley (1882); Wenham, Robinson (1875); Essex, Oakes; Nahant, Gilbert (1894).

AEGOPODIUM L. Goutweed.

A. podagraria L. Naturalized from Europe. Waste places and roadsides; rare. - Danvers, H.R. Towne (1912).

SIUM L. Water-Parsnip.

S. suave Walt. Meadows, wet thickets and muddy banks; common. - Amesbury, Bean (1950); Merrimac, H. 13680 (1957); Haverhill, H. 1830 (1935); Methuen, H. 18387 (1958); North Andover, Root (1945); Boxford, Alcott (1883); South Groveland, Pease 973 (1901); West Newbury, H. 10939 (1955); Newbury, White 213 (1913); Rowley, H. 13507 (1957); Ipswich, Oakes; Topsfield, Mrs. F. Peabody; Middleton, (1919); Danvers, E.L. Mackintosh (1924); Hamilton, H. 9410 (1953); Essex ponds, Lowell (1846); Gloucester, H. 13480 (1957); [Manchester, Hellquist (1971)] ; Beverly, Lowell (1847); Peabody, Mackintosh (1933).

S. suave, forma *carsonii* (Durand) Fassett - Methuen, Root (1949).

LILAEOPSIS Greene

L. chinensis (L.) Ktze. In mud of brackish marshes and tidal shores; occasional. - Newburyport, Moulton; Rowley, Fassett 2021 (1923); Ipswich, Sears (1886).

AETHUSA L. Fool's-Parsley.

A. cynapium L. Naturalized from Europe. Waste or cultivated ground; occasional. - Merrimac, H. 18314 (1958); Andover, Pease 1999 (1903); Ipswich, Oakes, Morong (1875); Salem, Robinson (1884); Marblehead, H. 18926 (1958).

FOENICULUM Mill. Fennel.

F. vulgare Mill. Naturalized from Europe. Dry fields and roadsides; rare. - Andover, Lowell Junction, A.P. Morse (1924).

LIGUSTICUM L. Lovage.

L. scothicum L. Scotch Lovage. Saline marshes and rocks along the coast; common. - Newbury, White 117 (1913); Newbury, Plum Island, Mackintosh (1924); Rowley, A.P. Morse (1925); Ipswich, Oakes; Ipswich, Plum Island, H. 12404 (1956); Essex, Hunnewell 3319 (1914); Gloucester, Magnolia, Clarke (1911); Rockport, Cushman, 3367 (1908); Beverly Farms, Hubbard (1913); South Salem, Sears (1886).

COELOPLEURUM Ledeb.

C. lucidum (L.) Fern. Rocky and gravelly coast; occasional. - Newburyport, Deane (1886); Plum Island, Mackintosh (1915); Gloucester, C.E. Faxon (1877); Rockport, Gates & Smith 1011 (1928); Beverly, Pickering (1847); South Salem, Sears (1886); Swampscott, Swan 1016 (1886).

ANGELICA L.

A. atropurpurea L. Rich thickets and swamps; common. - Amesbury, Robinson (1879); Merrimac, H. 10587 (1955); Haverhill, H. 625 (1932); Andover, Pease 947 (1900); North Andover, Morss (1884); Groveland, Horner; West Newbury, Bean & Pease (1940); Georgetown, Williams & Rearden (1906); Ipswich, Minns; Topsfield, O. Floyd; Danvers, M.E. Perley (1886); Gloucester, H. 10325 (1955); Rockport, Manning (1886); Peabody, Mackintosh (1933).

LEVISTICUM Hill Lovage.

L. officinale W.D.J. Koch. Introduced from Europe. Escape from cultivation; occasional. - Topsfield, Mrs. M. Gould; Wenham, Robinson (1875); Beverly, Robinson (1879).

PEUCEDANUM

P. palustre (L.) Moench. Adventive from Europe. - West Newbury, H. 18059 (1958). First record in North America.

PASTINACA L. Parsnip.

P. sativa L. Naturalized from Europe. Waste places and roadsides; frequent. - Salisbury, H. 12202 (1956); Haverhill, H. 910 (1933); Boxford, M.E. Perley (1882); West Newbury, H. 18057 (1958); Ipswich, Oakes; Danvers, H. 20760 (1959); Chebacco, Robinson (1879); Beverly, Lowell (1846); Peabody, Mackintosh (1933).

HERACLEUM L. Cow-Parsnip.

H. maximum Bartr. Low ground; occasional. - Ipswich, Oakes; Rockport, Gates & Smith 1012 (1928); Beverly, Lowell (1847); Salem, Cole (1902); Marblehead, F.W. Peabody (1893).

DAUCUS L. Carrot.

D. carota L. Queen Anne's-Lace. Naturalized from Europe. Dry fields and waste places; common. - Haverhill, H. 526 (1932); Methuen, H. 11012 (1955); Lawrence, H. 11226 (1955); Andover, H. 17737 (1958); North Andover, Root (1945); Boxford, M.E. Perley (1881); West Newbury, H. 10950 (1955); Newburyport, H. 10362 (1955); Georgetown, Horner; Ipswich, H. 13983 (1957); Topsfield, J.H. Towne; Hamilton, H. 10800 (1955); [Rockport, A & W (1969)]; West Gloucester, Robinson (1877); Manchester, H. 10718 (1955); Salem, Cole (1899); Peabody, Mackintosh (1933); Swampscott, H. 10673 (1955).

CORNACEAE (Dogwood Family).

CORNUS L. Cornel.

C. canadensis L. Bunchberry. Woods and thickets; frequent. - Salisbury, H. 18980 (1958); Amesbury, H. 18228 (1958); Haverhill, H. 441 (1932); Andover, Pease 978 (1902); Boxford, Frye (1881); Newburyport, E.M. Cole (1861); Georgetown, W.S. Horner (1880); Topsfield, B.B. Towne; Danvers, Frost; Chebacco, Robinson (1879); Gloucester, Magnolia, J.M. Hunnewell (1898); Rockport, Anderson, Shafer & Smith (1931); Manchester, Hubbard (1913); Beverly, Cole (1886); Saugus, Lund (1880).

C. florida L. Flowering Dogwood. Acidic woods; local. - Haverhill, H. 442 (1932); North Andover, Root (1951); Boxford, Oakes; Groveland, Mrs. Hall; Topsfield, P.E. Kneeland; Danvers, Frost; Wenham Swamp, Sears (1887); Manchester, Mrs. Bray (1875); Lynn, Cole (1900).

C. stolonifera Michx. Red Osier. Shores and thickets; occasional. - Amesbury, H. 18196 (1958); Methuen, H. 8906 (1953); North Andover, Root (1945); Newburyport, Cole (1902); Ipswich, St. John 11963 (1916); Topsfield, Oakes; Middleton, H. 8696 (1953); Danvers, Sears (1883).

C. rugosa Lam. Round-leaved Dogwood. Dry woods; occasional. - Methuen, H. 11002 (1955); North Andover, Root (1951); Boxford, H. 18552 (1958); Georgetown, Robinson (1876); Rowley, Sears (1883); Beverly, Robinson (1879).

C. amomum Mill. "Red Willow". Swamps and damp thickets; common. - Amesbury, Bean (1950); Haverhill, H. 527 (1932); Metheun, H. 9631 (1954); Andover, Pease 981 (1900); North Andover, Morss (1885); Boxford, M.E. Perley (1881); West Newbury, NEBC. (1951); Newbury, White 211 (1913); Georgetown, Horner; Ipswich, H.E. Parkhurst (1883); Topsfield, H. 9593 (1954); Danvers, Sears (1884); Wenham, Robinson (1874); Peabody, Mackintosh (1933); Lynnfield, Kennedy (1911).

C. obliqua Raf. Silky Dogwood. Swamps and damp thickets; common. - Lawrence, H. 11242 (1955); Andover, H. 17911 (1958); North Andover, H. 8664 (1953); Boxford, M.E. Perley (1885); Groveland, Bean & H. 8552 (1953); West Newbury, H. 11156 (1955); Rowley, Sears (1883); Ipswich, Oakes; Topsfield, H.R. Towne; Middleton, H. 18060 (1958); Danvers, Sears (1885); Salem, Cole (1888); Saugus, Bean & H. 9828 (1955).

C. racemosa Lam. Dry to moist open habitats; frequent. - Haverhill, H. 578 (1932); Boxford, M.E. Perley (1884); West Newbury, Bean & Pease (1940); Byfield, Sears (1875); Georgetown, Robinson (1876); Ipswich, Oakes; Topsfield, A.E. Kneeland; Danvers, Mackintosh (1933); Salem, Sears (1885); Peabody, Mackintosh (1890); Lynnfield, Kennedy (1911).

C. alternifolia L. Pagoda-Dogwood. Dry woods; common. - Andover, Pease 976 (1902); North Andover, Root (1957); Boxford, M.E. Perley (1885); Georgetown, M.E. Perley (1882); Ipswich, Oakes; Topsfield, M.E. Perley (1884); Danvers, Frost (1875); Swampscott, Buck (1896).

CLETHRACEAE (White Alder Family).

CLETHRA L. Summer-sweet.

C. alnifolia L. Sweet Pepperbush. Swamps and damp thickets; common. - Salisbury, White 339 (1913); Amesbury, H. 18239 (1958); Merrimac, H. 13717 (1957); Haverhill, H. (1932); Andover, Blake (1882); North Andover, Morss (1885); Boxford, M.E. Perley (1883); Newburyport, Cole (1900); Newbury, Cushing; Rowley, Oakes; Topsfield, Whipple; Rockport, H. 10294 (1955); Manchester, H. 10737 (1955); Beverly, Lowell (1846); Peabody, Mackintosh 555 (1933); Saugus, Bean (1936); Marblehead, H. 12600 (1956).

PYROLACEAE (Wintergreen Family).

CHIMAPHILA Pursh Pipsissewa.

C. umbellata (L.) Bart., var. *cisatlantica* Blake. Prince's Pine. Dry woods; common. - Amesbury, H. 18200 (1958); Haverhill, H. 194 (1931); Andover, Lowell (1847); Boxford, M.E. Perley (1882); West Newbury, Robinson (1879); Georgetown, Horner; Ipswich, Oakes; Topsfield, G.T. Dow; Middleton, Bean & H. 7761 (1951); Wenham, Cole (1911); Saugus, H. 7792 (1951).

C. maculata (L.) Pursh. Spotted Wintergreen. Dry woods; occasional. - Andover, Boutwell (1951); Boxford, H. 14198 (1957); Georgetown, W.S. Horner; Topsfield, M.I. Pierce; Wenham, Sears (1869); Beverly, Redmond (1878).

MONESES Salisb. One-flowered Pyrola.

M. uniflora (L.) Gray. Cool mossy woods; occasional. - Andover, Mrs. G.F. Moore & Prof. C.C. Torrey circa (1898); North Andover, Cole (1904); Boxford, Robinson & Sears (1877); Georgetown, Horner; Topsfield, M.I. Towne; Wenham, Oakes; Beverly, Cross & Street (1850).

PYROLA L. Shinleaf.

P. secunda L. One-sided Pyrola. Dry or moist woods; occasional. - Andover, Pease 1018 (1901); Boxford, M.E. Perley (1885); Middleton, Sears (1887); Wenham, Lowell (1847); Essex County, Oakes.

P. virens Schweigger. Dryish coniferous woods; rare. - Boxford, Hunnewell 8883 (1923); Essex County, Oakes.

P. virens, var. *convoluta* (Bart.) Fern. Dry woods; occasional. - Andover, Pease 1022 (1901); Boxford, M.P. Perley (1888); Georgetown, Horner; Middleton, (1887); Wenham, Lowell (1847); Beverly Farms, Hubbard (1913).

P. elliptica Nutt. Wild Lily-of-the-valley. Dry or moist woods; common. - Salisbury, White 81 (1913); Merrimac, H. 18326 (1958); Haverhill, H. 528 (1932); Lawrence, H. 10071 (1955); Andover, Pease 1021 (1901); Boxford, M.E. Perley (1881); Newbury, Palmer 46286 (1942);

Georgetown, Horner; Ipswich, H. 18133 (1958); Topsfield, A.M. Dow; Wenham, Lowell (1847); Chebacco Woods, Robinson (1879); Rockport, H. 19028 (1958); Manchester, Hubbard (1913); Beverly, Cole (1887); Swampscott, Sears (1885).

P. rotundifolia L., var. *americana* (Sweet) Fern. Wild Lily-of-the-valley. Woods and clearings; common. - Haverhill, Bradford, Cushman (1908); Andover, Pease 1024 (1901); North Andover, Morss (1879); Boxford, M.P. Perley (1883); Georgetown, Horner; Ipswich, H. 18434 (1958); Topsfield, Mrs. W. Herrick; Danvers, Lowell (1841); Manchester, Williams (1895); Beverly, Lowell (1845); Lynnfield, Young (1879); Lynn, Forbes (1909); Marblehead, H. 12594 (1956).

MONOTROPA L. Indian-pipe.

M. uniflora L. Indian-pipe. Woodland humus; common. - Merrimac, H. 18332 (1958); Haverhill, H. 78 (1931); Lawrence, H. 12281 (1956); Andover, Pease 1015 (1902); North Andover, Morss (1879); Boxford, M.P. Perley (1881); Georgetown, Horner; Ipswich, H. 18421 (1958); Topsfield, G.F. Dow; Danvers, Sears (1885); Essex, Lowell (1846); Gloucester, Magnolia, J.M. Hunnewell (1898); Beverly, Cole (1887); Marblehead, Cole (1944).

M. hypopithys L. Pine-sap. Woodland humus; frequent. - Merrimac, H. 18349 (1958); Andover, Pease 1017 (1896); Boxford, M.E. Perley (1881); Groveland, H. 911 (1933); Georgetown, Horner (1881); Ipswich, H. 18422 (1958); Topsfield, G.F. Dow; Danvers, Lowell (1847); Wenham, Oakes; Essex Woods, Lowell (1847); West Gloucester, Smith 364 (1925); Manchester, Williams (1895); Beverly, Cole (1899); Lynnfield, Cole (1898); Marblehead, Cole (1930).

ERICACEAE (Heath Family).

LEDUM L. Labrador-tea.

L. groenlandicum Oeder. Peaty soil; rare. - Gloucester, Magnolia Woods, Horner.

RHODODENDRON L. Rhododendron.

R. canadense (L.) Torr. Rhodora. Bogs and damp thickets; frequent. - Haverhill, H. 443 (1932); Andover, Hopkins & Pease 24033 (1934); North Andover, R.D. Morss (1901); Boxford, Robinson & Sears (1877); Newburyport, E.N. Cole (1861); Georgetown, Horner; Ipswich, H. 12001 (1956); Topsfield, G.W. Towne; Danvers, Robinson (1872); Rockport, Smith 942 (1928) [Gloucester A & W (1969)] ; Beverly, Saunders (1870); Peabody, Cole (1900); [MNS, Snyder (1950, 1971)] .

R. viscosum (L.) Torr. Swamp-Honeysuckle. Swamps and thickets; common. - Metheun, H. 18398 (1958); South Lawrence, Morss (1884); Andover, Ballardvale, Ashburn, Root & Zink (1945); North Andover, H. 17977 (1958); Boxford, M.E. Perley (1882); Groveland, Crane Pond, Horner; West Newbury, Schubert & Smith (1941); Ipswich, St. John 11889 (1916); Topsfield, A.M. Dow; Hamilton, H. 12806 (1957); Beverly, Montserrat-Prides, Hubbard (1913); Salem, Mackintosh (1933); Peabody, H. 9307 (1953); Marblehead Neck, H. 912 (1933); [MNS, Snyder (1950-1972)] .

R. viscosum, forma *glaucum* (Lam.) Voss - Haverhill, H. 529 (1932); Andover, Pease 1029 (1901); Ipswich, Oakes; Chebacco, Robinson (1879).

KALMIA L. Laurel.

K. latifolia L. Mountain-laurel. Rocky or gravelly woods; local. - Andover, Blake (1883); Boxford, H.N. Towne (1885); Newburyport, White 500 (1913); Newbury, Cushing; Rowley, W.S. Horner; Essex, H. 11980 (1955); Gloucester, Magnolia, Forbes 2757 (1902); Rockport, Sears (1885); Manchester, Oakes; Beverly, Cole (1902); Lynn, A.W. & C.A. Cheever (1900).

K. angustifolia L. Sheep-laurel. Old pastures and barrens; common. - Amesbury, H. 9436 (1953); Haverhill, Bradford, Cushman (1908); Andover, Pease 1007 (1902); North Andover, Horner (1880); Boxford, M.E. Perley (1881); Groveland, Bean & H. 8469 (1953); West Newbury, H. 9448 (1954); Georgetown, Horner; Ipswich, Oakes; Topsfield, E.B. Woodbury; Essex, Williams (1896); Gloucester, Annisquam, Ripley (1914); Rockport, H. 10285 (1955); Beverly, Lowell (1846); Salem, Sears (1885); Peabody, Mackintosh (1925); Saugus, Bean & H. 9777 (1955).

K. polifolia Wang. Pale-laurel. Peaty soils; rare. - Methuen, H. 7665 (1951); Peabody, Sears (1902); [Rockport, A & W (1966) PMSH].

ANDROMEDA L. Andromeda.

A. glaucophylla Link. Bog-rosemary. Bogs; rare. - Wenham, Oakes, Lowell, Sears (1886).

LYONIA Nutt.

L. ligustrina (L.) DC. Maleberry. Wet or dry thickets; common. - Amesbury, H. 12996 (1957); Merrimac, H. 13718 (1957); Haverhill, H. 530 (1932); Methuen, H. 9607 (1954); Lawrence, H. 11263 (1955); Andover, Pease 3779 (1904); North Andover, H. 8676 (1953); Boxford, H. 8087 (1952); Groveland, Bean & H. 8499 (1953); West Newbury, H. 10961 (1955); Georgetown, Horner; Ipswich, Oakes; Ipswich, Plum Island, St. John & White 537 (1913); Topsfield, H.R. Towne; Middleton, H. 8697 (1953); Danvers, Lowell (1846); Wenham, Robinson (1874); Hamilton, H. 9405 (1953); Essex, Hubbard (1913); Gloucester, Prince (1883); Rockport, Gates & Smith 1194 (1928); Manchester, Hubbard (1912); Beverly, Lowell; Salem, Cole (1901); Peabody, Sears (1885); Lynnfield, Bean (1919); [MNS, Snyder (1971)].

LEUCOTHOE D. Don Fetter-bush.

L. racemosa (L.) Gray. Moist thickets; rare. - Lynn, Cedar Pond, Sears (1886).

CHAMAEDAPHNE Moench Leather-leaf. Cassandra.

C. calyculata (L.) Moench, var. *angustifolia* (Ait) Rehd. Bogs and pond-margins; common. - Merrimac, H. 13681 (1957); Haverhill, H. 309 (1932); Methuen, Churchill (1909); South Lawrence, Morss (1885); Andover, Sutton 11 (1874); North Andover, H. 17969 (1958); Boxford, Duncan & H. 7657 (1951); Groveland, H. 9449 (1954); Georgetown, Horner (1883); Rowley, H. 13529 (1957); Ipswich, Oakes; Topsfield, Mrs. F. Peabody; Middleton, H. 18098 (1958); Danvers, Cole (1887); Hamilton, Hubbard (1913); Chebacco, Robinson (1872); Peabody, Sears (1885); Lynnfield, Cole (1904); Marblehead, Miss M.T. Saunders; [Lynn, Ingalls, (1971)].

EPIGAEA L. Trailing Arbutus.

E. repens L., var. *glabrifolia* Fern. Mayflower. Sandy to peaty woods and clearings; rare. - Collections seem to indicate that the plant was always local; it still persists in a few areas. - Andover,

Sutton 2 (1874), Root 10130 (1955); Boxford, Robinson (1884), Root 11876 (1955); Danvers, Mrs. N.P. Hayward (1860); Gloucester, [A & W (1969)] ; Magnolia, Sears (1884).

GAULTHERIA L. Aromatic Wintergreen.

G. procumbens L. Checkerberry. Sterile woods and clearings; common. - Salisbury, H. 18992 (1958); Amesbury, H. 18227 (1958); Haverhill, H. 1845 (1935); Andover, Pease 2036 (1903); North Andover, Root (1945); Boxford, G.B. Parkhurst (1882); West Newbury, H. 9446 (1954); Georgetown, Horner (1880); Topsfield, M. Clarke; Hamilton, H. 9399 (1953); Essex, H. 11983 (1955); Gloucester, Prince (1883); Rockport, H. 19048 (1958); Manchester, R.J. Eaton (1928); Beverly Farms, Hubbard (1913); Peabody, Sears (1885); Saugus, H. 7791 (1951); Marblehead, H. 12598 (1956).

G. hispidula (L.) Bigel. Creeping Snowberry. Mossy woods and bogs; occasional. - Andover, Lowell Junction, Hopkins & Pease 24040 (1934); Newburyport, Moulton (1887); Manchester, Oakes.

ARCTOSTAPHYLOS Adans. Bearberry.

A. uva-ursi (L.) Spreng., var. *coactilis* Fern. & Macbr. Exposed rocks and sands; local. - Andover, Pease 985 (1902); West Boxford, Root & Zink (1945); Rowley, H. 8291 (1952); Rowley, Plum Island, H. 12053 (1956); Danvers, Oakes; Wenham, Sears (1885); Gloucester, Magnolia, J.M. Hunnewell (1900); Manchester, Sears (1887); [W. Manchester, Burnett & Snyder (1971)] ; Beverly, Farlow; Peabody, Robinson (1872); Saugus, Lund; Marblehead, Robinson (1902).

A. uva-ursi, var. *coactilis,* forma *heterochroma* Fern. - Georgetown ?, Horner (1872).

CALLUNA Salisb. Heather.

C. vulgaris (L.) Hull. Introduced from Europe. Peaty or damp sandy spots; rare. - West Andover, Mitchell (1876), Ballardvale, Root & Zink (1951); Wenham, Bean (1937); Manchester, R.J. Eaton (1932).

GAYLUSSACIA HBK. Huckleberry.

G. dumosa (Andr.) T. & G., var. *bigeloviana* Fern. Dwarf Huckleberry. Sphagnous bogs; occasional. - Andover, Pease 1002 (1900); Danvers, Mrs. Kettle (1859); Essex, Oakes; Peabody, Sears (1885); Lynn, Sears (1885).

G. frondosa (L.) T. & G. Dangleberry. Dry woods and clearings; occasional. - Amesbury, H. 18201 (1958); Andover, Pease 1006 (1901); Gloucester, Oakes; Peabody, Cedar Pond, Mackintosh (1933).

G. baccata (Wang.) K. Koch. Black Huckleberry. Dry or moist thickets and clearings; common. - Salisbury, Bean & H. 8380 (1953); Merrimac, H. 18350 (1958); Methuen, H. 8914 (1953); Lawrence, H. 9877 (1955); Andover, H. 10042 (1955); North Andover, Root (1943); Boxford, H. 18541 (1958); Georgetown, Horner; Rowley, Bean & H. 8423 (1953); Topsfield, Sears (1899); Middleton, H. 18081 (1958); Danvers, Frost (1874); Gloucester, Oakes; Rockport, Smith 4334 (1945); West Manchester, Hubbard (1913); Beverly, Robinson (1875); Salem, Mackintosh (1925); Peabody, Sears (1887); Lynnfield, Bean (1916); Swampscott, Buck (1896).

G. baccata, forma *glaucocarpa* (Robins.) Mackenz. - Andover, Pease 2370 (1903).

VACCINIUM L. Blueberry. Cranberry.

V. vacillans Torr. Early Sweet Blueberry. Dry open woods and clearings; frequent. - Methuen, H. 18407 (1958); Andover, H. 17746 (1958); North Andover, H. 17962 (1958); Boxford, H. 7839 (1952); Groveland, Bean & H. 8509 (1953); Ipswich, Oakes; Topsfield, B. Woodberry; Middleton, Pease 27770 (1940); Danvers, Sears (1885); West Manchester, Hubbard (1913); Beverly, Lowell (1846); Saugus, Bean (1913).

V. angustifolium Ait., var. *laevifolium* House. Low Sweet Blueberry. Dry open barrens and rocks; our common low blueberry. - Amesbury, H. 18199 (1958); Haverhill, H. 444 (1932); Lawrence, H. 9880 (1955); Andover, Pease 1049 (1902); North Andover, Robinson (1884); Boxford, Robinson (1884); Ipswich, Oakes; Topsfield, Robinson (1875); Manchester, Hubbard (1913); Salem, Cole (1943); Saugus, Bean & H. 9816 (1955).

V. angustifolium, var. *laevifolium* X *V. corymbosum* - North Andover, H. 18006 (1958).

V. corymbosum L. Highbush-blueberry. Swamps and low woods; common. - Methuen, H. 12846 (1957); Andover, Hopkins & Pease 24042 (1934); North Andover, Robinson (1884); Boxford, H. 7913 (1952); Groveland, Bean & H. 8498 (1953); Georgetown, Williams (1906); Rowley, H. 10466 (1955); Ipswich, Oakes; Topsfield, Mackintosh (1925); Danvers, Lowell (1845); Rockport, H. 10289 (1955); West Manchester, Hubbard (1913); Beverly Farms, Hubbard (1913); Salem, Cole (1924); Swampscott, Buck (1896).

V. corymbosum, var. *albiflorum* (Hook) Fern. - North Andover, Root & Zink (1945); Rockport, Anderson, Shafer & Smith (1931); Lynnfield, Bean (1919).

V. corymbosum, var. *glabrum* Gray. - Salisbury, Bean & H. 8381 (1953); Amesbury, H. 13011 (1957); Metheun, H. 8949 (1953); Lawrence, H. 9879 (1955); Andover, H. 10038 (1955); North Andover, H. 8665 (1953); Rowley, Plum Island, H. 12072 (1956); Peabody, H. 12751 (1957); Saugus, Bean & H. 9807 (1955).

V. atrococcum (Gray) Heller. Black Highbush-blueberry. Swamps and low woods; frequent. - Merrimac, H. 18318 (1958); Haverhill, H. 445 (1932); Andover, Pease (1902); North Andover, Root & Zink (1945); Boxford, Cushman (1907); Newbury, Palmer 46285 (1942); Ipswich, Oakes; Topsfield, H.R. Towne; Middleton, H. 18092 (1958); Danvers, Pickman (1925); Wenham, Robinson (1891); Gloucester, H. 12155 (1956); Beverly, Robinson (1891); Saugus, Bean & H. 9814 (1955).

V. atrococcum, forma *leucocarpum* Deane - Andover, Pease 1034 (1901).

V. vitis-idaea L., var. *minus* Lodd. Mountain-cranberry. Rocks; rare. - Danvers, Oakes; Osgood, Lowell (1848), Putnamville, Sears (1897).

V. oxycoccos L. Small Cranberry. Bogs; occasional. - Methuen, H. 7666 (1951); Andover, Sutton 86 (1874); Danvers, Frost (1874); Peabody, Sears (1902),; Marblehead Neck, Sears (1882).

V. macrocarpon Ait. American Cranberry. Bogs and meadows; common. - Haverhill, H. 556 (1932); Methuen, H. 7667 (1951); North Andover, Morss (1885); Boxford, Sears (1883); Groveland, H. 9069 (1953); Newbury, Cushing; Newbury, Plum Island, White 99 (1913); Georgetown, Horner; Rowley, Plum Island, H. 12564 (1956); Ipswich, Prince; Ipswich, Plum Island, H., McGregor & Schweinfurth 12708 (1957); Topsfield, B.B. Towne; Middleton, Hopkins & Pease 24056 (1943); Danvers, Sears (1887); Gloucester, Prince (1883); Rockport, Forbes (1903); Beverly, Lowell (1846); Salem, Robinson (1879); Peabody, Mackintosh (1933); Lynnfield, Kennedy (1863).

DIAPENSIACEAE (Diapensia Family).

GALAX L. Wandflower.

G. aphylla L. Introduced from South. Garden escape; rare. - Swampscott, Mrs. M.E. Ward (1918); two clumps in oak woods.

PRIMULACEAE (Primrose Family).

PRIMULA L. Primrose.

P. officinalis Hill. Introduced from Europe. Garden escape. - Topsfield, B.B. Towne, cultivated?

ANDROSACE L.

A. occidentalis Pursh. Rare waif from West. - Georgetown, J.A. Bates.

LYSIMACHIA L. Loosestrife.

L. vulgaris L. Garden-Loosestrife. Introduced from Europe. Roadsides and waste places; occasional. - Topsfield, G.F. Dow (1916); Gloucester, H. Taylor (1944); Beverly, Robinson (1878).

L. punetata L. Garden-loosestrife. Introduced from Europe. Garden escape; rare. - Georgetown, Horner.

L. quadrifolia L. Whorled Loosestrife. Dry or moist open woods and clearings; common. - Amesbury, H. 10550 (1955); Merrimac, H. 18324 (1958); Haverhill, H. 579 (1932); Methuen, H. 9603 (1954); Andover, H. 10056 (1955); North Andover, Root (1945); Boxford, Frye (1881); Newbury, Oak Island, White 232 (1913); Georgetown, Horner, Rowley, H. 10441 (1955); Ipswich, Oakes; Ipswich, Plum Island, H., McGregor & Schweinfurth 12700 (1957); Topsfield, H.R. Towne; Hamilton, Hubbard (1913); Gloucester, Prince (1883); [Rockport, A & W (1969)] ; Manchester, Hubbard (1913); Beverly, Lowell (1846); Salem, Cole (1888); Peabody, Sears (1885); Lynn, Cushman (1908).

X *L. producta* (Gray) Fern. Damp thickets; occasional. - Merrimac, H. 18328 (1958); Lawrence, H. 11237 (1955); Boxford, H. 10221 (1955); A hybrid of *L. quadrifolia* and *L. terrestris.*

L. terrestris (L.) BSP. Swamp Loosestrife. Low grounds; common. - Amesbury, H. 9190 (1953); Merrimac, H. 13682 (1957); Haverhill, H. 79 (1931); Andover, Pease 1069 (1901); North Andover, Root (1945); Boxford, M.P. Perley (1882); Groveland, H. 9079 (1953); Newburyport, White 180 (1913); Newbury, Plum Island, White 103 (1913); Georgetown, Robinson (1876); Rowley, H. 8122 (1952); Ipswich, Hubbard (1913); Topsfield, H.R. Towne; Middleton, H. 18096 (1958); Danvers, Lowell (1846); Wenham, H. 13745 (1957); Hamilton, H. 9408 (1953); Essex, H. 13370 (1957); Gloucester, Prince (1883); Rockport, Cushman (1908); Manchester, Hubbard (1913); Salem, Sears (1884); Peabody, Sears (1896); Saugus, H. 10880 (1955); Lynn, Sears (1885).

L. nummularia L. Moneywort. Introduced from Europe. Damp roadsides and grasslands; occasional. - Andover, Pease 1071 (1901); Boxford, M.P. Perley (1882); Georgetown, Horner; Topsfield, H.R. Towne; Salem, Robinson (1877).

L. thyrsiflora L. Tufted Loosestrife. Springy marshes and bogs; occasional. - Andover, Blake (1880); North Andover, H. 9495 (1954); Boxford, M.E. Perley (1883); Georgetown, Horner; Ipswich, Oakes; Topsfield, U. Bailey; Danvers, Sears (1886); Wenham, Cole (1910); Hamilton, Hubbard (1913); Salem, Phippen (1858); Marblehead, Cole (1886).

L. ciliata L. Low grounds and shores; common in Merrimac valley, occasional elsewhere. - Salisbury, H. 12217 (1956); Amesbury, Bean (1950); Haverhill, H. 488 (1932); Lawrence, H. 12283 (1956); Andover, Pease 1074 (1901); Boxford, Williams (1899); Groveland, Sears (1883); West Newbury, H. 10936 (1955); Newbury, White 203 (1913); Georgetown, Horner; Ipswich, Oakes; Topsfield, Mrs. Conley.

L. hybrida Michx. Swamps and wet shores; more general through the county. - Andover, Pease 2373 (1903); North Andover, Root (1945); Boxford, M.P. Perley (1883); Groveland, H. 9074 (1953); West Newbury, Bean (1940); Georgetown, Horner; Ipswich, Oakes; Topsfield, Mrs. Conley; Middleton, Bean (1950); Essex, H. 13369 (1957); Beverly, Lowell (1846); Salem, Sears (1884); Peabody, Mackintosh (1933); Saugus, Fernald & Svenson 1016 (1928); Lynn, Sears (1885).

TRIENTALIS L. Chickweed-Wintergreen.

T. borealis Raf. Star-flower. Woodlands; common. Amesbury, H. 9518 (1954); Haverhill, H. 446 (1932); North Andover, Zink (1943); Boxford, Frye (1881); Newburyport, E.N. Cole (1861); Georgetown, Horner; Rowley, Plum Island, H., McGregor & Schweinfurth 12721 (1957); Topsfield, Mrs. L.T. Brown (1912); Middleton, H. 8327 (1953); Danvers, Frost; Wenham Swamp, Sears (1884); Rockport, Anderson, Shafer & Smith (1931); West Manchester, Hubbard (1913); Beverly, Phippen (1857); Lynnfield, Kennedy (1859); Lynn Woods, Phippen (1888); Marblehead, Cole (1898).

ANAGALLIS L. Pimpernel.

A. arvensis L. Scarlet Pimpernel. Naturalized from Europe. Waste sandy fields; chiefly coastal; frequent. - Salisbury, White 44 (1913); Ipswich, St. John & White 561 (1913); Topsfield, H. R. Towne; Gloucester, Hoop Pole Cove, Williams (1897); Rockport, Robinson (1874); Salem, Sears (1884); Marblehead, Kennedy (1887).

A. arvensis, forma *caerulea* (Schreb.) Baumg. - Under bird-feeders. - Boxford, H. 9598 (1954); Topsfield, H. 9588 (1954).

GLAUX L. Sea-Milkwort.

G. maritima Saline shores and sands; occasional. - Ipswich, H. 13883 (1957); Rockport, Sears (1886).

G. maritima, var. *obtusifolia* Fern. More frequent. - Newburyport, Moulton (1887); Newbury, Horner; Plum Island, Oakes; Ipswich, St. John (1911); Essex, Buxton (1926); Gloucester, Cushman (1908); Salem, Dr. Chas. Pickering.

SAMOLUS L. Water-Pimpernel.

S. parviflorus Raf. Shallow water and wet soil, chiefly in tidal eatuaries; occasional. - Methuen, Wm. Boott (1865); Newburyport, Moulton (1886); Newbury, Williams (1898); Rowley, Sears (1888); Ipswich, Oakes; Fassett 436 (1922); Gloucester, Sears (1887).

HOTTONIA L. Featherfoil.

H. inflata Ell. Slow streams and pools; frequent the year following a dry summer. - Amesbury, Huntington & Knowlton (1906); Boxford, M.E. Perley (1914); [and Stiles Pond, Snyder (1950)]; West Newbury, H. 915 (1934); Byfield, Eaton (1896); Topsfield; Manchester, Oakes; Beverly, Robinson (1872); Salem, Cole (1900).

PLUMBAGINACEAE (Leadwort Family).

LIMONIUM Mill. Sea-Lavender. Marsh Rosemary.

L. carolinianum (Walt.) Britt. Salisbury, Fernald, Fogg, Svenson (1928).

L. nashii Small. Salt-marshes; common. - Salisbury, White 59 (1913); Newbury, 914 (1933); Newbury, Plum Island, H. 12518 (1956); Rowley, Horner; Ipswich, Oakes; Danvers, Sears (1880); Essex, H. 9393 (1953); Gloucester, Prince (1883); Rockport, Smith 65 (1921); Salem, Phippen (1839), [Snyder (1950 on)]; Peabody, Mackintosh (1935), Marblehead, H. 11908 (1955).

OLEACEAE (Olive Family).

FRAXINUS L. Ash.

F. americana L. White Ash. Rich woods; frequent. - Haverhill, H. 739 (1932); Methuen, H. 9627 (1954); Lawrence, Pease 2398 (1903); Andover, Pease 3607 (1904); North Andover, Root (1949); Boxford, Robinson (1875); West Newbury, NEBC. (1951); Georgetown, Horner; Topsfield, H.R. Towne; Middleton, Cole (1919); Gloucester, Annisquam, Ripley (1914); Rockport, H. 10273 (1955); Swampscott, Cole (1924).

F. pennsylvanica Marsh. Red Ash. Low grounds; frequent. - Amesbury, Bean (1950); West Newbury, White 503 (1913); Rowley, H. 10422 (1955); Topsfield, P. Kneeland; Wenham, Robinson (1875); Beverly, Robinson (1875); Marblehead, Cole (1949).

F. pennsylvanica, var. *austini* Fern. - Haverhill, Bradford, Fernald 11411 (1913); North Andover, Pease 2664 (1903); Boxford, Fernald (1909); Topsfield, Sears (1880); Beverly, Robinson (1875); Swampscott, Buck (1895).

F. pennsylvanica, var. *subintegerrima* (Vahl) Fern. - Green Ash. - Ipswich, H. 13450 (1957).

F. nigra Marsh. Black Ash. Swamps and shores; rare. - Danvers, Robinson (1877); Wenham, Sears (1875).

SYRINGA L. Lilac.

S. vulgaris L. Common Lilac. Introduced from Europe. Escape from cultivation. - Boxford, H. 25466 (1962); Gloucester, Robinson (1877).

LIGUSTRUM L. Privet.

L. vulgare L. Privet. Introduced from Europe. Thickets and open woods; occasional. - Boxford, H. 14197 (1957); Newbury, H. 19062 (1959); Georgetown, H. 17663 (1958); Topsfield, H. Porter; Danvers, Sears (1889); Swampscott, Sears (1885).

L. amurense Carr. Amur Privet. Introduced from northeast Asia. Thickets an open woods. - Ipswich, H. 14019 (1957).

GENTIANACEA (Gentian Family).

Sabatia Adans. Sabatia.

S. angularis (L.) Pursh. Rose-Pink. Adventive from South. Fields; rare. - Salisbury, Eben True (1885), one plant in hayfield.

CENTAURIUM Hill. Centaury.

C. umbellatum Gilib. Naturalized from Europe. Fields and waste places; rare. - West Boxford, Byers 19054 (1958), J.M. Camp (1960).

GENTIANA L. Gentian.

G. crinita Froel. Fringed Gentian. Meadows and brooksides; now rare. - Haverhill, H. 654 (1932); Andover, Pease 1079 (1900); North Andover, Root & Zink (1945); Boxford, S.F. Blake (1918); West Newbury, H. & Wade 18802 (1958); Georgetown, Horner (1886); Ipswich, H. & Wade 18779 (1958); Topsfield, H. 11574 (1955); Danvers, Sears (1888); Beverly, Miss M.T. Saunders (1879); Peabody, Cole (1926); Lynnfield, Fernald (1917). Citations are most recent collections from each town.

G. crinita, forma *albina* Fern. - Boxford, J.P. Peabody (1886).

G. andrewsii Griseb. Closed Gentian. Meadows and low thickets; occasional, apparently more common along Merrimac valley. - Merrimac, H. 9296 (1953); Methuen, Root (1951); Lawrence, Sears (1885); North Andover, Bean (1916); Boxford, M.P. Perley (1881); West Newbury, H. 11118 (1955); Newburyport, White (1913); Newbury, Bean & Stockman (1954); Ipswich, H. & Wade 18798 (1958); Topsfield, G. Whipple; Danvers, Phippen (1857); Wenham, Sears (1892); Peabody, Mackintosh (1933).

G. andrewsii, forma *albiflora* Britt. - Newbury, Bean & Stockman (1954).

G. clausa Raf. Bottle-Gentian. Borders of rich woods and thickets; rare. - West Boxford, Horner; Groveland, Horner (1886).

G. cruciata L. Introduced from Europe. Garden escape. - Topsfield, Foye 9643 (1954).

BARTONIA Muhl.

B. virginica (L.) BSP. Dry or wet acid soil; occasional. - Andover, Pease 1553 (1902); Boxford, A.W. & C.A. Cheever (1913); Ipswich, Sears (1898); Danvers, Sears (1887); Wenham, Lund (1882); Gloucester, Lanesville, Williams (1897); Rockport, Smith 95 (1921); Manchester, Boott (1856).

MENYANTHES L. Buckbean.

M. trifoliata L., var. *minor* Raf. Bogbean. Quagmires and pond-margins; frequent. - Salisbury, White 273 (1913); Haverhill, H. 1300 (1934); Andover, Sutton 33 (1874); North Andover, Morss

(1884); Boxford, M.P. Perley (1882); Newburyport, Cushing (1819); Georgetown, Horner; Topsfield, G.T. Dow; Danvers, Robinson (1872); [Gloucester, A & W (1969)] ; North Beverly, Morong (1875).

NYMPHOIDES Hill Floating-heart.

N. cordata (Ell.) Fern. Ponds and slow streams; common. - Haverhill, Bradford, H. 13161 (1957); Andover, Pease 1084 (1902); Boxford, H. 918 (1933); Georgetown, H. 12962 (1957); Rowley, H. 13526 (1957); Topsfield, B.B. Towne; Wenham, Robinson (1875); Hamilton, H. 9396 (1953); Essex, H. 13372 (1957); Peabody, Sears (1885).

APOCYNACEAE (Dogbane Family).

VINCA L. Periwinkle.

V. minor L. "Myrtle". Introduced from Europe. Spread from cultivation, roadsides and open woods; occasional. - Haverhill, H. 447 (1932); North Andover, Root & Zink (1945); Boxford, H. 11903 (1953); Newburyport, Bean & H. 7726 (1951); Topsfield, H.R. Towne.

V. major L., var. *variegata* Loud. Introduced from Europe. Escape from cultivation, waste places; rare. - Saugus, H. 18690 (1958).

APOCYNUM L. Dogbane. Indian Hemp.

A. androsaemifolium L. Spreading Dogbane. Dry thickets and borders of woods; common. - Haverhill, H. 489 (1932); Andover, Pease 2454 (1903); North Andover, M.E. Perley (1885); Boxford, H. 8111 (1952); Georgetown, Horner (1885); Rowley, Plum Island, H. 12543 (1956); Ipswich, Oakes; Topsfield, H.R. Towne; Danvers, Sears (1889); Hamilton, H. 10786 (1955); Gloucester, H. 18878 (1958); Beverly, Prides Crossing, Hubbard (1913); Salem, Cole (1886); Peabody, Sears (1883); Lynn, Cushman (1908); Swampscott, Sears (1885).

A. cannabinum L. Indian Hemp. Open ground; rare. - Ipswich, Oakes.

A. sibiricum Jacq. Rocky or gravelly soil; occasional. - Haverhill, Ward Hill, H. 9890 (1955); Lawrence, Pease 1089 (1902); North Andover, H. 9902 (1955); Groveland, Paine (1885); Newburyport, Cole (1902); Georgetown, Sears (1905); Middleton, Oakes; Salem, Sears (1905).

ASCLEPIADACEAE (Milkweed Family).

ASCLEPIAS L. Milkweed.

A. tuberosa L. Butterfly-weed. Dry open soil; local and rare. - Haverhill, Horner (1874), H. 919 (1933); Danvers, Manning (1886); North Beverly, Horncliffe & Raymond (1902).

A. purpurascens L. Purple Milkweed. Dry to damp woods and openings; occasional. - Merrimac, H. 10582 (1955); Andover, Robinson (1878); North Andover, M.E. Perley (1887); West Boxford, G.S. Morse (1883); Georgetown, H. 12860 (1957); Middleton, Sears (1880); Hamilton, H. 10806 (1955).

A. incarnata L., var. *pulchra* (Ehrh.) Pers. Swamp-Milkweed. Swamps, wet thickets and shores; common. - Haverhill, H. 81 (1931); Methuen, H. 11848 (1957); Lawrence, H. 11248 (1955); Andover, Pease 1091 (1899); North Andover, Morss (1885); Boxford, M.E. Perley (1884); Groveland, H. 12868 (1957); Byfield, Horner; Georgetown, Horner; Rowley, H. 12230 (1956); Ipswich, Oakes; Topsfield, G.T. Dow; Middleton, Bean (1950); Danvers, Cushman 4829 (1908); Beverly, Lowell (1846); Salem, Russell (1859); Peabody, Sears (1896); Marblehead, H. 920 (1933).

A. quadrifolia Jacq. Dry woods; rare. - Topsfield, Oakes, Mrs. Conley; Danvers, Sears (1886); Swampscott, Young (1882).

A. exaltata L. Rich woods and clearings; occasional. - Merrimac, H. 12990 (1957); Haverhill, H. 531 (1932); North Andover, H. 17944 (1958); Boxford, M.E. Perley (1885); Ipswich, Oakes; Topsfield, Olive Floyd; Middleton, Pease 27863 (1940); Peabody, Sears (1896); Saugus, Young (1879).

A. syriaca L. Common Milkweed. Thickets, roadsides and dry fields; common. - Salisbury, H. 12205 (1956); Amesbury, H. 10549 (1955); Lawrence, H. 11258 (1955); North Andover, Morss (1885); Boxford, M.E. Perley (1884); West Newbury, H. 10949 (1955); Newburyport, Robinson (1875); Newburyport, Plum Island, H. 13056 (1957); Georgetown, Horner; Ipswich, Oakes; Ipswich, Plum Island, H., McGregor & Schweinfurth 12687 (1957); Topsfield, H.R. Towne; Middleton, Bean (1950); Hamilton, Robinson (1875); Rockport, H. 10263 (1955); Manchester, Hubbard (1913); Salem, Cole (1910); Swampscott, H. 10678 (1955); Nahant, Merrill (1850).

A. amplexicaulis Sm. Dry open soil; rare. - Andover, Young (1882); Pease 1095 (1900); Groveland, Horner (1884); H. & Wetherall 9555 (1954); Ipswich, Oakes.

A. verticillata L. Andover, "Prospect Hill on the Andover Turnpike - found by Mrs. Boott (1819)".

CYNANCHUM L.

C. nigrum (L.) Pers. Naturalized from Europe. Waste places and roadsides; occasionally escaping from cultivation. - Boxford, H. 23921 (1961); Newburyport, Sears (1886), Moulton; Topsfield, H. 9597 (1954); Lynnfield, Kennedy (1911); Bean (1916).

CONVOLVULACEAE (Convolvulus Family).

IPOMOEA L. Morning-glory.

I. coceinea L. Red Morning-glory. Naturalized from Tropical America. Waste places and cultivated ground; occasional. - Ipswich, H. 13981 (1957), spontaneous in garden for many years.

I. hederacea (L.) Jacq. Naturalized from Tropical America. Waste places; occasional. - Danvers, H. 20767 (1959); Salem, H. 11488 (1955); Saugus, H. 18700 (1958).

I. purpurea (L.) Roth. Common Morning-glory. Introduced from Tropical America. Waste places; occasional. - Haverhill, H. 532 (1932); Lawrence, H. 11789 (1955); Boxford, H. 30750 (1966); Newburyport, H. 20731 (1959); Gloucester, Magnolia, H. 18883 (1958); Beverly, H. 18895 (1958); Salem, Robinson (1877); Lynn, H. 11810 (1955); Marblehead, H. 11434 (1955).

CONVOLVULUS L. Bindweed.

C. sepium L. Hedge Bindweed. Thickets; occasional. - Merrimac, H. 18315 (1953); Methuen, H. 12280 (1956); Boxford, Root (1945); West Newbury, H. 18312 (1958); Plum Island, Knowlton (1913); Gloucester, Magnolia, H. 12335 (1956); Peabody, Mackintosh (1933).

C. sepium, forma *coloratus* Lange. The most common form. - Amesbury, H. 10533 (1955); Merrimac, Bean & H. 8589 (1953); Haverhill, H. 694 (1932); North Andover, Morss (1905); Boxford, M.E. Perley (1884); Newburyport, Cole (1902); Newbury, H. 10391 (1955); Rowley, H. 10426 (1955); Topsfield, H. 8866 (1953); Essex, H. 10249 (1955); Gloucester, Grigg & Sanford (1925); Beverly Farms, Hubbard (1913); Salem, Manning (1887); Swampscott, H. 10683 (1955); Marblehead, Cole (1930).

C. sepium, forma *malachophyllus* Fern. Chiefly coastal. - Salisbury, H. 10504 (1955); Newbury, Plum Island, Knowlton (1913); Ipswich, Plum Island, H., McGregor & Schweinfurth 12662 (1957); Gloucester, F.S. Collins (1915); Rockport, Cushman (1908); Beverly, Lowell (1846); Nahant, Merrill (1849).

C. arvensis L. Naturalized from Eurasia. Waste places and fields; occasional. - South Lawrence, Pease 1098 (1902); North Andover, H. & Root 14160 (1957); Ipswich, Oakes; Topsfield, H. 13327 (1957); Salem, Phippen; East Lynn, A.P. Morse (1924); Swampscott, Mrs. T.R. Beeber; Nahant, Rodman (1886).

C. arvensis, forma *cordifolius* Lasch. - Ipswich, Furber (1911); Peabody, Mackintosh (1933).

CUSCUTA L. Dodder. Love-vine.

C. pentagona Engelm. Dry open soil; rare. - Boxford, H. 8905 (1953).

C. gronovii Willd. Low ground, parasitic on many hosts; common. - Amesbury, H. 18967 (1958); Haverhill, H. 282 (1932); Lawrence, H. 11246 (1955); Andover, Pease 30385 (1943); North Andover, Morss (1885); Boxford, H.E. Parkhurst (1882); West Newbury, Pease 28529 (1940); Georgetown, Horner (1892); Ipswich, St. John (1917); Topsfield, H. 13606 (1957); Hamilton, Phippen; Gloucester, Magnolia, J.M. Hunnewell (1901); Rockport, Williams (1897); Manchester, Norong (1875); Salem, Cole (1908); Saugus, Robinson (1874); Lynn, H. 27048 (1964).

C. epilinum Weihe. Flax-Dodder. Adventive from Europe. No collections in the past century. - Topsfield, Lowell, flax field.

POLEMONIACEAE (Polemonium Family).

GILIA R. & S.

G. tricolor Benth., var. *longipedicellata* Greenm. Adventive from the West. Waste places; rare. - Lawrence, J.A. Collins, Jr. (1900).

PHLOX L. Phlox.

P. paniculata L. Fall-Phlox. Introduced from West. Garden escape; occasional. - Andover, Pease 1107 (1901); Topsfield, H.R. Towne; Beverly, Robinson (1875).

BORAGINACEAE (Borage Family).

BORAGO L. Borage.

B. officinalis L. Introduced from Europe. Escape from cultivation. - Danvers, Sears (1904).

SYMPHYTUM L. Comfrey.

S. asperum Lepechin. Prickly Comfrey. Introduced from the Caucasus. Waste places; occasional. - Andover, Pease 1119 (1901); Newburyport, E.N. Cole (1861); Salem, Sears (1904).

S. officinale L. Common Comfrey. Introduced from Europe. Waste places; occasional. - Andover, Pease 1121 (1901); Boxford, M.E. Perley (1885); Newburyport, Sears (1886); Newbury, Cushing, Horner; Marblehead, Sanford & Wies (1946).

LYCOPSIS L. Bugloss.

L. arvensis L. Small Bugloss. Naturalized from Europe. Dry or sandy fields and waste places; occasional. - Georgetown, Horner (1882, 1885); Ipswich, Oakes.

ECHIUM L. Viper's Bugloss.

E. vulgare L. Blueweed. Naturalized from Europe. Roadsides and waste places; occasional. - Methuen, Greene, H. 12259 (1956); North Andover, Kittredge (1897); Boxford, S. Perley (1884); Groveland, H. 12302 (1956); South Georgetown, Horner; Middleton, Phippen (1838); Rockport, Prybot 31784 (1967); Beverly, Robinson (1877).

E. plantagineum L. Introduced from Europe. Waste places; occasional. - Haverhill, Palfrey (1890); Boxford, H. 11561 (1955).

LITHOSPERMUM L. Gromwell.

L. arvense L. Corn-Cromwell. Naturalized from Europe. Sandy fields and roadsides; occasional. - Andover, Pease 1991 (1903); Georgetown, Horner; Salem, Sears (1886); Peabody, Mackintosh (1902).

CYNOGLOSSUM L. Hound's-tongue.

C. officinale L. Common Hound's-tongue. Naturalized from Europe. Roadsides and waste places; occasional. - Lawrence, Pease 1109 (1902); Byfield, Parker River (1886); Georgetown, Horner (1883); Topsfield, Horner. Danvers, Sears (1904); Salem, Mackintosh (1925).

MYOSOTIS L. Forget-me-not.

M. scorpioides L. True Forget-me-not. Naturalized from Europe. In quiet water or wet ground; common. - Merrimac, H. 10575 (1955); Haverhill, H. 83 (1931); Methuen, H. 18391 (1958); North Andover, Pease 24104 (1934); Groveland, Bean & H. 8551 (1953); Ipswich, H. 13456 (1957); Topsfield, H. 8838 (1953); Danvers, Sears (1886); Beverly Farms, Hubbard (1913); Saugus, Ripley 15523 (1915).

M. laxa Lehm. In water and wet ground; frequent. - Haverhill, H. 490 (1932); North Andover, Morss (1885); Boxford, H. 8637 (1953); West Newbury, H. 921 (1933); Danvers, Cushman (1908); Salem, Sears (1880); Peabody, Sears (1896).

M. verna Nutt. Rocky woods and openings; occasional. - Andover, Blake (1884); Boxford, Fernald (1909); Newbury, Cushing (1819); Ipswich; Danvers, Lowell (1847); Salem, Robinson (1872).

M. sylvatica Hoffm. Garden Forget-me-not. Introduced from Europe. Garden escape. - Andover, Pease 2045 (1903); Boxford, H. 7910 (1952).

M. arvensis (L.) Hill. Naturalized from Europe. Roadsides and waste places; occasional. - Andover, Pease 4242 (1904); Gloucester, Magnolia, H. 12324 (1956); Salem, Robinson (1885); Marblehead, Bean & H. 9771 (1955).

ASPERUGO L. Madwort.

A. procumbens L. Adventive from Europe. Waste places; occasional. - Newburyport, Moulton; Salem, Russell (1859); Sears (1896).

LAPPULA Moench Stickseed.

L. echinata Gilib. Naturalized from Europe. Roadsides and waste places; occasional. - North Andover, Root (1945); Byfield, Parker River Mills, Horner (1885); Topsfield, H.R. Towne; Salem, Mrs. A. Crosby (1876).

HACKELIA Opiz Stickseed.

H. virginiana (L.) I.M. Johnston. Rich woods and thickets; occasional. - Merrimac, H. 18313 (1958); Andover, Pease 1111 (1902); North Andover, Russell (1859); Morss (1885); Boxford, M. E. Perley (1885); Georgetown, Horner; Ipswich; Topsfield, Sears (1905); Middleton, Pease 1114 (1902).

VERBENACEAE (Vervain Family).

VERBENA L. Vervain.

V. officinalis L. European Vervain. Naturalized from Europe. Waste places; rare. - Rowley, Oakes.

V. urticifolia L. White Vervain. Rich thickets and borders of roads; common. - Haverhill, H. 557 (1932); Methuen, H. 11745 (1955); Lawrence, H. 11764 (1955); North Andover, Morss (1885); Boxford, M.E. Perley (1884); West Newbury, H. 922 (1933); Rowley, H. 12237 (1956); Ipswich, H. 14025 (1957); Topsfield, J. Thayer; Middleton, H. 8705 (1953); Danvers, H. 20754 (1959); Wenham, Robinson (1877); Hamilton, H. 12501 (1956); Salem, Cole (1899); Peabody, H. 8751 (1953).

V. hastata L. Blue Vervain. Swales and damp thickets; common. - Salisbury, White 254 (1913); Amesbury, Cole (1896); Haverhill, H. 626 (1932); Methuen, H. 12278 (1956); South Lawrence, Morss (1885); North Andover, Root (1945); Boxford, L.B. Twitchell; Groveland, H. 10969 (1955); West Newbury, H. 10952 (1955); Newbury, White 241 (1913); Rowley, H. 12231 (1956); Ipswich,

Furber (1911); Hamilton, H. 10840 (1955); Essex, Fernald, Hunnewell & Long 12064 (1913); Gloucester, Annisquam, Ripley (1914); Beverly, H. 11687 (1955); Peabody, Mackintosh (1933); Saugus, H. 10883 (1955); Lynn, Robinson (1875).

V. stricta Vent. Hoary Vervain. Adventive from West. Waste places; rare. - Lawrence, Gutterson (1902); Pease 1125 (1902).

V. bracteata Lag. & Rodr. Adventive from South. Waste places; rare. - Topsfield, M. Clarke; Salem, Hagar (1973).

LABIATAE (Mint Family).

TRICHOSTEMA L. Bluecurls.

T. dichotomum L. Dry open soil; common. - Salisbury, White 365 (1913); Amesbury, H. 9167 (1953); Merrimac, H. 9256 (1953); Haverhill, H. 707 (1932); Methuen, H. 11749 (1955); North Andover, H. 11050 (1955); Boxford, M.E. Perley (1883); Groveland, H. 11174 (1955); Newburyport, H. 11277 (1955); Newbury, H. 8244 (1952); Georgetown, Horner; Rowley, H. 11944 (1955); Rowley, Plum Island, H. 12467 (1956); Ipswich, Phippen (1857); Danvers, Sears (1887); Essex, Fernald, Hunnewell & Long 10275 (1913); Gloucester, Magnolia, J.M. Hunnewell (1900); Rockport, Smith 2713 (1933); Beverly, Lowell (1846); Salem, Cole (1888); Peabody, Mrs. F.R. Williams 552 (1937); Lynnfield, Bean (1913); Saugus, Robinson (1874); Marblehead, H. 11452 (1955).

AJUGA L. Bugleweed.

A. reptans L. Naturalized from Europe. Roadsides and fields; rare. - Topsfield, A.W. Conant.

A. genevensis L. Naturalized from Europe. Fields and waste places; rare. - Boxford, H. 9927 (1955).

TEUCRIUM L. Germander.

T. canadense L. Wood-Sage. Shores, thickets and woods; frequent. - North Andover, Russell (1859); Newburyport, Bean (1945); Newbury, White 433 (1913); Newbury, Plum Island, Pease 2642 (1903); Rowley, H. 8266 (1952); Ipswich, H. 14034 (1957); Ipswich, Plum Island, Furber (1911); Danvers, Mackintosh (1924); Essex, H. 9378 (1953); Gloucester, Robinson (1874); Rockport, Williams (1897); Beverly, Lowell (1846); Salem, Cole (1888).

T. occidentale Gray, var. *boreale* (Bickn.) Fern. Alluvial soil; rare. - Rowley, Bean & Stockman (1954); Rockport, Williams (1897); Salem, Cole (1908).

SCUTELLARIA L. Skullcap.

S. lateriflora L. Mad-dog Skullcap. Thickets, meadows and swampy woods; common. - Amesbury, H. 9196 (1953); Merrimac, H. 18330 (1958); Haverhill, H. 706 (1932); Andover, Pease 4270 (1904); North Andover, H. 11018 (1955); Boxford, H.E. Parkhurst (1882); Groveland, Horner; West Newbury, H. 11129 (1955); Newburyport, White 417 (1913); Topsfield, H.R. Towne; Danvers, Oakes?; Wenham, H. 13748 (1957); Essex, Lowell (1847); Manchester, H. 10724 (1955); Beverly, Lowell (1846); Peabody, Mackintosh (1933); Lynnfield, Robinson (1874); Lynn, H. 13175 (1957).

S. epilobiifolia A. Hamilton. Common Skullcap. Sandy shores, meadows and swampy thickets; common. - Amesbury, H. 13002 (1957); Merrimac, H. 13679 (1957); Haverhill, H. 84 (1931); Methuen, H. 8916 (1953); South Lawrence, Morss (1885); Boxford, M.E. Perley (1885); Newbury, White 207 (1913); South Georgetown, Horner; Ipswich, Forbes (1913); Topsfield, T. Conley; Middleton, H. 18089 (1958); Danvers, Sears (1885); Wenham, Lowell (1847); Gloucester, Prince (1883); Rockport, Rich (1898); Beverly, Lowell (1846); Salem, Phippen (1838).

S. epilobiifolia, forma *albiflora* (Millsp.) Fern. - North Andover, H. 12927 (1957).

MARRUBIUM L. Hoarhound.

M. vulgare L. Naturalized from Europe. Waste places; rare. - Danvers, Oakes.

NEPETA L. Catnip.

N. cataria L. Catnip. Naturalized from Europe. Dooryards and waste places; frequent. - Haverhill, H. 604 (1932); North Andover, Root (1955); Boxford, H. 12507 (1956); Georgetown, H. 13581 (1957); Rowley, Horner; Ipswich, Oakes; Topsfield, M.E. Perley (1883); Hamilton, Phippen (1857); Gloucester, Ripley 15886 (1915); Beverly, Lowell (1846); Saugus, H. 18644 (1958).

GLECHOMA L. Ground-Ivy.

G. hederacea L. Gill-over-the-ground. Naturalized from Europe. Roadsides, yards and damp shaded places; frequent. - Haverhill, H. 448 (1932); North Andover, Morss (1884); Boxford, M.E. Perley (1882); Georgetown, Horner; Ipswich, Cole (1890); Wenham, Robinson (1879); Hamilton, Cole (1899); East Gloucester, Deane (1901); West Manchester, Hubbard (1913); Beverly, H. 18904 (1958); Marblehead, Cole (1900).

G. hederacea, var. *micrantha* Moricand. Naturalized from Europe. Frequent. - Haverhill, H. 311 (1932); Metheun, Churchill (1909); Andover, Pease 1143 (1902); North Andover, Zink (1943); Boxford, H. 7849 (1952); West Newbury, H. 31889 (1968); Topsfield, H.R. Towne; Danvers, Cole (1952); Salem, Robinson (1879); Nahant, Bean & H. 9839 (1955).

DRACOCEPHALUM L. Dragonhead.

D. parviflorum Nutt. Adventive from West. Rocky or gravelly soils; rare. - Topsfield, H.R. Towne; Peabody, Mackintosh (1926).

PRUNELLA L. Selfheal.

P. vulgaris L. Heal-all. Naturalized from Europe. Roadsides and waste ground; occasional. - North Andover, Morss (1885); Lynnfield, Bean (1919).

P. vulgaris, var. *lanceolata* (Bart.) Fern. Native. Thickets and roadsides; Common. - Amesbury, H. 18213 (1958); Merrimac, H. 18348 (1958); Haverhill, H. 491 (1932); North Andover, Root (1945); Boxford, M.E. Perley (1882); Newburyport, White 475 (1913); Rowley, H. 10432 (1955); Ipswich, H. 18171 (1958); Topsfield, S.J. Towne; Danvers, Sears (1885); Manchester, Hubbard (1913); Salem, Sears (1885); Peabody, Mackintosh (1933); Lynn, Cushman 2998a (1908); Marblehead, Cole (1886).

P. vulgaris, var. *lanceolata,* forma *idocalyx* Fern. - Salisbury, White 65 (1913); Georgetown, Horner; Manchester, Hubbard (1913).

PHYSOSTEGIA Benth. False Dragonhead.

P. virginiana (L.) Benth. Introduced from the South. Garden escape; occasional. - Boxford, H. 13312 (1957); Middleton, Robinson (1874); Manchester, Farlow (1903).

P. virginiana, var. *speciosa* (Sweet) Gray. Garden escape. - Georgetown, Horner (1881); Topsfield, H.R. Towne.

BALLOTA L. Fetid Horehound.

B. nigra L. Black Horehound. Adventive from Europe. Waste places; rare. - Topsfield, H.R. Towne.

LEONURUS L. Motherwort.

L. cardiaca L. Naturalized from Europe. Waste places; common. - Merrimac, H. 10567 (1955); Haverhill, H. 606 (1932); Andover, Bullard (1880); North Andover, Morss (1885); Boxford, M.P. Perley (1882); Newburyport, White 481 (1913); Newbury, H. 20721 (1959); Georgetown, Horner; Rowley, Plum Island, H., McGregor & Schweinfurth 12729 (1957); Topsfield, H.R. Towne; Middleton, Bean & H. 7780 (1951); Wenham, Robinson (1875); Gloucester, Collins (1915); Rockport, L.B. & R.C. Smith 1137 (1930); Salem, Cole (1899).

GALEOPSIS L. Hemp-Nettle.

G. tetrahit L., var. *bifida* (Boenn.) Lej. & Court. Naturalized from Europe. Waste places; frequent. - Haverhill, H. 605 (1932); Andover, Pease 2447 (1903); Boxford, Horner; Groveland, Paine & M.E. Perley (1885); West Newbury, Robinson (1879); Byfield, M.E. Perley (1881); Georgetown, Horner (1874); Ipswich, Oakes; Topsfield, H.R. Towne; Manchester, Lowell (1848).

LAMIUM L. Dead-Nettle.

L. amplexicaule L. Henbit. Naturalized from Europe. Waste places and roadsides; occasional. - Haverhill, H. 8311 (1953); Andover, Pease 1131 (1902); Ipswich, H. 10228 (1955); Salem, Robinson (1886); Marblehead, Bean & H. 9734 (1955).

L. album L. Snowflake. Introduced from Europe. Waste places and lawns; occasional. - Newbury, Mrs. Gibbons (1959); Salem, Robinson (1877); Saugus, Collins (1890).

STACHYS L. Hedge-Nettle.

S. palustris L. Woundwort. Introduced from Europe. Ditches and waste places; occasional. [West Newbury, Cole (1900)]; Byfield, Horner; Ipswich, Oakes; Sears (1904); Rockport, Smith 27 (1921).

S. palustris, var. *nipigonensis* Jennings. Native. Low grounds; rare. - North Andover, Bean (1916).

S. palustris, var. *homotricha* Fern. Native. Low ground; rare. - "Danvers prob. Topsfield", Sears (1877).

S. tenuifolia Willd., var. *platyphylla* Fern. Low ground; rare. - West Newbury, Pease 28534 (1940).

S. tenuifolia, var. *hispida* (Pursh) Fern. Low ground; rare. - West Newbury, Bean (1940); Newburyport, White 501 (1913).

SALVIA L. Sage.

S. pratensis L. Naturalized from Europe. Waste places; rare. - North Andover, Root (1945), chicken yard.

S. splendens Ker. Introduced from Brazil. Escape from cultivation. - Lynn, H. 18837 (1958).

MONARDA L. Horsemint.

M. didyma L. Oswego-Tea. Introduced from South. Garden escape; occasional. - North Andover, Root (1945); Boxford, H. 10617 (1955); West Newbury, Horner; Topsfield, H.R. Towne.

M. media Willd. Introduced from South. Garden escape; rare. - Beverly, Robinson (1878); Essex County, Oakes.

M. fistulosa L., var. *mollis* (L.) Benth. Wild Bergamot. Dry thickets and borders of woods; occasional. - Haverhill, Hunnewell (1915); North Andover, Morss (1885); Root (1945); West Boxford, Horner; Groveland, M.E. Perley (1885); Georgetown, Horner (1874); Topsfield, B.B. Towne; Danvers, Lowell (1847), Sears (1887).

M. punctata L. Adventive from South? Dry sandy soil; rare. - Byfield, Parker River Mills (1885); Georgetown, Horner (1885).

HEDEOMA Pers. Mock Pennyroyal.

H. pulegioides (L.) Pers. Pudding-grass. Dry soil; occasional. - Haverhill, H. 607 (1932); Boxford, Robinson (1875), H. 8718 (1953); Georgetown, Horner; Ipswich, H. & Wade 18789 (1958); Topsfield, V.R. Dodge; Danvers, Lowell (1847); Salem, Cole (1901); Swampscott, Sears (1885).

MELISSA L. Balm.

M. officinalis L. Introduced from Europe. Escape from cultivation; rare. - West Boxford, Horner; Newbury, Horner.

SATUREJA L. Savory.

S. vulgaris (L.) Fritsch, var. *neogaea* Fern. Dogmint. Woods and thickets; rare. - North Andover, H. 17952 (1958); Salem, Oakes; Peabody, Sears (1884).

HYSSOPUS L. Hyssop.

H. officinalis L. Introduced from Europe. Dry pastures and roadsides, rare. - Georgetown, Horner (1877).

PYCNANTHEMUM Michx. Mountain-Mint.

P. tenuifolium Schrad. Dry to moist sandy ground and thickets; occasional. - Haverhill, Ward Hill, Bean (1916); Andover, Pease 1150 (1900); Boxford, Mackintosh (1924); Groveland, Horner; Newburyport, Currier (1896); Topsfield, Sears (1901); Hamilton, Hunnewell (1916).

P. virginianum (L.) Durand & Jackson. Dry to wet thickets; occasional. - Haverhill, Ward Hill, H. 608 (1932); Boxford, Williams (1899); Groveland, M.E. Perley (1885); Newburyport, Cole (1900); South Byfield, Sears (1888); Georgetown, Horner (1874); Topsfield, B.B. Towne.

P. verticillatum (Michx.) Pers. Dry to moist thickets; rare. - Haverhill, Ward Hill, Bean (1916); Andover, Pease 1148 (1902).

P. muticum (Michx.) Pers. Dry woods and clearings; frequent. - Merrimac, Bean & H. 7707 (1951); Haverhill, Ward Hill, Bean (1916); North Andover, Morss (1885); Boxford, H. 8152 (1952); Groveland, Horner; West Newbury, H. 10944 (1955); Newburyport, Cole (1900); South Georgetown, Horner; Ipswich, R.J. Eaton (1931); Topsfield, B.E. Towne; Danvers, Sears (1877).

P. incanum (L.) Michx. Dry woods and thickets; rare. - Georgetown, Horner.

ORIGANUM L. Wild Marjoram

O. vulgare L. Naturalized from Europe. Roadsides and old fields; rare. - Georgetown, Oakes; Gloucester, Magnolia, H. 12321 (1956).

THYMUS L. Thyme.

T. serpyllum L. Creeping Thyme. Naturalized from Europe. Old fields; rare. - Boxford, Oakes (1819); Newburyport, Pope (1898); Ipswich, Wade 18785 (1958); Nahant, Mrs. Wilson (1896).

T. vulgaris L. Naturalized from Europe. Garden escape; rare. - Topsfield, H.R. Towne.

LYCOPUS L. Bugleweed.

L. virginicus L. Rich moist soil; occasional. - Haverhill, Bradford, Fernald 11425 (1914); North Andover, Morss (1885); West Boxford, Root (1945); Groveland, H. 10968 (1955); Topsfield, Mrs. Conley; Wenham, H. 13749 (1957); Essex, Lowell (1847); Beverly, Lowell (1853).

L. uniflorus Michx. Low ground; common. - Amesbury, H. 9197 (1953); Merrimac, H. 13712 (1957); Haverhill, H. 611 (1932); Boxford, Williams (1906); Groveland, H. 9082 (1953); Newburyport, White 477 (1913); Newbury, Plum Island, White 313 (1913); Ipswich, Hubbard (1913); Danvers, Sears (1885); Manchester, Phillips (1877); Beverly, Phillips (1877); Salem, Cole (1898); Lynnfield, H. 14124 (1957).

L. americanus Muhl. Low ground; common. - Salisbury, White 262 (1913); Amesbury, Cole (1886); Haverhill, Bradford, Fernald 11428 (1914); North Andover, Morss (1885); Boxford, M.E. Perley (1881); West Newbury, H. 9130 (1953); Newburyport, White 476 (1913); Rowley, H. 8117 (1952); Ipswich, H. 13443 (1957); Ipswich, Plum Island, H. 12424 (1956); Topsfield, H.R. Towne; Danvers, Lowell (1853); Gloucester, H. 13834 (1957); Salem, Russell (1859); Peabody, H. 9315 (1953).

L. rubellus Moench. Damp soils; rare. - Newbury, White 314 (1913).

L. asper Greene. Adventive from West. - Rowley, Plum Island, H. 12463 (1956).

MENTHA L. Mint.

M. spicata L. Spearmint. Introduced from Europe. Wet places; occasional. - North Andover, Morss (1885); Boxford, M.P. Perley (1882); Rowley, Horner; Danvers, Gloucester, Magnolia, J.M. Hunnewell (1900); Marblehead, H. 18928 (1958); Nahant, Robinson (1877).

M. piperita L. Peppermint. Introduced from Europe. Wet meadows and brooksides; occasional. - North Andover, Morss (1885); Rowley, H. 13482 (1957); Ipswich, M.E. Perley (1886); Topsfield, Archie Pierce; Danvers, Cushman 4837 (1908); Essex, Sears (1892); West Gloucester, Robinson (1877); Rockport, Smith 84 (1921); Beverly, Lowell (1853); Peabody, Mackintosh (1933).

M. crispa L. Introduced from Europe. Waste places; rare. - Topsfield, H. 18305 (1958).

M. cardiaca Baker. Introduced from Europe. Wet meadows and waste places; rare. - North Andover, Root (1945).

M. gentilis L. Introduced from Europe. Rich damp soil; rare. - Lynn, H. 11840 (1955).

M. arvensis L. Both native and naturalized from Europe. Damp open soils; frequent. - Amesbury, H. 18214 (1958); Merrimac, H. 18369 (1958); Haverhill, H. 613 (1932); Boxford, M.P. Perley (1885); West Newbury, Cole (1900); Newburyport, White 189 (1913); Topsfield, F. Gould; Essex, Fernald, Hunnewell & Long 10333 (1913); Rockport, Smith 913 (1927); Salem, Cole; Peabody, Mackintosh (1933).

M. arvensis, forma *lanata* (Piper). S.R. Stewart - Amesbury, Cole (1886); Topsfield, H. 8867 (1953).

M. arvensis, forma *glabra* (Benth.) S.R. Stewart - Boxford, H. 13538 (1957); Hamilton, Robinson (1874).

M. arvensis, var. *villosa* (Benth.) S.R. Stewart - North Andover, Morss (1885); Boxford, A.W. & C.A. Cheever (1913); Newburyport, White 482 (1913); Byfield, Horner; Georgetown, Horner; Beverly, Lowell (1845).

ELSHOLTZIA Willd.

E. ciliata (Thunb.) Hylander. Naturalized from Asia. Roadsides and gardens; rare. - Rowley, H. & Wade 18799 (1958).

COLLINSONIA L. Horse-Balm.

C. canadensis L. Richweed. Rich moist woods; rare. - Haverhill, Robinson (1890); Georgetown, Horner; Ipswich, Oakes.

SOLANACEAE (Nightshade Family).

SOLANUM L. Nightshade.

S. dulcamara L. Nightshade, Bittersweet. Naturalized from Europe. Thickets and damp clearings; common. - Salisbury, D. & E.W. White 45 (1913); Haverhill, H. 533 (1932); Methuen, H. 11752 (1955); Lawrence, H. 9850 (1955); North Andover, Root (1945); Boxford, M.E. Perley

(1884); Groveland, H. 10979 (1955); Georgetown, Horner; Rowley, H. 10424 (1955); Ipswich, Oakes; Ipswich, Plum Island, H. 12081 (1956); Topsfield, Mrs. E.A. Perkins; Middleton, Bean & H. 7779 (1951); Hamilton, H. 10835 (1955); Essex, Fernald, Hunnewell & Long 10337 (1913); Rockport, Gates & Smith 967 (1928); West Manchester, Hubbard (1913); Salem, Sears (1884); Salem, Little Misery Island, H. 13301 (1957); Peabody, H. 12753 (1957); Lynnfield, H. 14127 (1957); Saugus, H. 10873 (1955); Swampscott, H. 10695 (1955); Nahant, Bean & H. 9834 (1955).

S. tuberosum L. Potato. Introduced from South America. Dumps and waste places; occasional. - Haverhill, Bradford, H. 656 (1932); Lawrence, H. 18740 (1958); Boxford, H. 8099 (1952); Newburyport, H. 20733 (1959); Newbury, H. 20718 (1959); Georgetown, H. 13582 (1957); Rowley, H. 18507 (1958); Danvers, H. 20757 (1959); Wenham, Robinson (1877); Salem, H. 11485 (1955); Saugus, H. 18672 (1958).

S. americanum Mill. Thickets, shores and openings; frequent. - Haverhill, H. 11198 (1955); Lawrence, H. 11779 (1955); Boxford, M.E. Perley (1885); Groveland, H. 11181 (1955); Newburyport, H. 20746 (1959); Gloucester, Robinson (1874); Saugus, H. 11869 (1955); Swampscott, H. 10691 (1955); Marblehead, H. 11455 (1955).

S. nigrum L. Naturalized from Europe. Waste places and other disturbed soils; frequent. - Easily confused with the previous species. - Haverhill, H. 655 (1932); Boxford, H. 9650 (1954); Newburyport, White 406 (1913); Ipswich, Forbes (1913); Danvers, H. 11640 (1955); West Gloucester, Robinson (1877); Beverly, Lowell (1847); Salem, Cole (1919); Salem, North Gooseberry Island, H. 13276 (1957); Peabody, H. 11545 (1955); Lynn, H. 11841 (1955).

S. sarachoides Sendtner. Adventive from South America. Waste places; occasional. - Boxford, H. 11569 (1955); Danvers, H. 18574 (1958); Saugus, H. 18714 (1958); Lynn, H. 11842 (1955); Marblehead, H. 18924 (1958).

S. carolinense L. Adventive from South. Waste places; rare. - North Andover, H. 18731 (1958).

S. rostratum Dunal. Naturalized from West. Waste places; occasional. - Methuen, H. 11732 (1955); Boxford, M.E. Perley (1883); Gloucester, Magnolia, H. 12316 (1956); Salem, Mr. Spencer (1882); Lynn, H. 11820 (1955); Marblehead, Mrs. Hopkinson.

S. melongena L., var. *esculentum* Nees. Eggplant. Introduced from . Waste places; rare. - Gloucester, Magnolia, H. 12333 (1956).

LYCOPERSICUM Tomato.

L. esculentum Mill. Introduced from South America. Dumps and waste places; frequent. - Haverhill, Bradford, H. 743 (1932); Methuen, H. 11736 (1955); Lawrence, H. 11788 (1955); Boxford, M.E. Perley (1884); Newburyport, H. 20729 (1959); Newbury, H. 20720 (1959); Rowley, H. 18505 (1958); Ipswich, E. 7691 (1951); Topsfield, H. 18293 (1958); Gloucester, Magnolia, H. 12381 (1956); Beverly, H. 11671 (1955); Salem, H. 11486 (1955); Peabody, H. 11535 (1955); Saugus, H. 11853 (1955); Lynn, H. 11837 (1955); Marblehead, H. 11454 (1955).

PHYSALIS L. Ground-Cherry.

P. pubescens L., var. *grisea* Waterfall. Introduced from Tropical America. Waste places; rare. - Haverhill, H. 741 (1932).

P. pruinosa L. Dry open soil; rare. - Lynn, Faxon (1879), H. 18830 (1958).

P. angulata L. Adventive from South. Waste places; rare. - Haverhill, H. 742 (1932).

P. virginiana Mill., var. *subglabrata* (Mackenz. & Bush) Waterfall. Adventive from South. Waste places; rare. - Salem, Mackintosh (1933).

P. heterophylla Nees. Dry open woods and waste places; occasional. - Andover, Pease 2138 (1903); Georgetown, Horner; Ipswich, H. 11604 (1955); Topsfield, Robinson (1897), H. 11381 (1955); Salem, Hall (1927).

P. alkekengi L. Chinese Lantern-plant. Introduced from Asia. Escape from cultivation. - Haverhill, H. 740 (1932); Saugus, H. 18645 (1958).

NICANDRA Adans. Apple-of-Peru.

N. physalodes (L.) Pers. Introduced from Peru. Old gardens and waste places; occasional. - Lawrence, Sears (1885); Boxford, M.E. Perley (1881); West Newbury, Wm. Merrill (1877); Rockport, Prybot 31785 (1967); Ipswich, Oakes; Saugus, H. 18681 (1958); Lynn, H. 11822 (1955).

LYCIUM L. Matrimony-vine.

L. halimifolium Mill. Introduced from Europe. Roadsides and waste ground; occasional. - Haverhill, H. 534 (1932); Boxford, M.E. Perley (1884); Newburyport, White (1913); Georgetown, H. 13575 (1957); [Wenham, L. Dodge (1950)]; Gloucester, Swan (1881); Beverly, Robinson (1877).

HYOSCYAMUS L. Henbane.

H. niger L. Naturalized from Europe. Waste places; rare. - Ipswich, Oakes; Essex County, Oakes.

DATURA L. Jimsonweed. Thorn-Apple.

D. stramonium L. Naturalized from Asia. Waste ground; frequent. - Haverhill, H. 535 (1932); Newburyport, White 93 (1913); Newbury, H. 20707 (1959); Ipswich, Oakes, H. 13864 (1957); Danvers, H. 11657 (1955); Salem, H. 10621 (1955); Salem, Misery Island, Cole (1898); Peabody, H. 11557 (1955); Lynn, H. 11846 (1955); Marblehead, Clifton, Ware; Nahant, Robinson (1877).

D. stramonium, var. *tatula* (L.) Torr. Purple Thorn-Apple. Naturalized from Asia. Waste places; frequent. - Lawrence, H. 18576 (1958); Boxford, H. 11572 (1955); Newbury, Plum Island, H. 18603 (1958); Ipswich, Furber (1911); Topsfield, H. 18283 (1958); Gloucester, Magnolia, H. 12332 (1956); Beverly, Lowell (1853); Salem, Cole (1901); Saugus, H. 18640 (1958); Lynn, H. 18819 (1958); Nahant, Robinson (1877).

D. innoxia Mill. Adventive from Tropical America. Waste ground; rare. - Swampscott, Swan (1888).

D. meteloides Dunal. Introduction from Tropical America. Waste places; occasional. - Lawrence, H. 11778 (1955); Boxford, H. 12514 (1956); Newburyport, Plum Island, H. 18602 (1958); Georgetown, H. 13570 (1957); Ipswich, H. 13863 (1957); Gloucester, Magnolia, H. 12331 (1956); Peabody, H. 27076 (1964); Saugus, H. 18628 (1958); Marblehead, H. 11436 (1955).

NICOTIANA L. Tobacco.

N. rustica L. Wild Tobacco. Of unknown nativity. Waste places; rare. - Lynn, Mackintosh (1929).

N. alata Link & Otto. Introduced from S. America. Waste places; occasional. - Rowley, H. 18479 (1958); Ipswich, H. 11581 (1955); Beverly, H. 18906 (1958); Peabody, H. 11546 (1955); Saugus, H. 18696 (1958); Lynn, H. 18834 (1958); Marblehead, H. 11419 (1955).

N. longiflora Cav. Introduced from Tropical America. Waste places; rare. - Boxford, H. 12513 (1956).

PETUNIA Juss. Petunia.

P. violacea Lindl. Introduced from South America. Waste places; frequent. - Haverhill, H. 536 (1932); Lawrence, H. 18578 (1958); Andover, Pease 1162 (1900); North Andover, H. 11062 (1955); Boxford, H. 10613 (1955); Newburyport, H. 11272 (1955); Rowley, H. 18502 (1958); Ipswich, H. 13858 (1957); Danvers, H. 11633 (1955); Essex, H. 18852 (1958); Gloucester, Magnolia, H. 12358 (1956); Beverly, H. 18899 (1958); Peabody, H. 11550 (1955); Saugus, H. 18669 (1958); Lynn, H. 11821 (1955); Marblehead, H. 11443 (1955); Nahant, Robinson (1877).

SCROPHULARIACEAE (Figwort Family).

VERBASCUM L. Mullein.

V. thapsus L. Common Mullein. Naturalized from Europe. Fields and waste places; common. - Haverhill, H. 1842 (1935); Andover, Pease 4249 (1904); North Andover, Root (1945); Boxford, M.E. Perley (1884); Georgetown, Horner; Rowley, H. 12228 (1956); Ipswich, H. 13999 (1957); Ipswich, Plum Island, H., McGregor & Schweinfurth 12645 (1957); Topsfield, H.R. Towne; Hamilton, H. 10790 (1955); Rockport, Gates & Smith 1010 (1928); Manchester, Hubbard (1913); Salem, Sears (1884); Saugus, Ripley 15541 (1915); Lynn, Robinson (1875).

V. phlomoides L. Naturalized from Europe. Roadsides and clearings; occasional. - Ipswich, Wade 14161 (1957); Gloucester, Collins (1915); Peabody, Mackintosh (1933).

V. blattaria L. Moth-Mullein. Naturalized from Europe. Fields and waste places; occasional. - Boxford, H. 17940 (1958); West Newbury, Merrill (1877); Georgetown, Horner (1877); Topsfield, Foye 14201 (1958); Gloucester, Magnolia, H. 12352 (1956); Peabody, Mackintosh (1933).

V. blattaria, forma *albiflora* (Don) House - North Andover, Root (1945); Boxford, H. 17998 (1958); Topsfield, H. 18051 (1958).

V. nigrum L. Naturalized from Europe. Waste places; occasional. - Danvers, Tapleyville, Sears (1879); Lynn, Bean (1953).

V. nigrum, forma (white-flowered) - Lynn, Bean (1953).

KICKXIA Dumort. Fluellin.

K. elatine (L.) Dumort. Canker-root. Naturalized from Europe. Waste places; rare. - Ipswich, Oakes.

LINARIA Mill. Toadflax.

L. vulgaris Hill. Butter-and-eggs. Naturalized from Europe. Roadsides, dry fields and waste places; common. - Salisbury, White 379 (1913); Haverhill, H. 558 (1932); Lawrence, H. 11222 (1955); North Andover, Morss (1885); Groveland, H. 10790 (1955); Newburyport, Plum Island, H. 13037 (1957); Newbury, Plum Island, H. 12516 (1956); Ipswich, Townsend (1904); Topsfield, Frye (1880); Danvers, Frye (1880); Hamilton, H. 10834 (1955); Gloucester, Annisquam, Ripley (1914); Manchester, Lane (1877); Beverly Farms, Hubbard (1913); Salem, Sears (1884); Salem, Little Misery Island, H. 13283 (1957); Peabody, Mackintosh (1933); Lynn, H. 11836 (1955); Swampscott, (1900); Marblehead, (1900).

L. canadensis (L.) Dumont. Old-field Toadflax. Dry sandy or sterile soil; common. - Salisbury, White 6 (1913); Haverhill, H. 581 (1932); Methuen, Wm. Boott (1854); North Andover, Root (1945); Boxford, M.E. Perley (1884); Groveland, Bean & H. 8514 (1953); Newbury, Plum Island, H. 10409 (1955); Georgetown, Horner; Rowley, H. 8121 (1952); Rowley, Plum Island, H. 12046 (1956); Ipswich, C.E. Brown (1912); Ipswich, Plum Island, H., McGregor & Schweinfurth 12702 (1957); Topsfield, M. Connolly; Essex, H. 10244 (1955); Gloucester, Prince (1883); Rockport, Forbes 3238 (1902); Manchester, Hubbard (1913); Beverly, Lowell (1848); Salem, Little Misery Island, H. 13306 (1957); Peabody Mackintosh (1933); Saugus, H. 10886 (1955); Lynn, H. 10915 (1955); Swampscott, H. 10690 (1955); Marblehead, Cole (1886).

L. dalmatica (L.) Mill. Introduced from southeastern Europe. Roadsides and fields; rare. - Essex, Mrs. E.S. Curtis (1945).

L. genistifolia (L.) Mill. Introduced from Eurasia. Garden escape; rare. - Georgetown, H. 12798 (1957).

ANTIRRHINUM L. Snapdragon.

A. majus L. Common Snapdragon. Introduced from Europe. Garden escape. - Newburyport, White 480 (1913); Beverly, H. 18910 (1958); Saugus, H. 18723 (1958).

SCROPHULARIA L. Figwort.

S. lanceolata Pursh. Thickets and borders of woods; frequent. - Haverhill, Ward Hill, H. 582 (1932); Andover, Fernald (1911); Boxford, H. 8022 (1952); Newbury, Pine Island, Farwell (1927); Rowley, J.H. Sears (1906); Ipswich, H. 14000 (1957); Topsfield, H.R. Towne; Wenham, Cole (1910); Essex, Mrs. J.I. Hopkins 9714 (1955); Gloucester, Annisquam, Smith 616 (1927); Beverly, Montserrat, Hubbard (1913); Lynn, Oakes.

CHELONE L. Turtlehead.

C. glabra L. Balmony. Low ground; common. - Haverhill, H. 1841 (1935); Lawrence, H. 11266 (1955); North Andover, Root (1945); Boxford, H.E. Parkhurst (1882); Groveland, H. 11083 (1955); West Newbury, White 509 (1913); Topsfield, H.R. Towne; Wenham, Robinson (1875); [Rockport, A & W (1969)] ; Salem, Cole (1908); Peabody, Mackintosh (1933).

C. glabra, forma *tomentosa* (Raf.) Pennell - Salisbury, White 333 (1913).

PENSTEMON Mitchell Beard-tongue.

P. digitalis Nutt. Introduced from the West. Fields and clearings; occasional. - Andover, Mrs. S.J. Bucklin (1898); Boxford, M.P. Perley (1881); Groveland, W.S. Horner; Rowley, Miss B. Dodge; Topsfield, U. Bailey; Foye 8046 (1952); Hamilton, Willowdale, St. John 11886 (1916); Rockport, Smith 41 (1921).

P. calycosus Small. Introduced from South and West. Fields; rare. - Haverhill, Palfrey (1892).

P. pallidus Small. Introduced from the West. Fields and roadsides, rare. - Boxford, H. 17614 (1958); Ipswich, Mackintosh (1915).

P. hirsutus (L.) Willd. Dry or rocky ground; rare. - Amesbury, Huntington and Knowlton (1906), Rich (1906).

MIMULUS L. Monkey-flower.

M. ringens L. Shores and wet places; frequent. - Amesbury, Cole (1886); Merrimac, Bean & H. 7703 (1951); Haverhill, H. 86 (1931); Lawrence, Robinson (1879); Andover, Pease 1188 (1901); North Andover, Morss (1885); Boxford, H.E. Parkhurst (1882); West Newbury, White 504 (1913); Ipswich, Oakes; Topsfield, H.R. Towne; Danvers, H. 11639 (1955); Manchester, H. 10758 (1955); Beverly, Lowell (1846); Peabody, Mackintosh (1933).

GRATIOLA L. Hedge-Hyssop.

G. neglecta Torr. Wet or muddy places; frequent. - Amesbury, Huntington and Knowlton (1906); Merrimac, Bean & H. 8564 (1953); Lawrence, Robinson & Sears (1877); North Andover, Morss (1885); Boxford, H. 13393 (1957); Groveland, H. 11190 (1955); West Newbury, H. 11127 (1955); Ipswich, H. 18156 (1958); Beverly, H. 10161 (1955); Peabody, Sears (1885); Lynn, Young (1879).

G. aurea Muhl. Golden-pert. Sandy shores and open swamps; common. - Salisbury, White 274 (1913); Amesbury, H. 9179 (1953); Merrimac, H. 13725 (1957); Haverhill, H. 614 (1932); Methuen, H. 18404 (1958); Andover, H. 14099 (1957); North Andover, Morss (1885); Boxford, Frye (1880); Groveland, H. 9099 (1953); West Newbury, White 527 (1913); Georgetown, Horner; Ipswich, Hubbard (1913); Topsfield, H.R. Towne; Danvers, Lowell (1846); Wenham, Deane (1886); Hamilton, Phippen (1857); Essex, H. 13362 (1957); Peabody, H. 9317 (1953); Lynnfield, Robinson (1874); Lynn, Robinson (1873).

G. aurea, forma *pusilla* Fassett - Merrimac, H. 13726 (1957).

G. aurea, forma *leucantha* Bartlett - Lynn, C.E. Faxon (1879).

LIMOSELLA L. Mudwort.

L. subulata Ives. Brackish sand and mud; rare. - Newburyport, Moulton (1886); B.L. Robinson (1903); Rowley, Fassett 2036 (1923); Ipswich, Sears (1886).

LINDERNIA All. False Pimpernel.

L. dubia (L.) Pennell. Shores and damp open spots; occasional. - Haverhill, H. 28386 (1965); Methuen, H. 12833 (1957); North Andover, H. 12919 (1957); Boxford, H. 14044 (1957); Rowley, H. 8124 (1952); Middleton, H. 9430 (1953); Manchester, H. 10716 (1955).

L. dubia, var. *riparia* (Raf.) Fern. - Boxford, Bean (1940).

L. dubia, var. *inundata* (L.) Pennell - North Andover, H. 12941 (1957); West Newbury, White 515 (1913).

L. anagallidea (Michx.) Pennell. Damp shores and sands; common. - Amesbury, H. 9180 (1953); Haverhill, H. 88 (1931); Andover, Pease 2646 (1903); North Andover, H. 14068 (1957); Boxford, H. 13789 (1957); Ipswich, Deane (1903); Danvers, Lowell (1848); Wenham Pond, Sears (1892); Essex, H. 13363 (1957); East Gloucester, Sears (1887); Lynn, Young (1879).

DIGITALIS L. Foxglove.

D. purpurea L. Introduced from Europe. Garden escape; occasional. - Boxford, H. 8039 (1952); West Manchester, Hubbard (1913).

D. lanata Ehrh. Introduced from southeastern Europe. Escape from cultivation; rare. - Rowley, H.R. Towne (1918).

VERONICA L. Speedwell.

V. longifolia L. Introduced from Europe. Roadsides and waste places; occasional. - Methuen, H. 12276 (1956); North Andover, Root (1945); Boxford, H. 12891 (1957); Georgetown, Horner; Topsfield, H. 18284 (1958).

V. spicata L. Introduced from Europe. Roadsides and waste places; rare. - Methuen, H. 11712 (1955).

V. serpyllifolia L. Thyme-leaved Speedwell. Introduced from Europe. Grassy places and damp open woods; frequent. - Amesbury, H. 9530 (1954); Haverhill, H. 791 (1931); Andover, Pease 1201 (1900); Boxford, M.E. Perley (1882); Georgetown, Horner (1885); Ipswich, H. 14015 (1957); Topsfield, H.R. Towne; Beverly, Robinson (1875); Salem, Cole (1902); Peabody, Mackintosh (1933).

V. officinalis L. Gypsyweed. Dry hills, open woods and fields; common. - Amesbury, H. 9516 (1954); Merrimac, Bean & H. 7723 (1951); North Andover, Zink (1945); Boxford, H. 17635 (1958); Newburyport, Moulton (1885); Newbury, H. 8230 (1952); Georgetown, H. 17679 (1958); Ipswich, Prince (1896); Topsfield, O. Floyd; Danvers, Hathorne, St. John 627 (1912); Rockport, H. 10297 (1955); West Manchester, Hubbard (1913); Salem, Cole (1901).

V. latifolia L. Introduced from Europe. Escape from cultivation; rare. - Topsfield, H. 10124 (1955).

V. chamaedrys L. Bird's-eye. Naturalized from Europe. Roadsides and fields; occasional. - Andover, Pease 1206 (1901); Newburyport, Moulton (1886); Topsfield, C.F. Dow (1915).

V. scutellata L. Marsh-Speedwell. Wet places; common. - Salisbury, White 358 (1913); Amesbury, H. 18190 (1958); Haverhill, H. 89 (1931); North Andover, Morss (1885); Boxford, H.E. Parkhurst (1882); Groveland, Bean & H. 8493 (1953); Newbury, Palmer (1942); Georgetown, Horner (1870); Byfield, M.E. Perley (1881); Topsfield, H.R. Towne; Middleton, H. 18071 (1958); Danvers, Sears (1885); Rockport, Smith 2721 (1933); West Manchester, Hubbard (1913); Beverly, Lowell; Salem, Cole (1898); Peabody, Mackintosh (1933); Lynnfield, Phippen (1837); Lynn, Lowell (1846); Marblehead, Sanford (1945).

V. americana (Raf.) Schwein. American Brooklime. Shallow water and swamps; occasional. - Boxford, H. 17711 (1958); Topsfield, A. Dodge; Danvers, Oakes; Peabody, Sears (1884).

V. anagallis-aquatica L. Water Speedwell. Ditches and shores; rare. - Ipswich, Oakes.

V. peregrina L. Neckweed. Damp open soil; frequent. - Haverhill, H. 449 (1932); Methuen, Wheeler (1922); North Andover, Root (1945); Georgetown, H. 17675a (1958); Ipswich, H. 10227 (1955); Topsfield, H.R. Towne; Danvers, H. 7978 (1952); Beverly, H. 10177 (1955); Salem, Robinson (1875).

V. peregrina, var. *xalapensis* (HBK.) St. John & Warren - Andover, Pease 3629 (1904); North Andover, H. 18023 (1958); Georgetown, Horner; Gloucester, Collins (1915); Rockport, Rouleau & Smith 4319 (1945).

V. arvensis L. Corn-Speedwell. Naturalized from Europe. Waste open ground; rare. - Andover, Blake (1882); Boxford, M.E. Perley (1882).

V. persica Poir. Bird's-eye. Naturalized from Europe. Waste places; occasional. - West Boxford, Mrs. Wilmarth; Newburyport, Moulton (1886); Georgetown, Horner (1886).

GERARDIA L. Gerardia.

G. maritima Raf. Salt marshes; frequent. - Newburyport, Horner; Newburyport, Plum Island, Cole (1901); Newbury, Horner; Newbury, Plum Island, Pease 2892 (1903); Rowley, H. 8292 (1952); Ipswich, Sears (1886); Ipswich, Plum Island, H. 12383 (1956); Essex, Bean (1938); Gloucester, Seymour 5158 (1938); Beverly, Lowell (1853); Salem, Russell (1859).

G. paupercula (Gray) Britt. Damp open ground; common. - Salisbury, White 350 (1913); Merrimac, H. 13707 (1957); Haverhill, H. (1932); Methuen, H. 8946 (1953); Lawrence, H. 11267 (1955); North Andover, Dudley (1957); Boxford, Frye (1880); Groveland, H. 13961 (1957); West Newbury, H. 11100 (1955); Newburyport, Horner; Newbury, Chute; Newbury, Plum Island, H. 13807 (1957); Georgetown, Horner (1881); Ipswich, Oakes; Ipswich, Plum Island, H. 12422 (1956); Topsfield, A.M. Dow; Danvers, Sears (1885); Hamilton, Bean (1938); Essex, Bean (1938); Gloucester, Magnolia, J.M. Hunnewell (1898); Rockport, Fernald, Hunnewell & Long 10394 (1913); Manchester, H. 11323 (1955); Beverly, Lowell (1846); Salem, Cole (1901); Peabody, Mackintosh (1933); Lynn, Sears (1885).

G. purpurea L. Damp soils; rare. - Newbury, Plum Island, Mackintosh (1924); Ipswich, H. 13897 (1957).

G. tenuifolia Vahl. Thickets and fields; occasional. - Haverhill, Ward Hill, Bean (1913); Andover, Pease 1178 (1901); North Andover, Morss (1884); Georgetown, Horner; Ipswich, Oakes; Topsfield, Mrs. Conley; Danvers, Lowell (1847); Beverly Hills, Phippen; Salem, Cole (1901).

G. virginica (L.) BSP. Downy False Foxglove. Dry open woods; occasional. - Lawrence, H. 11215 (1955); Boxford, H. 10699 (1955); South Georgetown, Horner; Danvers, Lowell (1847).

G. flava L. Deciduous woods; frequent. - Haverhill, H. 559 (1932); Andover, Pease 1179 (1901); Boxford, Williams (1906); Rowley, Horner; Ipswich, Oakes; Topsfield, H. Long; Danvers, Robinson (1875); Essex, R.H. Ture 44 (1932); Peabody, Mackintosh (1933); Lynnfield, Cole (1898).

G. pedicularia L. Dry deciduous woods and clearings; occasional. - Andover, Pease 1782 (1903); Georgetown, Horner (1874); Ipswich, M.E. Perley (1886); Topsfield, V.R. Dodge; Beverly, Lowell (1848); Salem, Phippen (1838, 9); Peabody, Mackintosh (1933); Lynnfield, Cole (1898).

CASTILLEJA Mutis Painted-cup.

C. coccinea (L.) Spreng. Scarlet Painted-cup. Peaty meadows and damp sands; occasional. - Haverhill, H. 492 (1932); Andover, Sutton 86 (1874), Pease 3635 (1904); North Andover, Robinson (1872); Topsfield, M.E. Perley (1885); Danvers, Oakes.

C. coccinea, forma *lutescens* Farw. - Danvers, Sears (1879).

MELAMPYRUM L. Cow-wheat.

M. lineare Desr., var. *americanum* (Michx.) Beauverd. Dry woods; common. - Amesbury, H. 18204 (1958); Merrimac, Bean & H. 8585 (1953); Haverhill, H. 678 (1932); Methuen, H. 8940 (1953); North Andover, Root (1945); Boxford, M.E. Perley (1882); Ipswich, H. 18134 (1958); Topsfield, S.W. Peabody; Danvers, Lowell (1848); Gloucester, H. 12170 (1956); Manchester, Hubbard (1913); Beverly, Cole (1899); Swampscott, Sears (1885).

ODONOTES Ludwig

O. serotina (Lam.) Dum. Red Bartsia. Naturalized from Europe. Roadsides and fields; rare. - Gloucester, Annisquam, Ripley 15875 (1915); F.E. & L.B. Smith 389 (1925).

RHINANTHUS L. Yellow-Rattle.

R. crista-galli L. Coastal thickets and openings; rare. - Ipswich, Plum Island, H., McGregor & Schweinfurth 12675 (1957); Rockport, Sears (1886).

R. crista-galli, var. *fallax* (Wimm. & Grab.) Druce. Occasional. - Salisbury, Bean & H. 8374 (1953); Gloucester, Mrs. Babson (1877); Rockport, Sears (1885); Smith 10 (1921).

PEDICULARIS L. Lousewort.

P. canadensis L. Wood-Betony. Woods and clearings; occasional. - Haverhill, H. 493 (1932); Andover, Pease 3532 (1904); North Andover, Morss (1885); Boxford, H. 7863 (1952); Rowley, Bean (1940); Middleton, Pease 27767 (1940); Danvers, Frost (1872); [Rockport, A & W (1969)]; Marblehead, Cole (1889).

P. canadensis, forma *praeclara* A.H. Moore - Essex County, Oakes.

BIGNONIACEAE (Bignonia Family).

GAMPSIS Lour Trumpet-flower.

C. radicans (L.) Seem. Trumpet-creeper. Introduced from the South. Escape from cultivation; occasional. - Boxford, H. 30743 (1966); Topsfield, H.R. Towne; Marblehead, H. 18935 (1958).

CATALPA Scop. Indian-Bean.

C. bignonioides Walt. Introduced from the South. Escape from cultivation; occasional. - Lawrence, H. 11241 (1955); West Newbury, H. 18055 (1958); Newburyport, H. 20743 (1959); Georgetown, H. 13591 (1957); Essex, H. 18860 (1958); Gloucester, H. 18880 (1958); Salem, Farrington (1877); Saugus, H. 18705 (1958); Lynn, H. 18841 (1958).

MARTYNIACEAE (Martynia Family).

PROBOSCIDEA Schmidel Unicorn-plant.

P. louisianica (Mill.) Thell. Proboscis-flower. Introduced from the West. Waste places; rare. - Salem, Robinson (1879), Cole (1907).

P. fragrans Decane. Introduced from South or West. Waste places; rare. - Lawrence, J.A. Collins, Jr. (1932).

PEDALIACEAE (Pedalium Family).

SESAMUM L.

S. indicum L. Sesame. Adventive from Old World. - Danvers, H. 20755 (1959).

OROBANCHACEAE (Broom-Rape Family).

EPIFAGUS Nutt. Beech-Drops.

E. virginiana (L.) Bart. Under beech-trees, parasitic or saprophytic; frequent. - Merrimac, Bean & H. 7722 (1951); South Georgetown, Horner; Hamilton, Hunnewell 4619 (1916); Essex, Phippen (1857); Gloucester, Annisquam, Ripley 15906 (1915); Rockport, Gates & Smith 1083 (1928); Beverly, Oakes.

OROBANCHE L. Broom-Rape.

O. uniflora L. One-flowered Cancer-root. Damp woods and thickets, parasitic on various plants; occasional. - Haverhill, H. 450 (1932); Andover, Pease 7725 (1905); North Andover, Root (1945); Boxford, M.E. Perley (1882); Georgetown, Horner; Ipswich, Oakes; Topsfield, H.S. Porter; Danvers, Frost (1874); Beverly, Mackintosh (1932); Salem, Ives (1910); Lynnfield, Lowell (1849); Saugus, Emerton (1868).

LENTIBULARIACEAE (Bladderwort Family).

UTRICULARIA L. Bladderwort.

U. purpurea Walt. Ponds and sluggish streams; occasional. - Haverhill, Bradford, H. 13141 (1957); Boxford, H. 13192 (1957); Hamilton, Phippen (1857); Essex, H. 13377 (1957); Peabody, H. 9339 (1953); Saugus, H. 13034 (1957).

U. inflata Walt., var. *minor* Chapm. Ponds and sluggish water; frequent. - Merrimac, H. 13690 (1957); Andover, Pease 1210 (1900); North Andover, Bean (1946); Boxford, H. 12801 (1957); Georgetown, H. 12975 (1957); Gloucester, H. 13816 (1957); Peabody, Sears (1884).

U. vulgaris L. Deep or shallow quiet water; common. - Amesbury, H. 13003 (1957); Haverhill, Bradford, Horner; Andover, Pease 1207 (1901); North Andover, H. 9498 (1954); Boxford, M.E. Perley (1882); Newbury, Oakes; Georgetown, H. 12958 (1957); Rowley, H. 13519 (1957); Ipswich, H. 13238 (1957); Middleton, H. 12902 (1957); Danvers, H. 9040 (1953); Wenham, H. 13771 (1957); Hamilton, Phippen (1875); Chebacco, Robinson (1874); Gloucester, H. 13817 (1957); Lynn, Robinson (1875); Marblehead, Cole (1886).

U. gibba L. Shallow pools and exposed mud; occasional. - Merrimac, H. 13736 (1957); Boxford, H. 13395 (1957); Georgetown, H. 12989 (1957); Ipswich, H. 13269 (1957); Wenham, H. 13777 (1957); Chebacco, Robinson (1876); Essex, H. 13330 (1957); Ipswich River, Oakes.

U. fibrosa Walt. Pools and wet peat; rare. - Haverhill, Bradford, Little Pond, "Chadwicks", Horner.

U. intermedia Hayne. Creeping at bottoms of shallow pools; frequent. - Amesbury, H. 13026 (1957); Merrimac, H. 13670 (1957); Haverhill, H. 1303 (1934); Andover, Pease 1214 (1901); Boxford, Horner; Groveland, H. 12881 (1957); West Newbury, Robinson (1879); Georgetown, H. 12597 (1957); Rowley, H. 13524 (1957); Ipswich, H. 18128 (1958); Wenham, H. 13782 (1957); Essex, H. 13378 (1957); Manchester, Oakes; Beverly, Lowell (1847); Lynn, Tracy.

U. cornuta Michx. Wet peaty shores or bogs; occasional. - Boxford, H. 12783 (1957); West Boxford, Horner; Chebacco Pond, Robinson (1874); Essex Ponds, Lowell (1846); Gloucester, Dogtown Common, Ames & Eaton (1903); Gloucester, Lanesville, Hoyt; Mancester, Oakes.

U. resupinata L. Pond and river shores; rare. - Haverhill, Bradford, Horner; Andover, C.E. Faxon; Boxford, Mrs. Wilmarth (1877).

PHRYMACEAE (Lopseed Family).

PHRYMA L. Lopseed.

P. leptostachya L. Rich thickets and woods; occasional. - Boxford, M.E. Perley (1884), H. 8764 (1953); Newbury, Cushing; Georgetown, Williams (1906); Byfield, Horner, Chute; Topsfield, H. Porter; Danvers, Robinson (1874).

PLANTAGINACEAE (Plantain Family).

PLANTAGO L. Plantain.

P. major L. Naturalized from Europe. Roadsides and dooryards; common. - Haverhill, H. 90 (1931); Methuen, H. 11750 (1955); Lawrence, H. 11799 (1955); North Andover, H. 13070 (1957); Boxford, Frye (1880); Groveland, H. 12862 (1957); Rowley, H. 11956 (1955); Ipswich, Plum Island, St. John & White 574 (1913); Topsfield, H.R. Towne; Wenham, Robinson (1875); Hamilton, H. 10847 (1955); Essex, H. 13336 (1957); Rockport, Fernald, Hunnewell & Long 10416 (1913); Manchester, Hubbard (1913); Beverly, Lowell (1847); Salem, Sears (1891); Peabody, H. 11547 (1955); Saugus, H. 10903 (1955).

P. major, forma *rosea* (Dene.) Prahl - Boxford, H. 8097 (1952); Ipswich, Oakes; Rockport, Robinson (1877); Lynn, H. 11818 (1955).

P. major, var. *scopulorum* Fries & Broberg. Brackish or saline shores; occasional. - Newbury, White 429 (1913); Rowley, Plum Island, H. 12562 (1956); Ipswich, Plum Island, H. 12398 (1956); Gloucester, Sears (1887); Rockport, Forbes (1902).

P. rugellii Dene. Damp shores, roadsides and waste places; common. - Merrimac, H. 9251 (1953); Haverhill, H. (1931); Methuen, H. 11010 (1955); Lawrence, H. 11223 (1955); Andover, Pease 1224 (1902); North Andover, Root (1945); Boxford, M.E. Perley (1884); Groveland, H. 8814 (1953); West Newbury, H. 10937 (1955); Ipswich, H. 11601 (1955); Middleton, Bean & H. 7775 (1951); Danvers, H. 11654 (1955); Gloucester, Lanesville, Collins (1915); Manchester, H. 10727 (1955); Salem, Robinson (1880); Lynnfield, H. 14121 (1957); Saugus, H. 10865 (1955).

P. juncoides Lam., var. *decipiens* (Barneoud) Fern. Seaside-Plantain. Seaside cliffs and dry beaches; occasional. - Gloucester, Magnolia, Higginson (1882); Rockport, Robinson (1874); Beverly, Lowell (1848); Peabody, Mackintosh (1933); Marblehead, Winslow (1910); Nahant, Hoar (1857).

P. oliganthos R. & S. Seaside-Plantain. Salt-marshes; frequent. - Salisbury. Amesbury, White 251 (1913); Newburyport, Bean (1940); Newbury, H. 927 (1933); Newbury, Plum Island, White 144 (1913); Rowley, H. 8288 (1952); Essex, H. 9380 (1953); West Gloucester, Bean (1945); Beverly, Lowell (1847); Marblehead Neck, H. 926 (1933); Nahant, Larrabee (1883).

P. lanceolata L. Naturalized from Europe. Grassland and waste places; common. - Amesbury, H. 10516 (1955); Haverhill, H. 451 (1932); Andover, Pease 1226 (1902); North Andover, Root (1945); Boxford, M.E. Perley (1882); West Newbury, H. 7898 (1952); Newbury, H. 10392 (1955); Newbury, Plum Island, White 130 (1913); Rowley, H. 18485 (1958); Ipswich, Oakes; Plum Island, H. 12094 (1956); Topsfield, H.R. Towne; Wenham, Robinson (1875); Hamilton, H. 10801 (1955); Gloucester, Annisquam, Ripley (1914); Rockport, Smith 719 (1926); Manchester, Hubbard (1913); Beverly, H. 10192 (1955); Lynn, 10925 (1955); Swampscott, H. 10689 (1955); Marblehead, Bean & H. 9728 (1955).

P. lanceolata, var. *sphaerostachya* Mert. & Koch - Haverhill, H. 92 (1931); Georgetown, Horner.

P. aristata Michx. Introduced from the West. Dry open soil; occasional. - Haverhill, H. 1838 (1935); Methuen, H. 11738 (1955); Boxford, H. 8627 (1953); Ipswich, Sears (1902); Topsfield, Mrs. F. Peabody; Gloucester, Prince (1883).

P. indica L. Naturalized from Europe. Waste places; rare. - Haverhill, H. 628 (1932); Rockport, Smith 2710 (1933).

RUBIACEAE (Madder Family).

SHERARDIA L. Field-Madder.

S. arvensis L. Adventive from Europe. Waste places; rare. - Newburyport, Sears (1887).

GALIUM L. Bedstraw. Cleavers.

G. aparine L. Cleavers. Thickets, seashores and waste places; occasional. - Amesbury, H. 10532 (1955); Merrimac, H. 10594 (1955); Rowley, Plum Island, H., McGregor & Schweinfurth 12720

(1957); Ipswich, Oakes; Topsfield, Mrs. Conley; Essex, H. 10235 (1955); Salem, Robinson (1879); Marblehead, Bean & H. 9726 (1955).

G. triflorum Michx. Sweet-scented Bedstraw. Woods and thickets; occasional. - Amesbury, H. 9522 (1954); Merrimac, H. 18362 (1958); Boxford, A.W. & C.A. Cheever (1914); Georgetown, Horner; Topsfield, Mrs. M. Peabody; Middleton, H. 18083 (1958); Wenham, Robinson (1875); Hamilton, Hunnewell 4629 (1916); Beverly, Robinson (1879); Lynn, Cushman 2922 (1908).

G. pilosum Ait. Dry woods; rare. - Andover, Blake (1882); Boxford, Robinson (1876); Hamilton, Purdie (1894).

G. circaezans Michx., var. *hypomalacum* Fern. Wild Licorice. Rich woods; occasional. - Andover, Pease 3222 (1900); Boxford, M.E. Perley (1884); Georgetown, Horner; Middleton, H. 18069 (1958); Danvers, Sears (1885); Beverly, Knowlton (1907); Salem, Cole (1899); Peabody, Sears (1896).

G. lanceolatum Torr. Wild Licorice. Dry woods; rare. - Boxford, Robinson (1877); Chebacco, Robinson (1874); North Beverly, Sears (1896).

G. boreale L., var. *intermedium* DC. Northern Bedstraw. Shores and gravelly or rocky banks; rare. - Topsfield, Mrs. Conley.

G. rubioides L. Naturalized from Europe. Roadsides; rare. - Boxford, H. 8047 (1953); Ipswich, H. 10223 (1955).

G. verum L. Yellow Bedstraw. Naturalized from Europe. Dry fields; rare. - Ipswich, H. & Wade 18793 (1958); Topsfield, Foye 13028 (1957); Danvers, Hitchings (1881).

G. mollugo L. Naturalized from Europe. Roadsides and fields; occasional. - Andover, Pease 3596 (1904); Groveland, H. 10599 (1955); Newbury, H. 10418 (1955); Danvers, Sears (1885).

G. erectum Huds. Naturalized from Europe. Fields; rare. - Georgetown, Sears (1905).

G. sylvaticum L. Scotch-Mist. Introduced from Europe. Escape from cultivation; rare. - North Andover, Pease 1240 (1901).

G. palustre L. Wet meadows and banks; common. - Haverhill, H. 695 (1932); Andover, Pease 1237 (1902); North Andover, H. 9906 (1955); West Boxford, Root (1945); Newbury, Schweinfurth (1950); Rowley, Plum Island, H., McGregor & Schweinfurth 12719 (1957); Topsfield, M.E. Perley (1881); Danvers, Sears (1878); Hamilton, Hubbard (1913); Manchester, Farlow (1915); Beverly, Knowlton (1907); Salem, Sears (1904); Saugus, Ripley 15528 (1915); Marblehead, Lindsay (1884).

G. trifidum L. Swamps; occasional. - Andover, H. 17901 (1958); North Andover, H. 17972 (1958); Boxford, H. 8044 (1952); Danvers, Sears (1885); Lynnfield, Fernald (1917).

G. tinctorium L. Swamps and damp places; common. - Amesbury, H. 9198 (1953); Merrimac, H. 13683 (1957); Haverhill, Bradford, H. 13139 (1957); Methuen, H. 12834 (1957); North Andover, H. 12932 (1957); Boxford, H.E. Parkhurst (1882); Groveland, H. 9092 (1953); Georgetown, Horner; Byfield, Sears (1904); Ipswich, Oakes; Ipswich, Plum Island, H., McGregor & Schweinfurth 12646 (1957); Topsfields, C.F. Dow; Danvers, Sears (1885); Wenham, Robinson (1875); Gloucester, Lanesville, Collins (1915); Manchester, H. 10739 (1955); Beverly, Lowell (1846); Lynnfield, Boott (1854).

G. obtusum Bigel. Low woods; rare. - North Andover, Morss (1884); Georgetown, Horner.

G. asprellum Michx. Rough Bedstraw. Low ground and damp thickets; frequent. - Andover, Pease 1239 (1901); North Andover, Morss (1885); West Boxford, Root (1945); Groveland, M.E. Perley (1885); Ipswich, Oakes; Topsfield, H.R. Towne; Rockport, Cushman 3360 (1908); Peabody, Mackintosh (1933); Lynnfield, H. 14129 (1957).

G. tricorne Stokes. Adventive from Europe. Waste places; rare. - Boxford, H. 9637 (1954).

ASPERULA L.

A. arvensis L. Adventive from Europe. Waste places; rare. - Boxford, H. 9599 (1954).

MITCHELLA L. Partridge-Berry.

M. repens L. Dry or moist knolls in woods; common. - Amesbury, Bean & H. 7744 (1952); Merrimac, H. 18325 (1958); Haverhill, H. 196 (1931); Methuen, H. 18389 (1958); Andover, Pease 1248 (1902); North Andover, Root (1945); Boxford, M.E. Perley (1882); Rowley, H. 11959 (1955); Topsfield, G.T. Dow; Danvers, Sears (1885); Hamilton, Hubbard (1913); Chebacco Woods, Robinson (1890); Gloucester, Smith 611 (1927); Manchester, Hubbard (1913); Beverly, Phippen (1859); Peabody, Mackintosh (1933); Saugus, Ripley 15524 (1915).

M. repens, forma *leucocarpa* Bissell - Middleton, A. Nichols, Jr. (1886); Beverly, Graves (1885).

CEPHALANTHUS L. Buttonbush.

C. occidentalis L. Swamps and borders of ponds and streams; common. - Salisbury, White (1913); Amesbury, H. 12999 (1957); Merrimac, H. 13689 (1957); Haverhill, H. 560 (1932); Andover, Pease 1230 (1901); North Andover, Morss (1885); Boxford, H.E. Parkhurst (1882); Groveland, H. 11097 (1955); Topsfield, B.B. Towne; Wenham, Cole; Gloucester, Williams (1897); Rockport, L.B. & R.C. Smith (1934); Manchester, H. 10782 (1955); Salem, Robinson (1879); West Peabody, Mackintosh (1933); Marblehead, Cole (1947).

HOUSTONIA L.

H. caerulea L. Bluets. Fields and open slopes; common. - Haverhill, Ward Hill, Cushman 2961 (1908); Lawrence, H. 9866 (1955); North Andover, Morss (1885); Boxford, M.P. Perley (1882); West Newbury, H. 31909 (1968); Georgetown, Horner; Rowley, H. 12250 (1956); Rowley, Plum Island, H. 12073 (1956); Ipswich, Oakes; Ipswich, Plum Island, H., McGregor & Schweinfurth 12652 (1957); Topsfield, C.E. Brown (1912); Danvers, Phippen (1857); Essex Woods, Lowell (1853); Gloucester, Prince (1883); Rockport, Sears (1885); Salem, Robinson (1875).

H. caerulea, forma *albiflora* Millsp. - Salem?, Sutton Road, Robinson (1872).

H. longifolia Gaertn. Rocky or gravelly soil; rare. - Boxford, Horner; Groveland, Paine.

CAPRIFOLIACEAE (Honeysuckle Family).

DIERVILLA Duham. Bush Honeysuckle.

D. lonicera Mill. Dry woods and clearings; common. - Haverhill, H. 494 (1932); Andover, Ballardvale, Zink (1945); Boxford, Frye (1881); West Newbury, H. 9535 (1954); Georgetown, Horner; Rowley, H. 10439 (1955); Topsfield, Mrs. L.T. Brown (1912); Danvers, Frost (1874); Hamilton, Hubbard (1913); Gloucester, Magnolia, J.M. Hunnewell (1898); [Rockport, A & W (1969)]; Salem, Cole (1907); Peabody, Mackintosh (1915); Saugus, Bean & H. 9797 (1955); Marblehead, Cole (1898); MNS, Snyder (1970).

WEIGELIA.

W. floribunda (Sieb. & Zucc.) C.A. Mey. Introduced from Asia. Escape from cultivation. - Essex, Hubbard (1913).

LONICERA L. Honeysuckle.

L. xylosteum L. Fly-Honeysuckle of Europe. Introduced from Eurasia. Roadsides and thickets; occasional. - Merrimac, Bean & H. 7704 (1951); Lawrence, H. 9851 (1955); Groveland, Bean & H. 8556 (1953); Rowley, Plum Island, H., McGregor & Schweinfurth 12718 (1957); Ipswich, H. 13887 (1957); Ipswich, Plum Island, H. 12106 (1956); Salem, Little Misery Island, H. 13294 (1957).

L. morrowi Gray. Introduced from Eurasia. Roadsides and thickets; occasional. - Lawrence, H. 9864 (1955); Wenham, Bean & H. 10027 (1955); Swampscott, H. 10649 (1955).

L. tatarica L. Tartarian Honeysuckle. Introduced from Eurasia. Thickets; an occasional escape from cultivation. - Haverhill, H. 679 (1932); Salem, Cole (1902); [MNS, Snyder (1970)].

L. canadensis Bartr. Fly-Honeysuckle. Cool woodlands; occasional. - Hamilton, Hunnewell 4621 (1916); Gloucester, Annisquam, Smith 855 (1928); Manchester, Oakes; Hubbard (1913); Beverly, Robinson (1876).

L. sempervirens L. Probably introduced from the South. Woods and thickets; occasional. - Andover, Root (1947); West Newbury, Horner; Georgetown, S. Noyes (1905); Topsfield, H.S. Porter, H. 18285 (1958); Salem, Whipple (1877); Peabody, Sears (1884).

L. dioica L. Woods and thickets; rare. - Topsfield, H.S. Porter.

SYMPHORICARPOS Duham. Snowberry.

S. albus (L.) Blake, var. *laevigatus* (Fern.) Blake. Introduced from the Pacific slope. Roadsides, escaped from cultivation, occasional. - Haverhill, H. 20963 (1959); Boxford, H. 11886 (1955); Salem, Robinson (1877).

LINNAEA Gronov. Twinflower.

L. borealis L., var. *americana* (Forbes) Rehd. Cool woods; local. - West Boxford; Horner; Rowley, Horner; Topsfield, A.F. Towne; Wenham, Morse (1919); Essex, Chute (1858); H. & Smith 2729 (1934); [John Kieran (1971)]; Manchester, Oakes; [Beverly Farms, Snyder (1950's)].

TRIOSTEUM L. Feverwort.

T. aurantiacum Bickn. Wild Coffee. Rich woods and thickets; frequent. - West Andover, Gutterson (1902); Boxford, M.E. Perley (1884); South Georgetown, Horner; Ipswich, Oakes; Topsfield, M. Clarke; Middleton, Pease 28055 (1940); Swampscott, Sears (1904).

VIBURNUM L. Viburnum.

V. alnifolium Marsh. Hobblebush. Woods and cool ravines; local. - Haverhill, Bradford, Fernald 11438 (1914); Danvers, Frost (1872); Essex, Buck (1897); Gloucester, Magnolia, Williams (1896); [Rockport, A & W (1969)] ; Manchester, Oakes; Beverly, Lowell (1849).

V. cassinoides L. Witherod. Thickets and clearings; frequent. - Andover, Pease 1258 (1902); Boxford, Root & Zink (1945); Topsfield, Mackintosh (1925); Gloucester, Oakes; Rockport, Smith 604 (1927); West Manchester, Hubbard (1913); Salem, Cole (1937).

V. lentago L. Sheepberry. Borders of woods; common. - Amesbury, H. 10538 (1955); Haverhill, H. 452 (1932); Andover, Pease 1261 (1901); North Andover, Root (1943); Boxford, S.P. Perley (1882); Groveland, H. 12585 (1956); West Newbury, H. 9113 (1953); Newbury, White 210 (1913); Ipswich, Oakes; Topsfield, Mrs. Conley; Middleton, H. 8707 (1953); Danvers, Robinson (1877); Hamilton, H. 10782 (1955); Salem, Cole (1889).

V. recognitum Fern. Arrow-wood. Damp thickets; common. - Amesbury, H. 10554 (1955); Haverhill, H. 584 (1932); Andover, Pease 1256 (1902); Boxford, M.E. Perley (1882); West Newbury, H. 9110 (1953); Newburyport, Schweinfurth (1950); Newbury, Rich (1898); Byfield, W.S. Horner; Ipswich, Oakes; Ipswich, Plum Island, H. 12103 (1956); Topsfield, A.M. Dow; Middleton, H. 8693 (1953); Hamilton, H. 10778 (1955); Essex, Hubbard (1913); Gloucester, Annisquam, Ripley (1914); Rockport, Gates & Smith 1008 (1928); Manchester, Williams (1895); Beverly, Sears (1885); Salem, Cole (1888); Peabody, Mackintosh (1933); Saugus, Ripley 15521 (1915); Swampscott, Buck (1896); Marblehead, Bean & H. 9733 (1955).

V. acerifolium L. Dockmackie. Dry or rocky woods; common. - Haverhill, Ward Hill, Cushman 2932 (1908); Andover, Pease 1257 (1900); North Andover, H. 8658 (1953); Boxford, M.E. Perley (1882); Newbury, Schweinfurth (1950); South Georgetown, Horner; Rowley, H. 7945 (1952); Topsfield, A. Dow; Wenham, Bean & H. 10021 (1955); Chebacco Woods; Robinson (1873); Essex, Hubbard (1913); Manchester, Oakes; Beverly, Lowell (1846); Peabody, Sears (1885); Saugus, Bean & H. 9802 (1955); Swampscott, Buck (1896).

V. acerifolium, forma *ovatum* Rehd. - Wenham, Cole (1911).

V. trilobum Marsh. Highbush Cranberry. Cool woods, occasional. - Merrimac, Bean & H. 7700 (1951); Andover, Pease 1264 (1902); North Andover, Pease 1263 (1902); West Boxford, Horner; Newbury, H. 12626 (1956); Rowley, Horner; Ipswich, Oakes; Topsfield, H. 26044 (1963); Danvers, Sears (1877); Salem, Robinson.

V. opulus L. Guelder-Rose. Introduced from Europe. Escape from cultivation; occasional. - Haverhill, Fernald 11439 (1914); North Andover, Zink (1945); Ipswich, H. 14005 (1957); Salem, Cole (1895).

SAMBUCUS L. Elder.

S. canadensis L. Common Elder. Damp or rich soils; common. - Amesbury, H. 10555 (1955); Merrimac, H. 10588 (1955); Haverhill, H. 629 (1932); Lawrence, H. 9870 (1955); Andover, Pease 1254 (1901); North Andover, Morss (1885); Boxford, M.E. Perley (1882); Groveland, Bean & H. 8525 (1953); Georgetown, Horner; Rowley, Plum Island, H. 12533 (1956); Ipswich, H. 13452 (1957); Ipswich, Plum Island, H., McGregor & Schweinfurth 12685 (1957); Topsfield, H.R. Towne; Hamilton, H. 10787 (1955); Rockport, Cushman 3359 (1908); West Manchester, Hubbard (1913); Salem, Robinson (1879); West Peabody, Mackintosh (1933); Saugus, Bean & H. 9798 (1955).

S. pubens Michx. Red-berried Elder. Woods and openings; occasional. - Salisbury, H. 12204 (1956); Wenham, Cole (1910); Essex Woods, Robinson (1872); Gloucester, Morss (1900); Rockport, Sears (1897); Manchester, Oakes; Swampscott, Sears (1897).

VALERIANACEAE (Valerian Family).

VALERIANA L. Valerian.

V. officinalis L. "Garden-Heliotrope." Introduced from Europe. Escape from cultivation; occasional. - Georgetown, H. 19310 (1959); Middleton, H. 17689 (1958); Rockport, Sanford (1946).

DIPSACACEAE (Teasel Family).

DIPSACUS Card. Teasel.

D. sylvestris Huds. Wild Teasel. Roadsides, old fields. Rare. [MNS, Snyder (1971)].

SUCCISA Neck. Devil's Bit.

S. pratensis Moench. Adventive from Europe. - Groveland, Lunt (1909).

KNAUTIA L.

K. arvensis (L.) Duby. Naturalized from Europe. Dry pastures and fields; rare. - Andover, Shaw (1945).

SCABIOSA Neck.

S. sp. Escape from cultivation. - Topsfield, B. Woodbury.

CUCURBITACEAE (Gourd Family).

CUCURBITA L.

C. pepo L. Pumpkin. Introduced from Tropical Africa, Orient. Waste places; occasional. - Newburyport, 1786 (1955); Newburyport, H. 20730 (1959); Newbury, H. 20714 (1959); Rowley, H. 18493 (1958); Danvers, H. 11663 (1955); Gloucester, Magnolia, H. 12380 (1956); Beverly, H. 11674 (1955); Salem, H. 11502 (1955); Peabody, H. 11553 (1955); Saugus, H. 18647 (1958); Marblehead, H. 11476 (1955).

C. pepo, var. *ovifera* Alef. - Boxford, H. 11885 (1955); Danvers, H. 11652 (1955).

CUCUMIS

C. anguria L. Bur Gherkin. Native of South. Escape from cultivation. - Lawrence, Gutterson (1909).

C. melo L. Muskmelon. Introduced from Tropical Old World. Waste places; occasional. - Haverhill, H. 586 (1932); Newburyport, H. 20739 (1959).

C. sativus L. Cucumber. Introduced from . Waste places; occasional. - Boxford, H. 11571 (1955); Gloucester, Magnolia, H. 12362 (1956); Peabody, H. 11539 (1955).

CITRULLUS

C. vulgaris Schrad. Watermelon. Introduced from Tropical Africa. Waste places; occasional. - Haverhill, H. 585 (1932); Lawrence, H. 11773 (1955); Boxford, H. 11570 (1955); Newburyport, H. 20734 (1959); Newbury, H. 20713 (1959); Rowley, H. 18494 (1958); Ipswich, H. 13866 (1957); Danvers, H. 11656 (1955); Essex, H. 11310 (1955); Salem, H. 11496 (1955); Peabody, H. 11532 (1955); Saugus, H. 18638 (1958); Marblehead, H. 11464 (1955).

SICYOS L. Bur-Cucumber.

S. angulatus L. River-banks and waste places; occasional. - Lawrence, Sears (1885); Andover, Pease 1268 (1900); Groveland, Seymour 4864 (1936); Newburyport, White 405 (1913); Newbury, H. 20715 (1959); Hamilton, Oakes; Gloucester, Jenks (1883); Salem, Robinson (1875); H. 11508 (1955); Peabody, H. 11530 (1955); Lynn, Young (1879); Marblehead, H. 18922 (1958).

ECHINOCYSTIS T. & G. Wild Balsam-Apple.

E. lobata (Michx.) T. & G. Wild Cucumber. Rich soil along streams and waste places; common. - Merrimac, Bean & H. 7732 (1951); Haverhill, Ward Hill, H. 561 (1932); Metheun, H. 11715 (1955); Lawrence, Sears (1885); North Andover, H. 11021 (1955); Groveland, M.E. Perley (1885); West Newbury, H. 11116 (1955); Newburyport, White 419 (1913); Georgetown, H. 13571 (1957); Ipswich, H. 13868 (1957); Topsfield, H.R. Towne; Danvers, H. 18588 (1958); Gloucester, Rand & B.L. Robinson (1896); Beverly, Robinson (1875); Salem, Robinson (1875); Saugus, H. 18616 (1958); Lynn, Young (1879); Marblehead, H. 11465 (1955).

CAMPANULACEAE (Bluebell Family).

SPECULARIA Fabricius Venus's Looking-glass.

S. perfoliata (L.) A. DC. Sterile open ground; occasional. - Salisbury, Cushing (1819); Haverhill, H. 1833 (1935); Andover, Pease 1274 (1902); North Andover, Root (1945); Boxford, H. 9600 (1954); Rowley, Horner; Danvers, Sears (1879).

CAMPANULA L. Bellflower.

S. glomerata L. Clustered Bellflower. Introduced from Europe. Roadsides and old fields; occasional. - West Boxford, Horner; Topsfield, Knowlton (1917); Danvers, Oakes, Sears (1902); Salem, Russell (1863).

C. rapunculoides L. Naturalized from Europe. Roadsides; frequent. - Andover, Pease 2392 (1903); North Andover, Root (1945); Groveland, H. 11085 (1955); Newburyport, H. 10934 (1955); Georgetown, H. 13587 (1957); Topsfield, H.R. Towne; Wenham, Robinson (1875); Gloucester, J.M. Hunnewell (1901); Salem, Hall (1925); Swampscott, H. 10697 (1955); Marblehead, H. 18943 (1958).

C. rotundifolia L. Harebell. Open or rocky banks and meadows; common. - Amesbury, Robinson (1879); Merrimac, H. 18367 (1958); Haverhill, Ward Hill, Bean (1916); Andover, Pease 2448 (1903); North Andover, Morss (1885); Boxford, H. 8959 (1953); Groveland, Greenough (1891); West Newbury, H. 11120 (1955); Newburyport, White 453 (1913); Georgetown, Horner (1881); Ipswich, H. & Wade 18786 (1958); Topsfield, H.R. Towne; Middleton, Frost (1870); Peabody, Mackintosh (1933).

C. aparinoides Pursh. Marsh-Bellflower. Meadows and swales; common. - Haverhill, H. 480 (1932); Methuen, H. 8913 (1953); North Andover, Morss (1885); Groveland, H. 8802 (1953); Ipswich, Linebrook, Sears (1883); Topsfield, B.B. Towne; Danvers, Hitchings (1881); Wenham, Cole (1911); Chebacco, L.P. Robinson (1879); Beverly, Lowell (1846).

C. uliginosa Rydb. Swales and wet thickets; rare. - Andover, Pease 2394 (1903); South Georgetown, Horner.

JASIONE L.

J. montana L. West Peabody, H. Lewis 31789 (1967). Native of Europe.

LOBELIA L. Lobelia.

L. cardinalis L. Cardinal-Flower. Damp shores and swamps; common. - Haverhill, H. 563 (1932); Methuen, H. 18381 (1958); Anodover, Ballardvale, Zink (1943); North Andover, Morss (1885); Boxford, H.E. Parkhurst (1882); Groveland, H. 11092 (1955); Newbury, White 217 (1913); Georgetown, Horner (1881); Ipswich, Oakes; H. 13453 (1957); Topsfield, B.B. Towne; Essex Woods, Lowell (1846); Rockport, Smith 66 (1921) [A & W (1969)]; Manchester, H. 11348 (1955); Salem, Robinson (1879); Peabody, Mackintosh (1933).

L. spicata Lam. "Highbelia." Rich meadows and thickets; common. - Amesbury, H. 10526 (1955); Haverhill, H. 1844 (1935); Andover, Pease 1279 (1902); North Andover, Morss (1885); Boxford, H. 8639 (1953); South Georgetown, Williams (1899); Ipswich, Horner; Topsfield, L. Dow; Danvers, Sears (1879); Gloucester, Magnolia, J.M. Hunnewell (1898); Beverly, Lowell (1846); Salem, Robinson (1879); Peabody, Mackintosh (1933); Swampscott, Sears (1896).

L. inflata L. Indian-Tobacco. Fields, roadsides and open woods, common. - Merrimac, Bean & H. 7724 (1951); Haverhill, H. 587 (1932); Methuen, H. 11011 (1955); North Andover, Morss (1885); Boxford, H. 13646 (1957); Groveland, H. 8829 (1953); West Newbury, Robinson (1879); Newburyport, White 491 (1913); Georgetown, Horner; Ipswich, H. 13252 (1957); Topsfield, Mrs. C. Floyd; Middleton, Bean & H. 7770 (1951); Wenham, Lowell (1847); Gloucester, Magnolia, J.M. Hunnewell (1898); Manchester, Hubbard (1913); Beverly, Lowell (1846).

L. dortmanna L. Water-Gladiole. Sandy or gravelly margins of ponds, emersed to deeply submersed; frequent. - Haverhill, Bradford, H. 564 (1932); Andover, H. 14098 (1957); Boxford, H. 8978 (1953); Georgetown, Horner; Ipswich, H. 13244 (1957); Topsfield, H. 13230 (1957); Hamilton, Oakes; Essex Ponds, Lowell (1846); Rockport, Sears (1890).

COMPOSITAE (Composite Family).

EUPATORIUM L. Thoroughwort.

E. dubium Willd. Joe-Pye-Weed. Swamps and damp thickets; common. - Amesbury, Bean (1950); Haverhill, H. 630 (1932); Lawrence, H. 11238 (1955); North Andover, Morss (1885); Boxford, M.E. Perley (1883); Groveland, H. 8796 (1953); West Newbury, Robinson (1879); Ipswich, Oakes; Topsfield, H.R. Towne; Middleton, H. 8711 (1953); Danvers, Lowell (1845); Rockport, H. 19022 (1958); Manchester, Prince; Salem, Robinson (1879); Saugus, H. 10896 (1955).

E. fistulosum Barratt. Trumpet-Weed. Damp thickets; rare. - Essex, Bean (1938).

E. pilosum Walt. Low or dry thickets; rare. - Hamilton, Phippen (1857); Essex, Purdie (1894).

E. pubescens Muhl. Moist to dry thickets; occasional. - Lawrence, H. 11221 (1955); Andover, Lowell Junction, Pease 30383 (1943); North Andover, M.E. Perley (1885); Ipswich, Oakes; Danvers, Hitchings (1880); Wenham Swamp, Sears (1885); Essex Woods, Robinson (1876); Lynnfield, Boott (1854).

E. rotundifolium L. Open woods and clearings; rare. - Georgetown, Horner; Saugus, Purdie (1902).

E. sessilifolium L., var. *brittonianum* Porter. Upland Boneset. Rich thickets; rare. - Newburyport, H. 11281 (1955); Essex County, Oakes.

E. perfoliatum L. Thoroughwort. Low woods and swales; common. - Merrimac, H. 9271 (1953); Haverhill, H. 631 (1931); Methuen, H. 8925 (1953); Lawrence, H. 11214 (1955); North Andover, Morss (1885); Boxford, M.E. Perley (1883); Groveland, H. 11089 (1955); West Newbury, H. 11149 (1955); Georgetown, Horner; Rowley, Plum Island, H. 12556 (1956); Ipswich, Oakes; Topsfield, C.T. Dow; Middleton, H. 8708 (1953); Danvers, Frost (1874); Gloucester, Magnolia, J.M. Hunnewell (1898); Rockport, Smith 638 (1927); Manchester, H. 11338 (1955); Salem, Robinson (1879); Peabody, Mackintosh (1933); Marblehead, H. 11448 (1955).

E. perfoliatum, forma *purpureum* Britt. - Haverhill, Blatchford (1894); West Newbury, Wade 25143 (1962).

E. rugosum Houtt. White Snakeroot. Rich woods and clearings; occasional. - Haverhill, Ward Hill, Bean (1916); Boxford, Bean (1940); Georgetown, Horner (1885); Ipswich, Oakes, H. 14010 (1957); Topsfield, Williams (1903).

MIKANIA Willd. Climbing Hempweed.

M. scandens (L.) Willd. Swamps and banks of streams; occasional. - Boxford, H. 8873 (1953); South Groveland, H. 13958 (1957); Georgetown, Horner (1874); Rowley, H. 13496 (1957); Ipswich, Oakes; Topsfield, Williams (1903); Middleton, Mackintosh (1924); Wenham, H. 13764 (1957); Beverly, Lowell (1848).

LIATRIS Schreb. Blazing-Star.

L. spicata (L.) Willd. Native of South. Escape from cultivation. - Lawrence, Sereno Watson (1877).

L. borealis Nutt. Dry open woods and thickets; local. - Haverhill, Ward Hill, Bean (1916); Andover, Pease 1437 (1894); North Andover, Morss (1884); Boxford, Williams (1899); Newburyport, Cole (1899); Georgetown, Horner (1874); Rowley, Boott (1856); Ipswich, Oakes; Topsfield, M.E. Perley (1884); Danvers, Lowell (1847); Salem, Cole (1885); Peabody, Mackintosh (1933); Lynnfield, Cole (1939).

L. borealis, forma *albiflora* (Shinners) Fern. - Boxford, Kennedy (1887) TYPE.

AGERATUM L.

A. houstonianum Mill. Introduced from Mexico. Escape from cultivation. - Lynn, E. & C.E. Faxon (1880).

SOLIDAGO L. Goldenrod.

S. caesia L. Blue-stem Goldenrod. Open woods and clearings; frequent. - Amesbury, Bean & H. 7735 (1951); Lawrence, H. 11217 (1955); North Andover, Root (1945); Boxford, H. 11705 (1955); Georgetown, Horner; Rowley, Bean & Stockman (1954); Ipswich, H. 11577 (1955); Topsfield, Mrs. F. Peabody; Danvers, Oakes; Essex, Knowlton (1908).

S. caesia, forma *axillaris* (Pursh) House - Boxford, M.E. Perley (1883); Ipswich, Oakes; Topsfield, H. 11892 (1955); Lynnfield, Phippen (1856).

S. flexicaulis L. Rich woods and thickets; occasional. - North Andover, Pease 1460 (1901); Boxford, M.E. Perley (1883); Georgetown, Horner; Ipswich, Oakes; Topsfield, H.R. Towne; Danvers, Sears (1884).

S. squarrosa Muhl. Rich woods, thickets and clearings; rare. - Rowley, Bean & Stockman (1954); Danvers, Oakes (1826); Lowell (1847); Beverly, Sears (1879).

S. bicolor L. Silver-rod. Dry open soil or thin woods; common. - Haverhill, H. (1932); North Andover, Root (1945); Boxford, M.E. Perley (1881); Groveland, H. 11171 (1955); Georgetown, Horner; Rowley, H. 8285 (1952); Ipswich, Oakes; Topsfield, P.E. Kneeland; Hamilton, Robinson (1875); Rockport, L.B. & R.C. Smith 4340 (1945); Salem, Cole (1888); Peabody, Sears (1884); Lynn, Russell (1859); Marblehead, H. 11420 (1955).

S. puberula Nutt. Dry sterile soils; frequent. - Merrimac, H. 9211 (1953); Haverhill, Lunt 789 (1911); Andover, Pease 9860 (1906); Boxford, M.E. Perley (1885); Groveland, H. 11170 (1955); Ipswich, Oakes; Middleton, Sears (1887); Danvers, Lowell (1853); Gloucester, Magnolia, H. 11994 (1955); Rockport, L.B. & R.C. Smith 4341 (1945); Manchester, Oakes; Salem, Cole (1886); Peabody, Sears (1884); Saugus, Bean (1936).

S. speciosa Nutt. Dry to moist thickets; occasional. - Rowley, Bean & Stockman (1954); Ipswich, Oakes; Danvers, Sears (1878); North Salem, Hall (1927).

S. sempervirens L. Seaside Goldenrod. Chiefly coastal; common. - Amesbury, H. 18960 (1958); Boxford, H. 28364 (1965); Newbury, H. 8220 (1952); Rowley, Horner; Rowley, Plum Island, H. 12458 (1956); Ipswich, Oakes; Ipswich, Plum Island, H. 12427 (1956); Danversport, Sears (1885); Essex, H. 9377 (1953); Rockport, Sears (1886); Manchester, H.E. Parkhurst (1885); Salem, Hall (1925); Salem, Little Misery Island, H. 13304 (1957); Salem, North Gooseberry Island, H. 13273 (1957); Peabody, Mackintosh (1933); Saugus, H. 8204 (1952); Lynn, H. 11825 (1955); Marblehead, Frost (1880); Nahant, Hoar (1857).

S. sempervirens, forma *citrea* Harris - Ipswich, H. 13874 (1957).

S. juncea Ait. Dry open soil; common. - Haverhill, Lunt 793 (1911); Lawrence, H. 12294 (1956); Andover, Pease 1483 (1901); North Andover, H. 11059 (1955); West Boxford, Root (1945); Groveland, H. 8831 (1953); West Newbury, H. 10947 (1955); Newburyport, Plum Island, H. 13055 (1957); Newbury, H. 18468 (1957); Georgetown, Horner; Ipswich, Oakes; Ipswich, Plum Island, H. 12448 (1956); Danvers, Sears (1887); Wenham, Robinson (1875); Hamilton, H. 12294 (1956); Rockport, L.B. & R.C. Smith 4342 (1945); Manchester, Stone (1902); Beverly, A.P. Morse (1925); Peabody, Mackintosh (1933); Lynn, H. 10919 (1955); Marblehead, Ives (1927).

S. uliginosa Nutt. Swamps, meadows and thickets; frequent. - North Andover, Root (1945); Groveland, Lunt 791 (1911); Georgetown, Horner; Ipswich, Oakes; Middleton, Sears (1887); Danvers, Mackintosh (1924); Wenham, Robinson (1875); Chebacco, Robinson (1876); Beverly, Lowell (1847); Peabody, Mackintosh (1933); Lynnfield, Bean (1932).

S. uliginosa, var. *linoides* (T. & G.) Fern. - Boxford, Williams (1899); Hamilton, H. 12497 (1956); Beverly, Knowlton (1907).

S. arguta Ait. Open woods and clearings; occasional. - Haverhill, H. 696 (1932); Groveland, Horner; Ipswich, Oakes, H. 11758 (1955); Beverly, Lowell (1849); Salem, Russell (1859).

S. nemoralis Ait. Dry open soil; common. - Haverhill, Lunt 795 (1911); Lawrence, H. 11261 (1955); North Andover, Root (1945); Boxford, H. 8163 (1952); Groveland, H. 11161 (1955); Newburyport, H. & McGregor 12306 (1956); Georgetown, Horner; Rowley, Wade 26809 (1964); Ipswich, Oakes, Topsfield, M. Clarke; Danvers, Oakes; Wenham, Robinson (1874); Hamilton, Oakes; West Gloucester, H. 19002 (1958); Rockport, Forbes (1902); Beverly, H. 11665 (1955); Salem, Forbes (1903); Peabody, Mackintosh; Lynnfield, H. 9341 (1953); Swampscott, A.P. Morse (1925); Marblehead, H. 11922 (1955); Nahant, Forbes (1912).

S. odora Ait. Sweet Goldenrod. Dry sterile open soil; occasional. - Andover, Gutterson (1905); Boxford, M.E. Perley (1881); Rowley, H. & Wade 18801 (1958); Ipswich, Oakes; Danvers, Sears (1883); Lynnfield, H. 9354 (1953).

S. elliottii T. & G., var. *ascendens* Fern. Swamps and low thickets; occasional. - Methuen, Bean & H. 7747 (1951); Lawrence, H. 11256 (1955); West Newbury, H. 11139 (1955); Topsfield, H. 11402 (1955).

S. ulmifolia Muhl. Dry rocky woods; rare. - Gloucester, Lanesville, Sears (1887).

S. rugosa Ait. Damp open soil and borders of woods; common. - Salisbury, H. 18979 (1958); Haverhill, H. 1276 (1932); Lawrence, H. 11259 (1955); North Andover, Root (1945); Boxford, M.E. Perley (1883); Groveland, H. 11072 (1955); Newbury, H. 8233 (1952); Georgetown, Horner; Rowley, H. 11957 (1955); Rowley, Plum Island, H. 12460 (1956); Ipswich, Oakes; Ipswich, Plum Island, H. 12428 (1956); Topsfield, H.R. Towne; Danvers, Cushman 4844 (1908); Hamilton, H. 12493 (1956); West Gloucester, H. 19004 (1958); Rockport, Forbes (1902); Beverly, Lowell (1847); Salem, Hall (1927); Peabody, Mackintosh (1926); Lynnfield, H. 14109 (1957); Swampscott, A.P. Morse (1925); Marblehead, H. 11418 (1955).

S. rugosa, var. *villosa* (Pursh) Fern. - West Newbury, H. 11153 (1955); Gloucester, Magnolia, H. 11993 (1955).

S. rugosa, var. *aspera* (Ait.) Fern. - Salisbury, Eaton (1899); Saugus, Bean (1936).

X *s. asperula* Desf. A hybrid of *S. sempervirens* and *S. rugosa.* Coastal areas; occasional. - Newburyport, Wade 26807 (1964); Rowley, Plum Island, H. 12459 (1956); Ipswich, Plum Island, H. 12429 (1956); Plum Island, St. John 860 (1913); Nahant, H. 12592 (1956).

S. canadensis L. Moist to dryish thickets and roadsides; common. - Salisbury, Eaton; Haverhill, Lunt 795 (1911); North Andover, Root (1945); Boxford, H.E. Parkhurst (1882); Georgetown, Horner; Ipswich, Oakes; Danvers, Sears (1884); Wenham, Robinson (1875); Hamilton, H. 12495 (1956); Manchester, H. 11332 (1955); Salem, Cole (1886); Peabody, Mackintosh (1933); Lynnfield, Bean (1932); Saugus, H. 10908 (1955).

S. altissima L. Dry to damp thickets and roadsides; occasional. - Georgetown, Wade 13669 (1957); Ipswich, Oakes; Topsfield, Mrs. Conley; Salem, Forbes (1903); Peabody, Mackintosh (1926); Swampscott, A.P. Morse (1925).

S. gigantea Ait., var. *leiophylla* Fern. Alluvial or rich soil, thickets; frequent. - Merrimac, H. 9263 (1953); Haverhill, Bradford, Lunt (1913); North Andover, Root (1945); Boxford, H.E. Parkhurst (1882); Groveland, Paine & M.E. Perley (1885); Newburyport, Moulton (1886); Newbury, H. 8232 (1953); Georgetown, Horner (1877); Rowley, O'Neil (1925); Ipswich, Oakes; Topsfield, H.R. Towne; Danvers, Robinson (1875); Lynn, C.E. & E. Faxon (1881).

S. graminifolia (L.) Salisb. Thickets; rare. - Georgetown, Horner; Rowley, H. 8270 (1952).

S. graminifolia, var. *nuttallii* (Greene) Fern. Thickets and roadsides; common. - Amesbury, Bean (1950); Merrimac, H. 9205 (1953); Haverhill, H. 632 (1932); Lawrence, H. 11239 (1955); North Andover, Morss (1885); Boxford, M.E. Perley (1883); Groveland, H. 8824 (1953); Ipswich, H. 14040 (1957); Ipswich, Plum Island, H. 12390 (1956); Topsfield, H.R. Towne; Danvers, H. 11625 (1955); Wenham, Robinson (1874); Gloucester, Sears (1887); Rockport, L.B. & R.C. Smith 4343 (1945); Beverly, H. 11680 (1955); Salem, Cole (1886); Peabody, H. 11549 (1955); Lynnfield, Rich (1892); Marblehead Neck, A.W. & C.A. Cheever (1900).

S. tenuifolia Pursh. Damp to dry sandy soil; rare. - North Andover, H. 14048 (1957); Newburyport, Moulton (1887); Ipswich, Oakes; Ipswich, Plum Island, Lewis 20775 (1959); Rockport, Sears (1887).

CALLISTEPHUS Cass.

C. chinensis Nees. Introduced from Asia. Garden escape; occasional. - Lawrence, H. 18751 (1958); Saugus, H. 18684 (1958).

BOLTONIA L'Her.

B. asteroides (L.) L'Her. Adventive from South; rare. - Danvers, Cushman 4836 (1908).

B. latisquama Gray. Introduced from West. Escaped from cultivation. - Haverhill, H. 12607 (1932).

ASTER L.

A. divaricatus L. Dry woods and clearings; frequent. - Merrimac, H. 9284 (1953); Boxford, Robinson (1877); West Newbury, H. 11112 (1955); Rowley, H. 11930 (1955); Ipswich, Oakes; Wenham, H. 13747 (1957); West Gloucester, H. 19005 (1958); Rockport, H. 19046 (1958); Salem, Cole (1899).

A. macrophyllus L. Dry to moist open woods and clearings; frequent. - Haverhill, Lunt (1912); Boxford, H. 8656 (1953); Georgetown, Horner; Rowley, Smith 714 (1927); Ipswich, Oakes; Hamilton, Oakes; Salem, Robinson (1879); Lynnfield, Rodman (1887).

A. macrophyllus, var. *ianthinus* (Burgess) Fern. - Boxford, M.E. Perley (1885).

A. cordifolius L. Open woods and clearings; common. - Amesbury, H. 18969 (1958); Merrimac, H. 9276 (1953); Haverhill, Ward Hill, Bean (1916); Methuen, H. 11716 (1955); Lawrence, H. 11796 (1955); North Andover, Root (1945); Boxford, Stevens (1884); West Newbury, H. 8212 (1952); Ipswich, Oakes; Topsfield, H. 11897 (1955); Danvers, Sears (1884); Hamilton, H. 12483 (1956); Essex, H. 18850 (1958); West Gloucester, H. 19008 (1958); Beverly, Lowell (1847); Salem, Sears (1884); Peabody, Mackintosh (1933); Marblehead, H. 11919 (1955); Nahant, H. 12590 (1956).

A. undulatus L. Dry open woods and clearings; frequent. - Andover, Pease 9861 (1906); Boxford, M.E. Perley (1884); Georgetown, Horner; Rowley, Plum Island, H. 12554 (1956); Ipswich, Oakes; Topsfield, P.E. Kneeland, Danvers, Sears (1885); Wenham, A.P. Morse (1925); Essex, Bean (1907); Rockport, H. 19042 (1958); Peabody, Sears (1884).

A. patens Ait. Dry open woods and clearings; frequent. - Andover, Pease 1358 (1900); North Andover, Bean (1916); Boxford, H. 9001 (1953); Groveland, M.E. Perley (1885); Georgetown, Horner; Topsfield, H.R. Towne; Danvers, Sears (1885); Essex, Fernald, Hunnewell & Long 10570 (1913); Beverly, Russell (1859).

A. novae-angliae L. New England Aster. Damp thickets and meadows; frequent. - Amesbury, H. 18956 (1958); Haverhill, H. 94 (1931); Methuen, H. 11714 (1955); Lawrence, H. 18761 (1958); Andover, Pease 9863 (1906); North Andover, H. 8997 (1953); Boxford, Horner; West Newbury, H. 8213 (1952); Newburyport, H. 18955 (1958); Newbury, H. 8215 (1952); Rowley, H. 18954 (1958); Ipswich, Hunnewell (1914); Topsfield, A. Jacobs; Danvers, H. 11662 (1955); Gloucester, H. 18998 (1958); Beverly, H. 11683 (1955); Salem, Robinson (1874); Peabody, Mackintosh (1933); Saugus, H. 18463 (1958); Lynn, H. 18823 (1958); Marblehead, H. 11412 (1955).

A. novae-angliae, forma *roseus* (Desf.) Britt. Both native and escaped from cultivation; frequent. - Amesbury, H. 18957 (1958); Haverhill, H. 12618 (1932); Lawrence, H. 11795 (1955); Andover, Pease 1353 (1900); North Andover, Root (1943); West Newbury, Pease 29530 (1940); Topsfield, Mrs. M. Pierce; Danvers, Fowler; Hamilton, H. 12494 (1956); Salem, Sears (1898).

A. puniceus L. Damp thickets and meadows; frequent. - Amesbury, Bean (1950); Haverhill, H. 12610 (1932); North Andover, Root (1945); Topsfield, Mary Rossi; Danvers, Cushman 7825 (1908); Gloucester, Magnolia, H. 12337 (1956); Beverly, Robinson (1874).

A. puniceus, var. *firmus* (Nees) T. & G. - Topsfield, H.R. Towne.

A. spectabilis Ait. Dry sandy open woods; rare. - Boxford, Horner (1884).

A. laevis L. Dry open habitats and borders of woods; common. - Haverhill, Lunt 809 (1911); Andover, Bean (1935); Boxford, M.E. Perley (1884); Newbury, Bean & Stockman (1954); Georgetown, Horner; Rowley, Bean & Stockman (1954); Ipswich, Oakes; Topsfield, M.E. Williams; Middleton, Sears (1887); Danvers, Lowell (1848); Beverly, Mackintosh (1933); Salem, Hall (1927); Peabody, Cole (1926); Saugus, H. 11874 (1955).

A. laevis, var. *latifolius* (Porter) Shinners - Salem, Hall (1925), Ives (1925); Essex County, Oakes.

A. radula Ait. Low woods, swamps and bogs; rare. - Methuen, H. 8918 (1953); Peabody, E. Faxon (1891).

A. pilosus Willd. var. *pilosus* Boxford, H. 28365 (1965).

A. pilosus Willd., var. *demotus* Blake. Dry thickets, fields and roadsides; rare. - Haverhill, H. 95 (1931); North Andover, Root 19057 (1958); Newbury, H. 18467 (1958).

A. pilosus, var. *pringlei* (Gray) Blake - Essex County, Oakes.

X *A. amethystinus* Nutt. A probable hybrid of *A. novae-angliae* and *A. ericoides.* Moist thickets and meadows; rare. - Boxford, Solbrig (1965); Beverly, Robinson (1880).

A. ericoides L. Dry open soil or thickets; common. - Haverhill, H. 12616 (1932); Lawrence, H. 11774 (1955); North Andover, Bean (1916); Boxford, H. 8178 (1952); Newbury, Bean & Stockman (1954); Georgetown, Horner; Rowley, A.P. Morse (1925); Ipswich, Ordway (1937); Ipswich, Plum Island, H. 12447 (1956); Topsfield, H.R. Towne; Danvers, H. 11646 (1955); Essex, Knowlton (1909); Gloucester, Robinson (1874); Rockport, Williams (1893); Manchester, H. 11360 (1955); Beverly, H. 11681 (1955); Salem, Sears (1874); Saugus, H. 11866 (1955); Lynn, H. 27050 (1964); Marblehead, H. 11241 (1955); Nahant, Blankinship (1895).

A. ericoides, forma *exiguus* Fern. - Beverly, Prides Crossing, Miss A.L. Jackson (1894).

A. ericoides X *A. novi-belgii* ? - Ipswich, Ordway (1937).

A. dumosus L. Dry to wet open ground, thickets and shores; common. - Haverhill, Ward Hill, Bean (1916); Lawrence, Robinson (1879); Andover, Pease 1326 (1900); North Andover, Morss (1885); Rowley, A.P. Morse (1925); Ipswich, Oakes; Middleton, Ives (1927); Danvers, Ives (1927); Chebacco, Robinson (1874); Essex Woods, Lowell (1848); Rockport, L.B. & R.C. Smith 4344 (1945); Beverly, (1925).

A. dumosus X *A. vimineus* - Peabody, Mackintosh (1933).

A. vimineus Lam. Dry to moist fields, meadows and shores; frequent. - Andover, Pease 2613 (1903); North Andover, Root 19058 (1958); Groveland, H. 9077 (1953); Ipswich, G.M. Allen (1912); Topsfield, M.I. Pierce; Danvers, Cushman 4839 (1908); Chebacco, Robinson (1874); Essex, Fernald, Hunnewell & Long 10598 (1913); Lynnfield, H. 11385 (1955); Saugus, Bean (1936).

A. vimineus X *A. lateriflorus* - North Andover, Bean (1935).

A. lateriflorus (L.) Britt. Dry to moist fields, clearings, thickets and shores; common. - Salisbury, H. 18971 (1958); Amesbury, Bean (1950); Merrimac, H. 9267 (1953); Haverhill, Ward Hill, Bean (1916); North Andover, Root (1945); West Newbury, H. 11142 (1955); Ipswich, Oakes; Topsfield, H.R. Towne; Danvers, A.P. Morse (1927); Hamilton, H. 12484 (1956); Beverly, A.P. Morse (1925); Peabody, Mackintosh (1933); Lynnfield, H. 9352 (1953).

A. praealtus Poir., var. *angustior* Wieg. Low woods and clearings; rare. - Haverhill, H. 12614 (1932); Ipswich, Oakes.

A. simplex Willd. Damp thickets, meadows and shores; occasional. - Haverhill, Ward Hill, Bean 16147 (1916); Lawrence, Pease 1366 (1901); Andover, Pease 1313 (1901); North Andover, Root (1945); Essex, H. 9370 (1953).

A. simplex, var. *ramosissimus* (T. & G.) Cronq. - Amesbury, Bean (1950); Haverhill, H. 12612 (1932); North Andover, Sargent 19055 (1958); Newbury, H. 8231 (1952); Topsfield, P.E. Kneeland; Danvers, Cushman (1908); Beverly, Lowell (1846); Lynn, H. 27045 (1964).

A. simplex X *A. novi-belgii* - Saugus, Bean (1936).

A. tradescanti L. Damp rocky or gravelly shores; rare. - Haverhill, H. 12619 (1932); Boxford, H. 8162 (1952); Danvers, Sears (1884).

A. johannensis Fern. Shores and damp clearings; occasional. - Amesbury, H. 18970 (1958); North Andover, Root 19056 (1958); West Newbury, Pease 28533 (1940); Georgetown, Horner.

A. novi-belgii L. Damp thickets, meadows and shores; very common and very variable. - Salisbury, H. 18982 (1958); Amesbury, H. 18962 (1958); Merrimac, H. 9217 (1953); Haverhill, H. 12617 (1932); Andover, Pease 8371½ (1906); Boxford, H.E. Parkhurst (1882); Newburyport, Forbes 3773 (1903); Newbury, Bean & Stockman (1954); Newbury, Plum Island, H. 12537 (1956); Rowley, A.P. Morse (1925); Rowley, Plum Island, H. 12560 (1956); Ipswich, Robinson (1875); Ipswich, Plum Island, H. 12432 (1956); Danvers, H. 11616 (1955); Wenham, A.P. Morse (1925); Chebacco, Robinson (1875); Essex, Winslow (1910); Gloucester, H. 13842 (1957); Rockport, H. 19020 (1958); Manchester, Hubbard (1913); Beverly, Russell (1859); Salem, Sears (1884); Peabody, H. 9319 (1953); Lynnfield, Bean (1913); Saugus, H. 18612 (1958); Lynn, Robinson (1890); Marblehead, H. 11920 (1955); Nahant, Blankinship (1895).

A. novi-belgii, forma *albiflorus* Rand & Redf. - Rockport, Forbes (1903); Lynn, H. 18828 (1958).

A. novi-belgii, var. *littoreus* Gray. Seashores and borders of saline marshes; occasional. - Essex, H. 9371 (1953); Rockport, Williams (1893); Salem, Robinson (1874); Saugus, H. 11867 (1955).

A. nemoralis Ait. Bog-Aster. Sphagnous bogs and peaty shores; occasional. - Andover, Pease 1344 (1901); Boxford, Robinson (1877); H. 13387 (1957); Lynn, Sears (1885); Essex County, Pickering.

A. blakei (Porter) House. Damp thickets and shores; occasional. - Hamilton, Hunnewell 4630 (1916); West Gloucester, Swan (1881); H. 19006 (1958); Lynn, Wheatland (1857), Sears (1885).

A. acuminatus Michx. Dry or moist woods and clearings; frequent. - Georgetown, Horner; Topsfield, M. Clark; Wenham, A.P. Morse (1925); Rockport, Gates & Smith 1088 (1928); Manchester, Oakes; Beverly, Robinson (1879); Lynn, Mackintosh (1933).

A. tenuifolius L. Saline marshes and shores; rare. - Ipswich, Morong (1873), Sears (1886).

A. umbellatus Mill. Damp to dryish thickets and meadows; common. - Haverhill, H. 12609 (1932); Lawrence, H. 11262 (1955); Boxford, M.E. Perley (1881); Groveland, H. 8826 (1953); West Newbury, H. 11140 (1955); Ipswich, H. 11583 (1955); Topsfield, H. Williams; Danvers, Mackintosh (1925); Rockport, Gates & Smith 1084 (1928); Beverly, Lowell (1847).

A. linariifolius L. Dry open soil; common. - Salisbury, H. 18983 (1958); Haverhill, H. 12620 (1932); Lawrence, H. 11219 (1955); Andover, Pease 9886 (1906); North Andover, Bean (1916); Boxford, H.E. Parkhurst (1882); Newbury, H. 18450 (1958); Georgetown, Horner; Rowley, Plum Island, H. 12582 (1956); Ipswich, Oakes; Ipswich, Plum Island, H. 12444 (1956); Topsfield, P.E.

Kneeland; Danvers, Sears (1885); Wenham, Robinson (1889); Essex, H. 11306 (1955); Rockport, Gates & Smith 1087 (1928); Lynnfield, H. 9342 (1953); Marblehead, H. 11473 (1955).

A. linariifolius, forma *leucactis* Benke - Plum Island, Snyder (1951); Manchester, Robinson (1875).

A. subulatus Michx. Saline marshes; common. Newbury, Plum Island, Pease 3077 (1903); Rowley, Horner; Rowley, Plum Island, H. 12450 (1956); Ipswich, Hunnewell 2788 (1914); Ipswich, Plum Island, H. 12402 (1956); Danvers, Mackintosh (1924); Essex, Bean (1938); Beverly, Lowell (1847); South Salem, A.P. Morse (1925); Lynn, Robinson (1874).

ERIGERON L. Fleabane.

E. pulchellus Michx. Robin's-Plantain. Copses and meadows; frequent. - Haverhill, H. 681 (1932); Andover, Robinson (1878); North Andover, Root (1944); West Newbury, Robinson (1879); Topsfield, A.M. Dow; Danvers, Oakes; Salem, Cole (1889); Peabody, Robinson (1877); Lynnfield, Bean (1916).

E. philadelphicus L. Rich thickets; rare. - Rowley, McGregor 17631 (1958); Middleton, Mrs. A.L. Downes (1890).

E. annuus (L.) Pers. Fields and waste places; common. - Salisbury, H. 10475 (1955); Amesbury, H. 10525 (1955); Merrimac, H. 10566 (1955); Lawrence, H. 9853 (1955); Andover, Pease 4245 (1904); North Andover, Root (1944); Boxford, M.E. Perley (1885); Groveland, Bean & H. 8529 (1953); West Newbury, H. 10329 (1955); Georgetown, Horner; Rowley, H. 12232 (1956); Ipswich, Oakes; Topsfield, H.R. Towne; Wenham, Robinson (1877); Hamiton, Robinson (1876); Essex, H. 10241 (1955); Gloucester, H. 12142 (1956); West Manchester, Hubbard (1913); Beverly, Lowell (1847); South Salem, Sears (1892); Lynnfield, Bean (1916); Marblehead, Sears (1882).

E. strigosus Muhl. Daisy-Fleabane. Dry open soil; common. - Haverhill, Ward Hill, Cushman 2927 (1908); North Andover, Morss (1885); Boxford, M.E. Perley (1881); Newburyport, H. 10361 (1955); Georgetown, Horner; Rowley, H. 11937 (1955); Ipswich, Plum Island, H., McGregor & Schweinfurth 12670 (1957); Topsfield, H.R. Towne; Middleton, Bean & H. 7778 (1952); Wenham, Robinson (1875); Gloucester, H. 10141 (1955); Rockport, Gates & Smith 1002 (1928); Manchester, H. 11333 (1955); Salem, Cole (1945); Marblehead, Bean & H. 9769 (1955).

E. strigosus, forma *discoideus* (Robbins.) Fern. - Groveland, H. 12859 (1957); Topsfield, O. Floyd.

E. canadensis L. Horse-weed. Waste places; common. - Merrimac, H. 9262 (1953); Haverhill, H. 933 (1933); Lawrence, H. 11265 (1955); North Andover, Root (1945); Boxford, H. 8780 (1953); West Newbury, H. 10945 (1955); Georgetown, Horner; Rowley, H. 8284 (1952); Rowley, Plum Island, H. 12470 (1956); Ipswich, Oakes; Topsfield, H.R. Towne; Hamilton, H. 10829 (1955); Gloucester, H. 12345 (1956); Rockport, L.B. & R.C. Smith 4345 (1945); Salem, Robinson (1875); Salem, Little Misery Island, H. 13285 (1957); Peabody, Mackintosh (1933); Lynnfield, H. 14120 (1957); Lynn, Lowell (1847).

SERICOCARPUS Nees. White-topped Aster.

S. asteroides (L.) BSP. Dry woods and clearings; common. - Haverhill, H. 96 (1931); Andover, Pease 1457 (1901); North Andover, H. 8666 (1953); Boxford, M.E. Perley (1885); West Newbury, H. 10951 (1955); Georgetown, Horner; Rowley, Williams (1906); Ipswich, Oakes; Essex, Sears

(1887); Gloucester, Eastern Point, Prince (1883); Rockport, Smith 1126 (1930); Beverly, Phippen (1857); [MNS Snyder (1971)] ; Lynn, E.L. Mackintosh (1924).

S. linifolius (L.) BSP. Dry woods and clearings; occasional. - Boxford, H. 25141 (1962); Georgetown, Horner; Ipswich, Oakes; Topsfield, Oakes, H. 20962 (1959); Middleton, Ives (1927); Peabody, Mackintosh (1933).

PLUCHEA Cass. Marsh-Fleabane.

P. purpurascens (Sw.) DC., var. *succulenta* Fern. Salt-marsh Fleabane. Saline marshes; frequent. - Salisbury, A.P. Morse (1925); Newbury, Plum Island, H. 12308 (1956); Rowley, Pingree Island, Horner (1878); Ipswich, A.P. Morse (1921); Ipswich, Plum Island, H., McGregor & Schweinfurth 12667 (1957); Salem, Robinson (1874); Nahant, Williams (1894).

ANTENNARIA Gaertn. Ladies'-Tobacco.

A. canadensis Greene. Open woods, pastures and fields; occasional. - Andover, Pease 3089 (1903); Ipswich Neck, Sears (1901); North Beverly, Sears (1901).

A. neglecta Greene. Sterile fields and pastures; occasional. - Haverhill, H. 99 (1931); Andover, Pease 1300 (1902); North Andover, Zink (1943); Boxford, Forbes (1909); Lynnfield, Bean (1916); Swampscott, Robinson (1879).

A. neodioica Greene. Dry fields, pastures and open woods; common. - Andover, Sutton 10 (1874); North Andover, Root & Zink (1945); Boxford, Pease & St. John 11946 (1917); Ipswich, G.M. Allen (1912); Topsfield, H. 17648 (1958); Gloucester, H. 12164 (1956); Rockport, Smith 927 (1928); North Beverly, Sears (1901); Salem, Winter Island, Sears (1901); Lynnfield, Bean (1916); Lynn, Cushman 2994 (1908).

A. necdioica, var. *attenuata* Fern. - Rowley, Bean & H. 8421 (1953); Topsfield, Bean & H. 9958 (1955).

A. petaloidea Fern. Fields, pastures and open woods; rare. - Salem, Winter Island, Sears (1901).

A. munda Fern. Dry sandy or gravelly ground; rare. - Hamilton, Hubbard (1913).

A. fallax Greene. Fields and clearings; occasional. - Andover, Pease 1296 (1902); Danvers, Putnamville, Sears (1901); Salem, Winter Island, Sears (1901).

A. parlinii Fern. Dry open woods and fields; rare. - Andover, Pease 2171 (1903); Boxford, Fernald (1909).

A. plantaginifolia (L.) Hook. Dry open woods, fields and pastures; common. - Haverhill, H. 98 (1931); Andover, Pease 1299 (1902); North Andover, Morss (1885); Boxford, Cushman 773 (1907); Newbury, Robinson (1873); Rowley, Bean & H. 8420 (1953); Rowley, Plum Island, H. 12074 (1956); Ipswich, Bean & H. 9979 (1955); Topsfield, H.R. Towne; Danvers, Dr. Nichols or Oakes (1838); Wenham, Sears (1899); Beverly, Robinson (1879); Salem, Winter Island, Sears (1901); Marblehead, H. 9675 (1955); Nahant, S.S. Clark 9 (1916).

A. plantaginifolia, var. *petiolata* (Fern.) Heller - West Boxford, Horner; Georgetown, Horner.

ANAPHALIS DC. Everlasting.

A. margaritacea (L.) C.B. Clarke, var. *intercedens* Hara. Pearly Everlasting. Gravelly or sandy soils; common. - Amesbury, Robinson (1879); Haverhill, H. 565 (1932); North Andover, Morss (1885); Boxford, M.E. Perley (1884); Newbury, C. Cushing; South Georgetown, Horner; Ipswich, Oakes; Topsfield, H.R. Towne; Rockport, Gates & Smith 2395 (1928); Beverly, Lowell (1847); Salem, Cole (1886); Peabody, Mackintosh (1933).

GNAPHALIUM L. Cudweed.

G. maccounii Greene. Clearings and borders of woods; occasional. - Boxford, M.E. Perley (1885); Topsfield, H.R. Towne; Danvers, Russell (1859); Essex County, Oakes.

G. obtusifolium L. Catfoot. Borders of woods and fields; common. - Haverhill, H. (1932); Methuen, Bean & H. 7748 (1951); Lawrence, H. 11216 (1955); North Andover, Root (1945); Boxford, M.E. Perley (1885); Groveland, H. 11088 (1955); West Newbury, Robinson (1879); Georgetown, Horner; Rowley, H. 3283 (1952); Rowley, Plum Island, H. 12465 (1956); Ipswich, Furber (1911); Ipswich, Plum Island, H. 12430 (1956); Topsfield, H.R. Towne; Middleton, Bean & H. 7777 (1951); Chebacco, Robinson (1874); Rockport, Forbes 3851 (1902); Beverly, H. 11695 (1955); Salem, Cole (1887); Peabody, Mackintosh (1933); Marblehead, H. 11445 (1955).

G. uliginosum L. Low Catfoot. Damp clearings and waste places; common. - Haverhill, H. 660 (1932); Methuen, H. 10995 (1955); North Andover, H. 11055 (1955); Boxford, M.E. Perley (1885); Groveland, H. 935 (1933); Newburyport, Plum Island, H. 13052 (1957); Newbury, Plum Island, Pease 2641 (1902); Byfield, Robinson (1879); Georgetown, Horner; Ipswich, Forbes (1913); Topsfield, P.M. Gould; Danvers, Mackintosh (1933); Wenham, Sears (1891); Gloucester, Magnolia, J.M. Hunnewell (1898); Manchester, Hubbard (1913); Salem, Cole (1928); Peabody, Mackintosh (1933).

ELICHRYSUM

E. bracteatum Ands., var. *atrococcum* Hort. Introduced from Australia. Escape from cultivation. - Haverhill, H. 145 (1932).

INULA L.

I. helenium L. Elecampane. Naturalized from Europe. Roadsides and fencerows; occasional. - West Andover, Morss (1885); Boxford, Williams (1899); West Newbury, Cole (1900); Newbury, H. 12201 (1956); Georgetown, Horner; Ipswich, M.E. Perley (1883); Topsfield, E.M. Dow; Danvers, Oakes; Hamilton, Robinson (1875); Beverly, Lowell (1847).

I. salicina L. Adventive from Old World. - Danvers, Sears (1879), Station destroyed in 1885.

IVA L. Marsh-Elder.

I. frutescens L., var. *oraria* (Bartlett) Fern. & Grisc. Saline marshes; common. - Newburyport, H. 11270 (1955); Newbury, Horner; Rowley, H. 8250 (1952); Ipswich, Oakes; Ipswich, Plum Island, H. 12386 (1956); Gloucester, Seymour 5174 (1938); Beverly, Lowell (1847); South Salem, Sears (1892); Marblehead, H. 11916 (1955).

I. xanthifolia Nutt. Adventive from West. - Danvers, Tapleyville, Robinson & Sears (1880).

AMBROSIA L. Ragweed.

A. trifida L. Great Ragweed. Alluvium and waste places; occasional. - Haverhill, Bean & H. 20852 (1959); North Andover, H. 11063 (1955); Boxford, H. 31781 (1967); Topsfield, H.R. Towne; Middleton, H. 26414 (1963); Danvers, Hathorne, Sears (1904); Danversport, Mackintosh (1924).

A. trifida, forma *integrifolia* (Muhl.) Fern. Boxford, H. 31782 (1967).

A. artemisiifolia L. Common Ragweed. Sea-beaches and waste land; occasional. - Lawrence, H. 18759 (1958); Newburyport, H. 20736 (1959); Rowley, H. 11962 (1955); Rowley, Plum Island, H. 12469 (1956); Ipswich, H. 11594 (1955); Danvers, H. 20758 (1959); Gloucester, Magnolia, H. 12355 (1956); Saugus, H. 18703 (1958).

A. artemisiifolia var. *elatior* (L.) Descourtils. Roadsides and waste places; too common. - Haverhill, H. 633 (1932); Methuen, H. 11009 (1955); North Andover, H. Morss (1884); Boxford, M.E. Perley (1883); Newburyport, McGregor 11269 (1955); heads abnormal; Newbury, Plum Island, H. 12532 (1956); Georgetown, Horner; Rowley, Plum Island, H. 22468 (1956); Ipswich, H. & Wade 18795 (1958); Topsfield, H.R. Towne; Danvers, H. 11624 (1955); Essex, H. 11316 (1955); Gloucester, Magnolia, Deane (1901); Rockport, Anderson & Smith 1144 (1930); Manchester, Williams (1895); Beverly, H. 11698 (1955); South Salem, Sears; Salem, North Gooseberry Island, H. 13274 (1957); Peabody, Mackintosh (1933); Lynnfield, H. 14143 (1957); Saugus, H. 10861 (1955); Lynn, Robinson (1879).

A. artemisiifolia, var. *elatior,* forma *villosa* Fern. & Grisc. Waste places; frequent. - Merrimac, H. 9277 (1953); North Andover, Root (1945); Groveland, H. 11079 (1955); West Newbury, H. 10946 (1955); Hamilton, H. 12503 (1956); Gloucester, Magnolia, H. 12354 (1956); Beverly, H. 11689 (1955); Salem, Cole (1886).

A. artemisiifolia, var. *paniculata* (Michx.) Blankinship. Waste places; occasional. - Lawrence, H. 11232 (1955); Boxford, H. 30772 (1966); Groveland, H. 11182 (1955); Ipswich, Plum Island, H. 12407 (1956); Essex, H. 11304 (1955).

XANTHIUM L. Cocklebur.

X. spinosum L. Naturalized from Europe. Waste places; occasional. - Lawrence, Pease 1498 (1902); Beverly, Oakes; South Salem, Sears (1884).

X. chinense Mill. Low ground and waste places; occasional. - Methuen, H. 11721 (1955); North Andover, H. 14046 (1957); Newburyport, Knowlton (1912).

X. pensylvanicum Wallr. Low ground and waste places; occasional. - Methuen, H. 11756 (1955); Lawrence, Robinson (1877); Newbury, Bean & Stockman (1954); Essex, H. 18853 (1958); Gloucester, Magnolia, Deane (1901); Saugus, H. 18706 (1958); Marblehead, H. 18920 (1958).

X. italicum Moretti. Low ground and waste places; occasional. - Lawrence, H. 11797 (1955); Newbury, Bean & Stockman (1954); Lynn, H. 11845 (1955).

X. echinatum Murr. Sea-Burdock. Beaches, dune-hollows and borders of saltmarsh; frequent. - Newburyport, H. 11276 (1955); Rowley, Oakes; Ipswich Beach, M.E. Perley (1903); Manchester, Robinson (1876); Beverly, Lowell (1846); Lynn, Robinson (1875); Marblehead, Sears; Nahant, Bankinship (1895).

ACANTHOSPERMUM Schrank

A. australe (Loef.) Ktze. Adventive from Tropical America. Waste places; rare. - Lawrence, Mrs. E.S. Schnelder (1907).

SILPHIUM L. Rosinweed.

S. perfoliatum L. Cup-plant. Introduced from West. Waste places; occasional. - West Newbury, Lewis 24032 (1961); Danvers, Sears (1880).

HELIOPSIS Pers. Ox-eye.

H. helianthoides (L.) Sweet, var. *scabra* (Dunal) Fern. Introduced from South. Escape from cultivation. - Boxford, H. 9640 (1954).

RUDBECKIA L. Coneflower.

R. laciniata L. Rich low ground; common. - Amesbury, Bean (1950); Merrimac, H. 9261 (1953); Methuen, H. 12273 (1956); Lawrence, H. 12284 (1956); Andover, Pease 1448 (1900); North Andover, Morss (1885); Boxford, Hunnewell (1915); Groveland, M.E. Perley (1885); West Newbury, H. 11131 (1955); Ipswich, Oakes; Topsfield, G.F. Dow; Middleton, H. 8706 (1953); Danvers, Sears (1882); Salem, Cole (1888); Peabody, Mackintosh (1933).

R. bicolor Nutt. Introduced from West. Waste places; occasional. - Haverhill, H. 746 (1932); North Andover, H. 18725 (1958).

R. serotina Nutt. Black-eyed Susan. Introduced from the West. Fields and clearings; common. - Haverhill, H. 634 (1932); Methuen, H. 11001 (1955); Andover, Ballardvale, Zink (1943); North Andover, Morss (1884); Boxford, Frye (1880); Georgetown, Horner; Topsfield, A.M. Dow; Beverly, Montserrat, Hubbard (1913); Peabody, Mackintosh (1933).

R. serotina, forma *tubuliformis* (Burnham) Fern. & Schub. - Essex County, Miss C. Goodrich (1881).

R. serotina, forma *rubra* (Clute) Fern. - Beverly, Hall.

R. serotina, forma *pulcherrima* (Farw.) Fern. - Byfield, Robinson (1875); Topsfield, Foye 8746 (1953).

R. serotina, forma *pleniflora* (Moldenke) Fern. & Schub. - Essex County, Miss C. Goodrich.

HELIANTHUS L. Sunflower.

H. annuus L. Common Sunflower. Introduced from the West. Escaping from cultivation to roadsides and waste places; common. - Haverhill, H. 12623 (1932); Methuen, H. 11762 (1955); Lawrence, H. 11781 (1955); North Andover, Root (1945); Boxford, H. 8988 (1953); Newburyport, H. 11278 (1955); Newbury, H. 20712 (1959); Newbury, Plum Island, H. 12530 (1956); Rowley, H. 18523 (1958); Rowley, Plum Island, H. 12578 (1956); Ipswich, Plum Island, H. 12393 (1956); Topsfield, H. 11387 (1955); Danvers, Sears (1879); Beverly, H. 11703 (1955); South Salem, Sears (1884); Peabody, H. 11559 (1955); Saugus, H. 18620 (1958); Lynn, H. 11827 (1955); Marblehead, H. 11460 (1955); Nahant, Kennedy (1894).

H. laetiflorus Pers. Introduced from the West. Adventive in waste places; occasional. - Beverly, H. 18898 (1958); Marblehead, H. 18945 (1958).

H. laetiflorus, var. *rigidus* (Cass.) Fern. - Boxford, H. 9018 (1953); Salem, Phippen (1839).

H. divaricatus L. Thin woods and thickets; common. - Haverhill, H. 697 (1932); Andover, Pease 2449 (1903); Boxford, M.E. Perley (1883); Byfield, Horner; Georgetown, Horner (1884); Rowley, Smith 713 (1927); Ipswich, Oakes; Topsfield, H.M. Goodale; Danvers, Sears (1885); Hamilton, Robinson (1875); Gloucester, Magnolia, Deane (1901); Salem, Cole (1888); Peabody, Mackintosh (1933); Marblehead, Cole (1944).

H. strumosus L. Open woods and clearings; common. - Salisbury, A.P. Morse (1924); Haverhill, Hunnewell 3833 (1915); Methuen, H. 8909 (1953); North Andover, H. 11036 (1955); Ipswich, Oakes; Topsfield, H.R. Towne; Danvers, Sears (1885); Essex, H. 11318 (1955); Gloucester, Robinson (1874); Rockport, Churchill (1887); Beverly, Hunnewell 3315 (1914); Salem, Cole (1907); Peabody, H. 9293 (1935); Marblehead, H. 11415 (1955).

H. decapetalus L. Open woods and thickets; occasional. - Amesbury, Bean (1950); Haverhill, Hunnewell 3817 (1915); North Andover, H. 11022 (1955); Georgetown, Horner (1876).

H. tuberosus L. Jerusalem Artichoke. Introduced from the West. Rich or damp thickets; frequent. - Haverhill, H. 12622 (1932); Methuen, H. 11707 (1955); Lawrence, Pease 1408 (1901); North Andover, Root (1945); Boxford, M.E. Perley (1881); Ipswich, H. 12586 (1956); Topsfield, P.E. Kneeland; Hamilton, H. 10788 (1955); Salem, Sears (1886); Saugus, H. 18722 (1958); Lynn, H. 11843 (1955); Marblehead, H. 11416 (1955).

VERBESINA L. Crown-Beard.

V. encelioides (Cav.) B. & H., var. *exauriculata* Robins. & Greenm. Adventive from the West. - Georgetown, Horner.

GUIZOTIA Cass.

G. abyssinica (L. f.) Cass. Adventive from tropical Africa. - Marblehead, H. 11453 (1955).

COREOPSIS L. Tickseed.

C. tinctoria Nutt. Introduced from the West. Escape from cultivation; occasional. - Haverhill, H. 100 (1931); Wenham, Kimball (1877).

C. lanceolata L. Introduced from the West. Escape from cultivation; occasional. - Methuen, H. 12265 (1956); Middleton, H. & Wetherall 9554 (1954); Marblehead, Bean & H. 9743 (1955).

C. grandiflora Hogg. Introduced from the South. Escape from cultivation. - Danvers, H. 18567 (1958).

C. major Walt., var. *stellata* (Nutt.) Robins. Introduced from the South. - Boxford, H. 8791 (1953).

BIDENS L. Bur-Marigold.

B. laevis (L.) BSP. Marshes and margins of pools and streams; occasional. - Andover, Young (1882); Rowley, Williams (1906); Topsfield, H.S. Porter; Danvers, M.E. Perley (1880); Cushman 4841 (1908); Wenham, Fernald, Hunnewell & Long 10689 (1913).

B. cernua L. Sloughs, pools and shores; common. - Amesbury, H. 9191 (1953); Merrimac, H. 9244 (1953); Andover, Williams (1899); North Andover, Andrade (1957); Boxford, H. 9009 (1953); Groveland, H. 9068 (1953); West Newbury, H. 9126 (1953); Ipswich, Forbes (1913); Ipswich, Plum Island, H. 12391 (1956); Danvers, Sears (1902); Gloucester, H. 13845 (1957); Beverly, Lowell (1853); Saugus, Bean (1936).

B. cernua, var. *integra* Wieg. - Danvers, Sears (1902).

B. cernua, var. *elliptica* Wieg. - Haverhill, H. 101 (1931); Groveland, H. 28366 (1965); Danvers, Sears (1902).

B. cernua, var. *oligodonta* Fern. & St. John - Amesbury, H. 9154 (1953).

B. cernua, var. *minima* (Huds.) Pursh - Haverhill, H. 102 (1931); North Andover, H. 14062 (1957).

B. hyperborea Greene, var. *colpophila* (Fern. & St. John) Fern. Tidal mud of estuaries; occasional. - Newburyport, Eaton & Fernald (1902); Rowley, Fassett 787 (1923).

B. connata Muhl. Beggar-Ticks. Wet shores and swamps; occasional. - Salisbury, Eaton & Fernald (1902); Haverhill, H. 591 (1932); Andover, Williams (1899); Danvers, Kennedy (1907); Wenham, Conant (1877); North Salem, A.P. Morse (1925).

B. connata, var. *gracilipes* Fern. - Peabody, H. 9322 (1953).

B. connata, var. *petiolata* (Nutt.) Farw. Common. - Amesbury, H. 9161 (1953); Haverhill, H. 12625 (1932); North Andover, Root (1943); Boxford, H. 8202 (1952); Groveland, H. 28372 (1965); West Newbury, Conant (1876); Rowley, Plum Island, H. 12472 (1956); Topsfield, Mrs. Conley; Middleton, H. 9424 (1953); Danvers, H. 20770 (1959); Lynnfield, H. 14107 (1957).

B. eatoni Fern. Tidal shores of Merrimac River; occasional. - Salisbury, Eaton & Fernald (1902); Amesbury, S.F. Blake 10784A (1928); Haverhill, H. 662 (1932); Newburyport, Eaton & Fernald (1902).

B. eatoni, var. *fallax* Fern. - Newburyport, Eaton & Fernald (1902).

B. eatoni, var. *kennebecensis* Fern. - Amesbury, Fassett 2300 (1924).

B. comosa (Gray) Wieg. Rich moist or fairly dry soil; occasional. - Salisbury, Eaton (1898); Gloucester, Magnolia, J.M. Hunnewell (1899).

B. vulgata Greene. Rich low ground, roadsides and waste places; occasional. - Haverhill, H. 664 (1932); North Andover, Root (1945); Georgetown, Horner; Wenham, A.P. Morse (1925); Essex, Fernald, Hunnewell & Long 10676 (1913).

B. vulgata, forma *puberula* (Wieg.) Fern. - North Salem, A.P. Morse (1925).

B. frondosa L. Damp open habitats and waste ground; common. - Salisbury, Fernald (1902); Merrimac, H. 9283 (1953); Haverhill, H. 663 (1932); Andover, Williams (1899); Boxford, M.E. Perley (1881); Groveland, H. 9061 (1953); Newburyport, Fernald (1902); Ipswich, Plum Island, H. 12414 (1956); Topsfield, H.R. Towne; Wenham, A.P. Morse (1925); Hamilton, Phippen (1857); Gloucester, Magnolia, J.M. Hunnewell (1899); Manchester, H. 11358 (1955); Beverly, H. 11685 (1955); Lynnfield, H. 14147 (1957); Marblehead, H. 11459 (1955).

B. frondosa, forma *anomala* (Porter) Fern. - Amesbury, H. 9195 (1953); Boxford, H. 9010 (1953).

B. discoidea (T. & G.) Britt. Peaty or wet sandy swamps and shores; occasional. - Merrimac, H. 13710 (1957); Haverhill, H. 590 (1932); Andover, Pease 1371 (1901); Groveland, H. 28371 (1965).

B. coronata (L.) Britt., var. *brachyodonta* Fern. Bogs and peaty meadows; locally common. - Ipswich, Alcott (1913); Ipswich, Plum Island, H. 12394 (1956); Topsfield, H.S. Porter; Wenham, Oakes; Hamilton, Cole (1908); Essex, Bean (1938); Manchester, Oakes; Beverly, Lowell (1846); Lynnfield, Fernald (1917); Lynn, Young (1879).

COSMOS Cav.

C. bipinnatus Cav. Introduced from Mexico. Garden escape; occasional about dumps. - Haverhill, H. 665 (1932); Lawrence, H. 18742 (1958); Danvers, H. 18577 (1958); Beverly, H. 11699 (1955); Salem, H. 11510 (1955); Peabody, H. 11524 (1955); Saugus, H. 18656 (1958); Lynn, H. 11828 (1955); Marblehead, H. 18923 (1958).

C. parviflorus HBK. Introduced from Western U.S. Garden escape. - South Lawrence, Pease 1500 (1902).

MEGALODONTA Greene Water-Marigold.

M. beckii (Torr.) Greene. Ponds and slow streams; occasional. - Haverhill, H. 566 (1932); Boxford, H. 13566 (1957); West Newbury, Robinson (1879); Ipswich, H. 13239 (1957); Topsfield, H. 13235 (1957).

GALINSOGA R. & P.

G. ciliata (Raf.) Blake. Naturalized from Tropical America. Gardens and waste places; common. - Merrimac, H. 9260 (1953); Methuen, H. 11711 (1955); North Andover, Root (1945); Boxford, H. 8014 (1952); West Newbury, H. 10942 (1955); Rowley, H. 18516 (1958); Topsfield, Bean & H. 9972 (1955); Danvers, Mackintosh (1924); Wenham, H. 13746 (1957); Essex, H. 18859 (1958); Gloucester, Magnolia, H. 12360 (1956); Beverly, H. 11701 (1955); Salem, Files (1939); Peabody, Mackintosh (1933); Saugus, H. 10872 (1955); Lynn, H. 11838 (1955); Marblehead, H. 11449 (1955).

G. parviflora Cav. - Salem, Hall (1925).

HELENIUM L. Sneezeweed.

H. tenuifolium Nutt. Introduced from the South. Open ground; rare. - Georgetown, Byfield, Horner (1885).

H. nudiflorum Nutt. Introduced from the South. Openings and roadsides; occasional. - Andover, Miss Twitchell (1903); North Andover, Bragdon (1943); Boxford, H. 9648 (1954); Ipswich, H. & Wade 18787 (1958); Peabody, Mackintosh (1933).

H. autumnale L. Introduced from the South. Rich thickets and meadows; rare. - Meramec River, Prince (1877); Danvers, Sears (1880).

TAGETES L. Marigold.

T. patula L. Introduced from Tropical America. Escape from cultivation; occasional. - Peabody, H. 11522 (1955); Marblehead, H. 11466 (1955).

ACHILLEA L. Yarrow.

A. ptarmica L. Introduced and naturalized from Europe. Damp fields and roadsides; occasional. - Danvers, Oakes; Kennedy (1907); Salem, Russell; Saugus, H. 18687 (1958); Lynn, Tracy (1859).

A. millefolium L. Naturalized from Europe. Fields and roadsides; common. - Salisbury, H. 10500 (1955); Amesbury, H. 10564 (1955); Haverhill, Ward Hill, Cushman 2946 (1908); Methuen, H. 9602 (1954); Lawrence, H. 11211 (1955); Andover, Root (1943); Boxford, M.E. Perley (1884); Groveland, Bean & H. 8543 (1953); Newbury, Plum Island, H. 12015 (1956); Georgetown, Horner; Rowley, H. 8362 (1952); Ipswich, Plum Island, H. 12044 (1956); Ipswich, Plum Island, H. 12091 (1956); Topsfield, H.R. Towne; Wenham, H. 9418 (1953); Hamilton, H. 10797 (1955); Essex, H. 10234 (1955); Rockport, Smith 957 (1928); Salem, Phippen (1857); Salem, Little Misery Island, H. 13292 (1957); Peabody, Mackintosh (1933); Saugus, H. 10856 (1955); Lynn, H. 10926 (1955); Swampscott, H. 10644 (1955); Marblehead, Bean & H. 9762 (1955).

A. millefolium, forma *rosea* Rand & Redf. - Andover, Pease 2933 (1903); North Andover, A. Russell (1945); Georgetown, Horner (1875); North Salem, Hall (1925).

A. lanulosa Nutt. Gravelly or sandy shores; rare. - Newbury, Plum Island, Pease 13671 (1912).

ANTHEMIS L. Chamomile.

A. cotula L. Mayweed. Naturalized from Europe. Roadsides and waste places; common. - Haverhill, H. 666 (1932); Methuen, H. 10994 (1955); North Andover, Morss (1885); Boxford, M.E. Perley (1882); Groveland, H. 936 (1933); Newburyport, H. 10376 (1955); Rowley, H. 18486 (1958); Ipswich, H. 11608 (1955); Topsfield, H.R. Towne; Essex, H. 11303 (1955); Gloucester, Lanesville, Collins (1915); Rockport, Smith 613 (1927); Salem, H. 11495 (1955); Peabody, Sears (1883); Saugus, H. 18646 (1958).

A. arvensis L., var. *agrestis* (Wallr.) DC. Corn-Chamomile. Naturalized from Europe. Fields and roadsides; rare. - Danvers, Oakes, Dr. Geo. Osgood (1820).

A. tinctoria L. Yellow Chamomile. Naturalized from Europe. Fields and roadsides; occasional. - Haverhill, H. 747 (1932); North Andover, H. 18026 (1958); Boxford, H. 8114 (1952); Georgetown, H. 12800 (1957); Gloucester, Magnolia, Clarke (1906); Rockport, Fernald, Hunnewell & Long 10696 (1913); Beverly, H. 18893 (1958).

MATRICARIA L. Wild Chamomile.

M. maritima L., var. *agrestis* (Knaf) Wilmott. Naturalized from Europe. Fields and waste places; rare. - Andover, Pease 1992 (1902); Marblehead, Robinson (1890), Rand (1910).

M. chamomilla L. - Boxford, H. 30769 (1966).

M. matricarioides (Less.) Porter. Naturalized from Pacific States. Roadsides and waste places; common. - Haverhill, Ward Hill, H. 496 (1932); North Andover, Pease 2166 (1903); West Boxford, Root & Zink (1945); Newburyport, H. 10375 (1955); Newbury, H. 10393 (1955); Rowley, Plum Island, H. 12060 (1956); Ipswich, H. 10225 (1955); Topsfield, H. 8861 (1953); Danvers, H. 7975 (1952); Wenham, Bean & H. 9998 (1955); Essex, H. 18855 (1958); Gloucester, Knowlton (1908); Rockport, Smith 4335 (1945); Salem, North Gooseberry Island, H. 13272 (1957); Peabody, Mackintosh (1933); Marblehead, A.P. Morse (1924).

CHRYSANTHEMUM L. Chrysanthemum.

C. leucanthemum L., var. *pinnatifidum* Lecoq & Lamotte. Ox-Eye-Daisy. Naturalized from Europe. Fields and roadsides; common. - Salisbury, H. 10491 (1955); Merrimac, H. 10593 (1955); Haverhill, H. 497 (1932); Andover, Pease 2141 (1903); North Andover, M.E. Perley (1882); Boxford, H.E. Parkhurst (1882); Groveland, Bean & H. 8485 (1953); West Newbury, H. 10332 (1955); Newbury, H. 10411 (1955); Rowley, H. 10448 (1955); Rowley, Plum Island, H. 12043 (1956); Ipswich, Plum Island, H. 12121 (1956); Topsfield, H.R. Towne; Danvers, Nettie Sears (1877); Wenham, L.G. Kimball (1875); Hamilton, H. 10827 (1955); Essex, H. 10257 (1955); Gloucester, H. 10153 (1955); Rockport, Forbes (1904); Manchester, Hubbard (1913); Beverly, Robinson (1879); Salem, H. 9707 (1955); Swampscott, H. 10688 (1955); Marblehead, H. 9697 (1955).

C. leucanthemum, var. *pinnatifidum* forma *tubuliflorum* Henrikssen. Naturalized from Europe; rare. - Danvers, Miss A.L. Page (1890); Salem Neck, A.J. Whipple (1883).

C. coronarium L. Garland-Chrysanthemum. Introduced from Europe. Escape from cultivation. - Methuen, H. 11747 (1955).

C. parthenium (L.) Bernh. Feverfew. Introduced from Europe. Waste places, and roadsides; occasional. - Boxford, H. 12506 (1956); Byfield, Robinson (1879); Wenham, Robinson (1875); Essex, H. 18864 (1958); Saugus, H. 18654 (1958).

C. balsamita L. Mint-Geranium. Introduced from Europe. Escape from cultivation; occasional. - North Andover, Pease 1380 (1900); Salem, Misery Island, Robinson (1877).

C. uliginosum Pers. Introduced from Europe. Escape from cultivation. - Gloucester, Magnolia, J.M. Hunnewell (1900).

TANACETUM L. Tansy.

T. vulgare L. Introduced from Europe. Roadsides and waste places; common. - Haverhill, H. 592 (1932); Methuen, H. 11717 (1955); North Andover, Thomas (1952); Boxford, M.P. Perley (1883); West Newbury, H. 10956 (1955); Newburyport, H. 11274 (1955); Newbury, H. 18443 (1958); Georgetown, Horner; Rowley, H. 11939 (1955); Ipswich, Oakes; Ipswich, Plum Island, H., McGregor & Schweinfurth 12655 (1957); Topsfield, P.E. Kneeland; Danvers, Robinson (1876); Wenham, Robinson (1875); Hamilton, H. 10794 (1955); Essex, H. 11308 (1955); Gloucester,

Magnolia, H. 11990 (1955); Rockport, L.B. & R.C. Smith (1934); Lynn, H. 10931 (1955); Swampscott, H. 10686 (1955); Marblehead, H. 11432 (1955).

T. vulgare, forma *crispum* (L.) Hayek - Boxford, M.E. Perley (1885); Topsfield, H.R. Towne; Wenham, Robinson (1876).

ARTEMISIA L. Wormwood.

A. caudata Michx. Sands near coast; frequent. - Salisbury Beach, Horner; Newburyport, Plum Island, Cole (1900); Newbury, Plum Island, Pease 2940 (1903); Georgetown, Horner; Ipswich Beach, Sears (1880); Ipswich, Plum Island, Cole (1898); Rockport, Williams (1898); Lynn, Lowell (1847); Nahant, Robinson (1875).

A. dracunculus L. Dry ground; rare. - Danvers, Sears (1904).

A. biennis Willd. Open ground and waste places; rare. - Salem, Sears (1889); Marblehead, Sanford (1946).

A. annua L. Introduced from Asia. Waste places and roadsides; rare. - South Salem, E.S. Rogers (1876).

A. pontica L. Roman Wormwood. Introduced from southeastern Europe. Waste places and roadsides; rare. - Newbury, Kents Island, Sears (1904).

A. stelleriana Bess. Dusty Miller. Naturalized from northeastern Asia. Sandy beaches and dunes; common. - Salisbury, H. 10482 (1955); Newburyport, Margaret Cushing (1927); Newburyport, Plum Island, H. 13053 (1957); Rowley, Plum Island, H. 12570 (1956); Ipswich, Wiegand 2120 (1911); Ipswich, Plum Island, H. 12443 (1956); Lynn, Farlow (1879); Nahant, Robinson (1877).

A. vulgaris L. Common Mugwort. Roadsides and waste places; common. - Lawrence, H. 11224 (1955); North Andover, H. 18729 (1958); Boxford, H. 11070 (1955); Newburyport, H. 10347 (1955); Rowley, H. & Wetherall 9557 (1954); Ipswich, Oakes; Topsfield, H.R. Towne; Wenham, Dr. Osgood & Pitman (1857); Gloucester, Magnolia, H. 12346 (1956); North Salem, Sears (1885); Peabody, H. 11558 (1955); Saugus, Bean & H. 9822 (1955); Lynn, H. 10923 (1955); Marblehead, H. 11429 (1955).

A. vulgaris, var. *glabra* Ledeb. - Haverhill, H. 635 (1932).

A. ludoviciana Nutt., var. *gnaphalodes* (Nutt.) T. & G. Naturalized from West. Waste places, rare. - Newbury, Williams (1899).

A. ludoviciana, var. *latifolia* (Bess.) T. & G. Naturalized from West. Waste places; rare. - Newbury, Williams (1899).

A. absinthium L. Wormwood. Introduced from Europe. Escape from cultivation; rare. - Danvers, Sears (1904).

TUSSILAGO L. Coltsfoot.

T. farfara L. Introduced from Europe. Damp sandy soils and brooksides; common. - Amesbury, Eaton (1899); Boxford, H. 9923 (1955); Newburyport, H. 10351 (1955); Georgetown, Horner; Rowley, H. 9543 (1954); Topsfield, H.S. Porter; Middleton, H. 8301 (1953); Danvers, Frost (1874);

Wenham, H. 13738 (1957); Manchester, Robinson (1876); Beverly, Sears (1887); Salem, Ropes (1919); Peabody, H. 11551 (1955); Lynnfield, H. 8300 (1953); Saugus, H. 9653 (1955); Swampscott, H. 10684 (1955); Marblehead, Bean & H. 9764 (1955).

PETASITES Mill. Sweet Coltsfoot.

P. hybridus (L.) Gaertn. Butterbur. Introduced from Europe. Waste places; rare. - Essex County, Miss M.W. Brooks (1923).

ERECHTITES Raf. Fireweed.

E. hieracifolia (L.) Raf. Damp thickets and clearings; common. - Salisbury, Knowlton (1935); Haverhill, H. 708 (1932); Andover, Root (1947); Boxford, M.E. Perely (1881); Georgetown, H. 13583 (1957); Ipswich, H. 10230 (1955); Ipswich, Plum Island, H. 12405 (1956); Gloucester, Magnolia, H. 11996 (1955); Salem, Little Misery Island, H. 13303 (1957).

E. hieracifolia, var. *intermedia* Fern. - Topsfield, H. 8848 (1953); Salem, Cole (1908); Marblehead, Cole (1942).

E. hieracifolia, var. *praealta* (Raf.) Fern. - Haverhill, Bradford, H. 748 (1932); Boxford, M.E. Perley (1881); West Newbury, H. 11136 (1955); Topsfield, H.R. Towne; Gloucester, Magnolia, J.M. Hunnewell (1898); Lynnfield, H. 14138 (1957).

SENECIO L. Ragwort.

S. vulgaris L. Common Groundsel. Naturalized from Europe. Cultivated and waste land; occasional. - Newbury, W.S. Horner; Rockport, Forbes 4093 (1904); Salem, Robinson (1877).

S. jacobaea L. Stinking Willie. Naturalized from Europe. Fields and roadsides; rare. - Ipswich, Wade 11803 (1955); [Ipswich, Plum Island, S.S. Goodale (1968)] ; Danvers, Mrs. J.M. Kennedy (1910); Essex County, Mackintosh (1924).

S. pauperculus Michx., var. *praelongus* (Greenm.) House. Rocky open woods; rare. - Andover, Pease 1451 (1901); Lynnfield, Young (1879).

S. aureus L., var. *intercursus* Fern. Golden Ragwort. Meadows, swamps and low woods; common. - Amesbury, H. 9521 (1954); Merrimac, Bean & H. 8583 (1953); Haverhill, Bradford, Cushman 2950 (1908); Andover, Pease 1449 (1902); North Andover, Morss (1884); Boxford, M.E. Perley (1882); Groveland, Bean & H. 8472 (1953); West Newbury, H. 7900 (1952); Georgetown, Horner; Topsfield, H. 17651 (1958); Danvers, Sears (1886); Hamilton, Oakes; Manchester, Sears (1892); Salem, Cole (1886); Lynnfield, Fernald (1914); Swampscott, Phippen.

S. obovatus Muhl. Rich wooded banks; occasional. - Boxford, M.E. Perley (1885), Pease 13667 (1912); Ipswich, Oakes; Topsfield, Oakes.

S. macdougallii Heller. Adventive from the West. - North Andover, Pease 2666 (1904).

CALENDULA L.

C. officinalis L. Introduced from Europe. Escape from cultivation. - Peabody, Mackintosh (1925); Saugus, H. 18698 (1958); Marblehead, H. 18918 (1958).

ARCTIUM L. Burdock.

A. lappa L. Great Burdock. Naturalized from Europe. Waste places; rare. - Topsfield, H.R. Towne.

A. minus (Hill) Bernh. Common Burdock. Naturalized from Europe. Waste land; common. - Haverhill, H. 667 (1932); Lawrence, H. 11780 (1955); North Andover, Root (1945); Boxford, M.E. Perley (1885); Newburyport, H. 11284 (1955); Georgetown, Horner; Rowley, H. 11929 (1955); Ipswich, Oakes; Ipswich, Plum Island, H. 12396 (1956); Hamilton, H. 12486 (1956); Gloucester, Magnolia, H. 12326 (1956); Salem, Robinson (1875); Peabody, Mackintosh (1935); Marblehead, Cole (1930).

CIRSIUM Mill. Common Thistle.

C. vulgare (Savi) Tenore. Bull-Thistle. Naturalized from Europe. Clearings, pastures and roadsides; common. - South Lawrence, Morss (1885); North Andover, Root (1945); Newbury, H. 18454 (1958); Georgetown, M.E. Perley (1886); Rowley, H. 11934 (1955); Ipswich, Forbes (1913); Ipswich, Plum Island, H., McGregor & Schweinfurth 12664 (1957); Topsfield, H. 11887 (1955); Gloucester, Magnolia, J.M. Hunnewell (1901); Beverly, Lowell (1849); Salem, Cole (1887); Saugus, H. 10862 (1955); Nahant, Blankinship (1895).

C. horridulum Michx. Yellow Thistle. Sandy ground near coast; occasional. - Ipswich, Oakes; Ipswich, Plum Island, H., McGregor & Schweinfurth 12663 (1957); Gloucester, Jenks (1884); Rockport, Robinson (1875); East Saugus, Forbes 4069 (1903).

C. discolor (Muhl.) Spreng. Thickets and river-banks; occasional. - Ipswich, Oakes; Danvers, Robinson (1875); Essex, Knowlton (1909); Beverly, Lowell (1847); Peabody, Mackintosh (1925); Swampscott, Kidder (1886).

C. muticum Michx. Swamp-Thistle. Low woods, thickets and swamps; occasional. - Groveland, H. 18561 (1958); Ipswich, Oakes; Wenham Swamp, Robinson (1877); Essex County, Oakes.

C. pumilum (Nutt.) Spreng. Pasture-Thistle. Dry open soil; frequent. - North Andover, H. 11043 (1955); Boxford, M.E. Perley (1885); Ipswich, Plum Island, H. 12420 (1956); Topsfield, H.R. Towne; Middleton, H. 8679 (1953); Danvers, Lowell (1848); Gloucester, Magnolia, J.M. Hunnewell (1899); North Beverly, Robinson (1875); Salem, Cole (1887).

C. pumilum, forma *candidum* Fern. - Salem, A. Smith (1884).

C. arvense (L.) Scop. Canada Thistle. Naturalized from Europe. Weed of cultivated and waste ground; common. - Haverhill, H. 636 (1932); North Andover, Morss (1885); Boxford, M.E. Perley (1882); Newbury, H. 18463 (1958); Plum Island, Robinson (1876); Topsfield, B. Woodbury; East Danvers, Sears (1880); Gloucester, Magnolia, J.M. Hunnewell (1898); Manchester, Hubbard (1913); Beverly, Robinson (1875); Salem, Cole (1910).

ONOPORDUM L. Scotch Thistle.

O. acanthium L. Naturalized from Europe. Roadsides and waste places; occasional. - Haverhill, Palfrey (1892); Andover, Manning (1886); North Andover, Morss (1885); West Boxford, Bean & Stockman (1954); Georgetown, Pease 1439 (1902); Salem, Cole (1899); Nahant, Robinson (1875).

CENTAUREA L. Star-Thistle.

C. solstitialis L. Naturalized from Mediterranean region. Waste places; rare. - Swampscott, L.A. Wentworth (1892).

C. cyanus L. Bachelor's-button. Introduced from Europe. Escape from cultivation to roadsides and waste places; occasional. - Boxford, H. 8349 (1953); Georgetown, Horner; Rowley, H. 9547 (1954); Ipswich, H. 13872 (1957); Gloucester, Magnolia, J.M. Hunnewell (1901); Beverly, Stone (1879); Salem, Robinson (1879); Marblehead, H. 11440 (1955).

C. jacea L. Naturalized from Europe. Roadsides and waste places; occasional. - Danvers, Cole (1916); Gloucester, Magnolia, H. 12315 (1956); Saugus, Cliftondale, Collins (1890); Marblehead, M.W. Harris (1906).

C. nigra L. Knapweed. Naturalized from Europe. Fields and roadsides; common. - Andover, Pease 1378 (1901); Boxford, H. 8153 (1952); Topsfield, M.E. Perley (1885); Danvers, Oakes; Beverly Farms, Hubbard (1913); Peabody, Mackintosh (1933); Marblehead, Clifton, A.P. Morse (1924).

C. nigra, var. *radiata* DC. - Andover, Pease 1376 (1900); Newbury, H. 12463 (1957).

C. vochinensis Bernh. Naturalized from Europe. Waste places; rare. - Gloucester, Magnolia, Clarke (1911).

LAPSANA L. Nipplewort.

L. communis L. Naturalized from Europe. Roadsides and waste places; rare. - Newburyport, Fernald 2570 (1932); Salem, Russell, Robinson (1877).

CICHORIUM L. Chicory.

C. intybus L. Blue Sailors. Naturalized from Europe. Fields and roadsides; common. - Haverhill, H. 1829 (1935); Methuen, H. 12258 (1956); Lawrence, H. 11767 (1955); North Andover, Root (1945); Boxford, M.E. Perley (1884); Newburyport, Cole (1899); Newburyport, Plum Island, H. 13045 (1957); Rowley, H. 18481 (1958); Ipswich, H. 13984 (1957); Topsfield, Mrs. W. Poor; Danvers, H. 11658 (1955); Essex, H. 18862 (1958); Gloucester, Magnolia, H. 12349 (1956); South Salem, Sears (1884); Peabody, Mackintosh (1933); Saugus, H. 11849 (1955); Lynn. Phippen (1857); Marblehead, H. 11425 (1955).

C. intybus, forma *album* Neum. - Haverhill, Bradford, H. 13124 (1957); Ipswich, H. 13328 (1957).

KRIGIA Schreb. Dwarf Dandelion.

K. virginica (L.) Willd. Dry sterile soil; common. - Haverhill, H. 1828 (1935); Methuen, H. 7993 (1952); Lawrence, H. 9857 (1955); Andover, Sutton 99 (1874); Boxford, M.P. Perley (1882); Newbury, Horner; Topsfield, Mrs. Dodge; Middleton, Bean & H. 7759 (1951); Wenham, Robinson (1875); Chebacco Pond, C.E. Brown (1912); Gloucester, Morss (1900); Rockport, Smith 4336 (1945); Manchester, Hubbard (1913); Salem, Cole (1937); Saugus, Bean & H. 9806 (1955).

LEONTODON L. Hawkbit.

L. autumnalis L. Fall-Dandelion. Naturalized from Europe. Fields and roadsides; common. - Haverhill, H. 593 (1932); North Andover, Morss (1885); Boxford, M.E. Perley (1881); Groveland, H. 11071 (1955); West Newbury, H. 11119 (1955); Newbury, Schweinfurth (1950); Georgetown, Horner; Topsfield, H.R. Towne; Wenham, Sears (1905); Gloucester, Lanesville, Collins (1915); Rockport, Sears (1888); Beverly, Prides Crossing, Robinson & Sears (1890); Salem, Sears (1884).

L. autumnalis, var. *pratensis* (Link) W.D.J. Koch. - Andover, Pease 3730 (1904); Boxford, M.E. Perley (1883).

PICRIS

P. hieracioides L. Adventive from Europe. Fields, roadsides and waste places [MNS, Snyder (1973)]. Local, spreading rapidly. Det. by S.K. Svenson & R.J. Eaton.

TRAGOPOGON L. Goat's-beard.

T. porrifolius L. Salsify. Introduced from Europe. Escape from cultivation; rare. - Salem, Cole (1901).

T. pratensis L. Naturalized from Europe. Fields and roadsides; occasional. - Andover, Pease 2113 (1903); North Andover, Root (1945); Newbury, H. 10406 (1955); Byfield, Horner; Ipswich, H. 10222 (1955); Middleton, H. 17943 (1958); Salem, Sears (1888).

TARAXACUM Zinn Dandelion.

T. latilobum DC. Rich slopes; rare. - Beverly Farms, Hubbard (1913).

T. erythrospermum Andrz. Red-seeded Dandelion. Naturalized from Europe. Thin dry soil; common. - Haverhill, H. 103 (1931); Andover, Pease 1494 (1902); Boxford, M.E. Perley (1882); Newbury, H. 10413 (1955); Ipswich, Oakes; Topsfield, C.E. Brown; Essex, H. 18865 (1958); Gloucester, Rand & Robinson (1896); Salem, A.P. Morse (1925); Peabody, Mackintosh (1925).

T. officinale Weber. Common Dandelion. Naturalized from Europe. Lawns, grasslands and open grounds; common. - Haverhill, H. 498 (1932); Andover, Pease 2004 (1903); North Andover, Morss (1885); Boxford, M.E. Perley (1882); Georgetown, Horner; Topsfield, Mrs. G. Gleason (1906); Gloucester, H. 11997 (1955); Rockport, Smith 2633 (1932); Salem, Robinson (1875); Peabody, Mackintosh (1927); Saugus, Bean (1940).

T. officinale, var. *palustre* (Sm.) Blytt. Naturalized from Europe. - Andover, Pease 1950 (1903).

SONCHUS L. Sow-Thistle.

S. arvensis L. Field-Sow-Thistle. Naturalized from Europe. Roadsides, shores and waste places; common. - Haverhill, Ward Hill, H. 937 (1933); Lawrence, Pease 1489 (1902); Danvers, Oakes; Salem, Robinson (1877); Peabody, Mackintosh (1925); Saugus, H. 11854 (1955); Marblehead, Cole (1934).

S. arvensis, var. *glabrescens* Guenth, Grab. & Wimm. Naturalized from Europe; occasional. - North Andover, Root (1945); Peabody, Mackintosh (1933); Lynn, H. 11826 (1955).

S. uliginosus Bieb. Salem, A.F. Hall (1925).

S. oleraceus L. Common Sow-Thistle. Naturalized from Europe. Cultivated fields and waste places; frequent. - Salisbury, Eaton (1899); Andover, Pease 1491 (1900); North Andover, Root (1945); Boxford, H. 8985 (1953); Rowley, H. 18480 (1958); Topsfield, H.R. Towne; Danvers, H. 18568 (1958); Gloucester, Lanesville, Collins (1915); Lynn, H. 18840 (1958); Nahant, Deane (1884).

S. asper (L.) Hill. Spiny-leaved Sow-Thistle. Naturalized from Europe. Waste places and roadsides; common. - Merrimac, H. 9233 (1953); Georgetown, Horner (1879); Ipswich, H. 11602 (1955); Topsfield, H.R. Towne; Gloucester, H. 18873 (1958); Salem, Robinson (1874); Salem, North Gooseberry Island, H. 13275 (1957); Marblehead, H. 11447 (1955); Nahant, Deane (1884).

S. asper, forma *glandulosus* Beckh. North Andover, Pease 7488 (1902); Rockport, Forbes 3536 (1902).

S. asper, forma *inermis* (Bisch.) G. Beck. Naturalized from Europe; occasional. - Groveland, H. 8803 (1953); Georgetown, Horner (1883); Essex, J.M. Hunnewell (1901); Manchester, Hubbard (1913).

LACTUCA L. Lettuce.

L. scariola L. Prickly Lettuce. Naturalized from Europe. Roadsides and waste places; occasional. - Newbury, Plum Island, H. 12517 (1956); Topsfield, H. 18275 (1958); Danvers, H. 18590 (1958); Marblehead, H. 18938 (1958).

L. scariola, forma *integrifolia* (Bogenh.) G. Beck. Naturalized from Europe; occasional. - Lawrence, Gutterson (1901); North Andover, Root (1945); Rowley, H. 18477 (1958); East Gloucester, W.W. Bailey (1900); Salem, Cole (1910); Peabody, Mackintosh (1925); Marblehead Neck, H. 9702 (1933).

L. sativa L. Garden Lettuce. Introduced from Europe. Dumps and garden-refuse; occasional. - Gloucester, Magnolia, H. 12341 (1956); North Beverly, Robinson (1875); Saugus, H. 18694 (1958); Lynn, H. 11816 (1955).

L. canadensis L. Thickets and clearings; rare. - West Newbury, H. 10959 (1955).

L. canadensis, var. *obovata* Wieg. Occasional. - Ipswich, Oakes; Topsfield, H. 18287 (1958); Gloucester, Magnolia, J.M. Hunnewell (1899); Rockport, Forbes 3541 (1902).

L. canadensis, var. *obovata,* forma *stenopoda* Wieg. Occasional. - Salisbury, Eaton (1899); Haverhill, H. (1933).

L. canadensis, var. *longifolia* (Michx.) Farw. Occasional. - Merrimac, Rich (1906); Methuen, H. 11003 (1955); Boxford, H. 8147 (1952); Topsfield, H.R. Towne; Hamilton, H. 10772 (1955); Salem, Cole (1910).

L. canadensis, var. *latifolia* Ktze. Frequent. - North Andover, Root (1945); Boxford, C.E. Faxon & Robinson (1877); Newbury, H. 18466 (1958); Georgetown, Horner; Ipswich, H. 13444 (1957); Topsfield, H. 18286 (1958); Wenham, Robinson (1875); Rockport, H. 19018 (1958); Manchester, Hubbard (1913); Marblehead, Cole (1943).

L. hirsuta Muhl., var. *sanguinea* (Bigel.) Fern. Dry open woods and clearings; occasional. - Haverhill, H. 537 (1932); Gloucester, Smith 630 (1927).

L. biennis (Moench) Fern. Damp thickets and openings; frequent. - North Andover, H. 11015 (1955); Groveland, M.E. Perley (1885); Georgetown, H. 13574 (1957); Ipswich, Plum Island, H. 12397 (1956); Wenham, Oakes; West Gloucester, H. 19011 (1958); Rockport, Smith 2711 (1933); Manchester, Morris (1860).

L. biennis, forma *integrifolia* (T. & G.) Fern. Occasional. - Topsfield, H. 11900 (1955); Swampscott, Phippen (1857).

CREPIS L. Hawk's-beard.

C. capillaris (L.) Wallr. Adventive from Europe. Fields and waste places; rare. - Ipswich, H. 13996 (1957); Topsfield, H. 18296 (1958).

PRENANTHES L. Rattlesnake-root.

P. alba L. White Lettuce. Rich woods and thickets; occasional. - Georgetown, Horner; Beverly, Robinson (1874); Salem, Russell (1859).

P. serpentaria Pursh. Lion's-foot. Dry thickets and open woods; rare. - Andover, Bean (1936).

P. trifoliolata (Cass.) Fern. Gall-of-the-earth. Dry slopes, thickets and clearings; common. - Salisbury, H. 18973 (1958); Haverhill, H. 567 (1932); Lawrence, H. 11244 (1955); North Andover, Morss (1885); West Boxford, Horner; Groveland, M.E. Perley (1885); Topsfield, H.R. Towne; Hamilton, H. 12490 (1956); Gloucester, Magnolia, Deane (1901); Rockport, L.B. & R.C. Smith 2705 (1933); Beverly, Robinson (1875); Salem, Cole (1889); Peabody, Mackintosh (1933); Lynnfield, Bean (1949).

P. altissima L. Moist woods; occasional. - Boxford, Robinson (1875); Ipswich, Oakes; Chebacco, Robinson (1874).

P. altissima, forma *hispidula* Fern. - Hamilton, Hunnewell 4628 (1916).

HIERACIUM L. Hawkweed.

H. pilosella L. Mouse-ear. Naturalized from Europe. Pastures, fields and lawns; frequent. - Andover, H. 17745 (1958); Boxford, H. 17640 (1958); Groveland, H. 19626 (1959); Rowley, H. 17662 (1958); Ipswich, Bean & H. 8430 (1953); Beverly, H. 10215 (1955).

H. flagellare Willd. Naturalized from Europe. Fields and roadsides; frequent. - Haverhill, H. 19194 (1959); Boxford, H. 8351 (1953); West Newbury, Bean & H. 19467 (1959); Newbury, Bean & H. 8416 (1953); Georgetown, H. 17678 (1958); Ipswich, Plum Island, H. 12122 (1956); Danvers, Bean & H. 9997b (1955); Wenham, Bean & H. 10014 (1955).

H. aurantiacum L. Devil's Paint-brush. Naturalized from Europe. Fields and roadsides; frequent. - Amesbury, H. 9515 (1954); Haverhill, Bradford, H. 10456 (1955); Lawrence, H. 9845 (1955); Andover, Pease 1413 (1902); North Andover, Root (1943); Boxford, H. 7915 (1952); Groveland, Bean & H. (1953); West Newbury, H. 9531 (1954); Newburyport, H. 10346 (1955); Newbury, Bean & H. 8417 (1953); Georgetown, Horner (1890); Rowley, H. & Wetherall 9562 (1954); Ipswich, Bean & H. 8432 (1953); Topsfield, H. 8006 (1952); Danvers, Miss M. Nichols (1885) cult.;

Gloucester, Magnolia, H. 10147 (1955); Rockport, Pigeon Cove, Rev. Mr. Putnam (1884); Peabody, H. 10618 (1955); Lynnfield, H. 9720 (1955).

H. floribundum Wimm. & Grab. Naturalized from Europe. Fields and clearings; frequent. - Salisbury, H. 10469 (1955); Andover, H. 10060 (1955); Newburyport, Knowlton (1938); Newbury, H. 10412 (1955); Rowley, H. 7936 (1952); Topsfield, Bean & H. 9943 (1955); Wenham, Bean & H. 10032 (1955); Hamilton, H. 10808 (1955); Marblehead, H. 9757 (1955).

H. pratense Tausch. King Devil. Naturalized from Europe. Pastures and roadsides; occasional. - Andover, H. 17755 (1958); North Andover, H. 9487 (1954); Boxford, H. 7906 (1952); Ipswich, Plum Island, H., McGregor & Schweinfurth 12661 (1957); Topsfield, Bean & H. 9957 (1955); Beverly, H. 10206 (1955); Marblehead, Bean & H. 9763 (1955).

H. praealtum Gochnat, var. *decipiens* W.D.J. Koch. Naturalized from Europe. Grasslands and pastures; occasional. - Andover, Pease 1423 (1901); North Andover, Root (1943); Georgetown, H. 17664 (1958); Topsfield, Bean & H. 9961 (1955); Middleton, H. 9468 (1954); Wenham, H. 12131 (1956); Marblehead, H. 9682 (1955).

H. florentinum All. King Devil. Naturalized from Europe. Fields and roadsides; common. - Salisbury, Bean & H. 8393 (1953); Merrimac, Bean & H. 8565 (1953); Haverhill, H. 8462 (1953); Lawrence, H. 9883 (1955); Boxford, Hunnewell 8886 (1923): Groveland, Bean & H. 8533 (1953); West Newbury, H. 9532 (1954); Newburyport, Bean & H. 8395 (1953); Newbury, Bean & H. 8409 (1953); Rowley, Bean & H. 8428 (1953); Rowley, Plum Island, H. 12064 (1956); Ipswich, Bean & H. 8431 (1953); Ipswich, Plum Island, H., McGregor & Schweinfurth 12671 (1957); Topsfield, H. 8057 (1952); Danvers, Bean & H. 9997a (1955); Wenham, Bean & H. 10005 (1955); Essex, H. 10250 (1955); Gloucester, H. 10140 (1955); Rockport, H. 10307 (1955); Beverly, H. 10205 (1955); Saugus, Bean & H. 9785 (1955).

H. vulgatum Fries. Naturalized from Europe. Roadsides and waste places; rare. - Andover, H. 12301 (1956).

H. canadense Michx., var. *fasciculatum* (Pursh) Fern. Thickets and borders of woods; frequent. - Andover, Pease 1420 (1900); North Andover, Root (1945); Boxford, H. 24033 (1961), Rowley, Plum Island, H. 12480 (1956); Ipswich, Oakes; Topsfield, Hunnewell 3810 (1915); Rockport, Smith 4347 (1945); Manchester, Phippen; Beverly, Lowell (1846); [MNS, Snyder (1973)].

H. sabaudum L. Naturalized from Europe. Waste open ground; occasional. - North Andover, Bean (1945); West Gloucester, H. 19012 (1958); Lynn, H. 27061 (1964).

H. venosum L., var. *nudicaule* (Michx.) Farw. Open woods and clearings; common. - Haverhill, H. 499 (1932); Lawrence, H. 9865 (1955); North Andover, H. 9491 (1954); Boxford, M.P. Perley (1882); Groveland, Horner; Newbury, Cushing (1819); Topsfield, Oakes; Peabody, Mackintosh (1877); Saugus, Bean & H. 9812 (1955); Lynn, Cole (1900); Swampscott, Cole (1889).

H. paniculatum L. Open woods; occasional. - Haverhill, Ward Hill, Cushman 2942 (1908); Boxford, H. 18556 (1958); Groveland, Horner; Ipswich, Oakes; Topsfield, A. Jacobs; Rockport, H. 19025 (1958); Manchester, Robinson (1874).

H. paniculatum forma *glandulosum* R. Hoffm. - Boxford, H. 8159 (1952); Peabody, Mackintosh (1933).

H. scabrum Michx. Dry open woods; frequent. - Salisbury, H. 18972 (1958); Haverhill, H. 594 (1932); Boxford, Robinson (1876); Newbury, H. 8234 (1952); Ipswich, Plum Island, H. 11296 (1955); Topsfield, H.R. Towne; Danvers, Sears (1887); Rockport, H. 19016 (1958).

H. marianum Willd. - Ipswich, Oakes.

APPENDIX

The Museum has recently published a book entitled *Weeds and Wildflowers of Eastern North America* containing 114 colored plates of flowering plants from watercolors by T. Merrill Prentice (Salem: Peabody Museum of Salem and Barre Publishers, 1973). Most of those plants occur in Essex County and the book can be used as a companion volume of illustrations to this work. For the convenience of readers who may own both works, we here include the index of the Prentice volume for cross-reference.

Index

The numbers refer to plates